Genome Informatics 2008

GENOME INFORMATICS SERIES (GIS)

ISSN: 0919-9454

The Genome Informatics Series publishes peer-reviewed papers presented at the International Conference on Genome Informatics (GIW) and some conferences on bioinformatics. The Genome Informatics Series is indexed in MEDLINE.

No.	Title	Year	ISBN Cl./Pa.
1	Genome Informatics Workshop I	1990	(in Japanese)
2	Genome Informatics Workshop II	1991	(in Japanese)
3	Genome Informatics Workshop III	1992	(in Japanese)
4	Genome Informatics Workshop IV	1993	4-946443-20-7
5	Genome Informatics Workshop 1994	1994	4-946443-24-X
6	Genome Informatics Workship 1995	1995	4-946443-33-9
7	Genome Informatics 1996	1996	4-946443-37-1
8	Genome Informatics 1997	1997	4-946443-47-9
9	Genome Informatics 1998	1998	4-946443-52-5
10	Genome Informatics 1999	1999	4-946443-59-2
11	Genome Informatics 2000	2000	4-946443-65-7
12	Genome Informatics 2001	2001	4-946443-72-X
13	Genome Informatics 2002	2002	4-946443-79-7
14	Genome Informatics 2003	2003	4-946443-82-7
15	Genome Informatics 2004 Vol. 15, No. 1	2004	4-946443-88-6
16	Genome Informatics 2004 Vol. 15, No. 2	2004	4-946443-91-6
17	Genome Informatics 2005 Vol. 16, No. 1	2005	4-946443-93-2
18	Genome Informatics 2005 Vol. 16, No. 2	2005	4-946443-96-7
19	Genome Informatics 2006 Vol. 17, No. 1	2006	4-946443-97-5
20	Genome Informatics 2006 Vol. 17, No. 2	2006	4-946443-99-1
21	Genome Informatics 2007 Vol. 18	2007	978-1-86094-991-3
22	Genome Informatics 2007 Vol. 19	2007	978-1-86094-984-5
23	Genome Informatics 2008 Vol. 20	2008	978-1-84816-299-0
24	Genome Informatics 2008 Vol. 21	2008	978-1-84816-331-7

Genome Informatics Series Vol. 21 ISSN: 0919-9454

Genome Informatics 2008

Proceedings of the 19th International Conference

Gold Coast, Queensland, Australia 1 – 3 December 2008

Editors

Jonathan Arthur
University of Sydney, Australia

See-Kiong Ng
Institute for Infocomm Research, Singapore

Imperial College Press

Published by

Imperial College Press
57 Shelton Street
Covent Garden
London WC2H 9HE

Distributed by

World Scientific Publishing Co. Pte. Ltd.
5 Toh Tuck Link, Singapore 596224
USA office: 27 Warren Street, Suite 401-402, Hackensack, NJ 07601
UK office: 57 Shelton Street, Covent Garden, London WC2H 9HE

British Library Cataloguing-in-Publication Data
A catalogue record for this book is available from the British Library.

GENOME INFORMATICS 2008
Proceedings of the 19th International Conference (GIW 2008)

ISBN-13 978-1-84816-331-7
ISBN-10 1-84816-331-2

CONTENTS

PREFACE

This book contains papers presented at the Ninteenth International Conference on Genome Informatics (GIW 2008) held on the Gold Coast, Queensland, Australia on December 1st to 3rd, 2008.

The GIW series provides an international forum for presentation and discussion of original research papers on all aspects of bioinformatics, computational biology, and systems biology. Its scope includes biological sequence analysis, protein structure prediction, gene regulatory networks, clustering algorithms, comparative genomics, text mining, and many other areas. GIW has a history of 19 years and is the longest running international bioinformatics conference. The first GIW was held at Kikai Shinko Kaikan, Tokyo during December 3-4, 1990 as an open workshop just before the Japanese Human Genome Project started in 1991. GIW 2008 was the first time the conference has been held in Australia. This year it was hosted by Bioinformatics Australia, representing the bioinformatics community in Australia, and incorporated the annual Bioinformatics Australia conference. Bioinformatics Australia is organized within AusBiotech, the national peak body for biotechnology in Australia.

The Program Committee of GIW 2008 received a total of 55 submissions from authors in 16 different countries around the world. Each submitted paper was peer-reviewed by at least three members of the Program Committee. Based on their reports, 18 papers were accepted (33%) for presentation at the conference. These 18 papers appear in this book and are indexed in Medline. In addition, this book contains abstracts from the six invited speakers: Sean Grimmond, University of Queensland (Australia), Eugene Koonin, National Centre for Biotechnology Information (USA), Ming Li, University of Waterloo (Canada), Yixue Li, Shanghai Jiaotong University (China), John Mattick, University of Queensland (Australia), and Eric Schadt, Rosetta Inpharmatics (USA).

The electronic versions of all the papers in this issue are also publicly available from the website of the Japanese Society for Bioinformatics (JSBi) (`http://www.jsbi.org/journal.html`).

Jonathan Arthur
See-Kiong Ng
GIW 2008 Program Committee Co-Chairs

Mark Ragan
GIW 2008 Conference Chair

ACKNOWLEDGMENTS

We thank all the authors for their efforts in preparing their manuscripts. We also appreciate the great efforts made by the Program Committee members in rigourously reviewing the manuscripts. The high quality of the papers presented by the authors provided a challenging task in selecting the very best for acceptance. We greatly appreciate the time and effort of both the authors and the Program Committee, in their respective contributions, to continuing the GIW tradition of a high quality, engaging scientific program.

We also acknowledge Bioinformatics Australia (within AusBiotech Ltd) for hosting GIW 2008 as well as the assistance of the National Organizing Committee, the Local Organizing Committee, and the Conference Organisers (Martin Lack and Associates) for the coordination of the conference.

We are grateful for the support of the Department of Innovation, Industry, Science and Research, the Queensland State Government, and:

AIST Computational Biology Research Center
ARC Research Network in Enterprise Information Infrastructure
Australian Centre for Plant Functional Genomics
Australian Genome Research Facility
CSIRO
NICTA
Queensland Cyber Infrastructure Foundation
SGI
Sydney Bioinformatics
University of Queensland

Finally, we give special thanks to those who presented papers or posters at GIW 2008, and those who attended the conference. GIW 2008 would not be a complete success without their enthusiastic participation.

PROGRAM COMMITTEE

Jonathan Arthur	– University of Sydney, Australia; Co-Chair
See-Kiong Ng	– Institute for Infocomm Research, Singapore; Co-Chair
Cathy Abbott	– Flinders University, Australia
Gary Bader	– University of Toronto, Canada
Vladimir Bajic	– South African National Bioinformatics Institute, South Africa
Christopher Baker	– Institute for Infocomm Research, Singapore
Guillaume Bourque	– Genome Institute of Singapore, Singapore
Jung-Hsien Chiang	– National Cheng Kung University, Taiwan
Francis YL Chin	– University of Hong Kong, Hong Kong
Peter Clote	– Boston College, USA
Aaron Darling	– University of Queensland, Australia
Bhaskar DasGupta	– University of Illinois, USA
Colin Dewey	– University of Wisconsin, USA
Chris Ding	– University of Texas at Arlington, USA
Roland Dunbrack	– Fox Chase Cancer Center, USA
Jenny Graves	– Australian National University, Australia
Win Hide	– South African National Bioinformatics Institute, South Africa
Tamás Horváth	– University of Bonn and Fraunhofer IAIS, Germany
Wen-Lian Hsu	– Academia Sinica, Taiwan
Seiya Imoto	– University of Tokyo, Japan
Lars Jermiin	– University of Sydney, Australia
Minoru Kanehisa	– Kyoto University, Japan
George Karypis	– University of Minnesota, USA
Uri Keich	– Cornell University, USA
Daisuke Kihara	– Purdue University, USA
Edda Klipp	– Max Planck Institute for Molecular Genetics, Germany
Stefen Kramer	– Technische Universität München, Germany
Dong-Yup Lee	– Bioprocessing Institute & National University of Singapore, Singapore
Sang Yup Lee	– KAIST, Korea

Ming Li	– University of Waterloo, Canada
Frédérique Lisacek	– Swiss Institute of Bioinformatics, Switzerland
Hiroshi Mamitsuka	– Kyoto University, Japan
Aleksandar Milosavljevic	– Baylor College of Medicine, USA
Satoru Miyano	– University of Tokyo, Japan
Bernard Moret	– Swiss Federal Institute of Technology, Switzerland
Shin-ichi Morishita	– University of Tokyo, Japan
Pablo Moscato	– University of Newcastle, Australia
William Stafford Noble	– University of Washington, USA
Laxmi Parida	– IBM T. J. Watson Research Center, USA
Ron Pinter	– Technion, Israel
Shoba Ranganathan	– Macquarie University, Australia
Allen Rodrigo	– University of Auckland, New Zealand
Rintaro Saito	– Keio University, Japan
Yasubumi Sakakibara	– Keio University, Japan
Christian Schönbach	– Nanyang Technological University, Singapore
Tetsuo Shibuya	– University of Tokyo, Japan
Mona Singh	– Princeton University, USA
Wing Kin Sung	– National University of Singapore, Singapore
Koji Tsuda	– Max Planck Institute for Biological Cybernetics, Germany
Alfonso Valencia	– Universidad Autonoma, Spain
Gabriel Valiente	– Technical University of Catalonia, Spain
Jean-Philippe Vert	– École des Mines de Paris, France
Lusheng Wang	– The City University of Hong Kong, Hong Kong
Marc Wilkins	– University of New South Wales, Australia
Michael Wise	– University of Western Australia, Australia
Ying Xu	– University of Georgia, USA
Gwan-Su Yi	– Information & Communications University, Korea
Mohammed J. Zaki	– Rensselaer Polytechnic Institute, USA

CO-REVIEWERS

Satya Arjunan	Hong-Jie Dai	Kevin DeRonne
Jun-tao Guo	Kosuke Hashimoto	Rajaraman Kanagasabai
Chris Kauffman	Ian Menz	Nini Rao
Tadahiko Sakiyama	Michael Shmoish	Michihiro Tanaka
Haibao Tang	Katsuyuki Yugi	

STEERING COMMITTEE

Minoru Kanehisa	– Kyoto University, Japan
Satoru Miyano	– University of Tokyo, Japan
Mark Ragan	– University of Queensland, Australia
Toshihisa Takagi	– University of Tokyo, Japan
Limsoon Wong	– National University of Singapore, Singapore

CONFERENCE CHAIR

Mark Ragan – University of Queensland, Australia

NATIONAL ORGANIZING COMMITTEE

Cathy Abbott	– Flinders University, Australia
Jonathan Arthur	– University of Sydney, Australia
Tim Bailey	– University of Queensland, Australia
Mark Baker	– Australian Proteome Analysis Facility, Australia
Jeremy Barker	– Queensland Facility for Advanced Bioinformatics, Australia
Matthew Bellgard	– Murdoch University, Australia
Kevin Burrage	– University of Queensland, Australia
Phoebe Chen	– Deakin University, Australia
Ross Coppel	– Monash University, Australia
Brian Dalrymple	– CSIRO Livestock Industries, Australia
Simon Easteal	– Australian National University, Australia
Dave Edwards	– Australian Centre for Plant Functional Genomics, Australia
Sue Forrest	– Australian Genome Research Facility, Australia
Bruno Gaëta	– University of New South Wales, Australia
Jenny Graves	– Australian National University, Australia
David Hansen	– Australian e-Health Research Centre, Australia
James Hogan	– Queensland University of Technology, Australia
Jonathan Keith	– Queensland University of Technology, Australia
Vladimir Likic	– University of Melbourne & Bio21, Australia

Tim Littlejohn	– IBM Australia, Australia
John Mattick	– University of Queensland, Australia
Geoff McLachlan	– University of Queensland, Australia
Annette McGrath	– Australian Genome Research Facility, Australia
David Mitchell	– CSIRO CMIS, Australia
Pablo Moscato	– University of Newcastle, Australia
Tuan Pham	– James Cook University, Australia
Michael Poidinger	– Johnson & Johnson, Australia
Mark Ragan	– University of Queensland, Australia
Shoba Ranganathan	– Macquarie University, Australia
Allen Rodrigo	– University of Auckland, New Zealand
Rohan Teasdale	– University of Queensland, Australia
Mervyn Thomas	– Emphron Informatics, Australia
Matthew Wakefield	– Walter & Eliza Hall Institute of Medical Research, Australia
Marc Wilkins	– University of New South Wales, Australia
Sue Wilson	– Australian National University, Australia
Michael Wise	– University of Western Australia, Australia
Xiaofang Zhou	– University of Queensland, Australia
Albert Zomaya	– University of Sydney, Australia

LOCAL ORGANIZING COMMITTEE

Mark Ragan	– University of Queensland, Australia
Tim Bailey	– University of Queensland, Australia
Mikael Bodén	– University of Queensland, Australia
Brian Dalrymple	– CSIRO Livestock Industries, Australia
Dave Edwards	– Australian Centre for Plant Functional Genomics, Australia
James Hogan	– Queensland University of Technology, Australia
Rohan Teasdale	– University of Queensland, Australia

PART A

Full Papers

AN APPROACH TO TRANSCRIPTOME ANALYSIS OF NON-MODEL ORGANISMS USING SHORT-READ SEQUENCES

LESLEY J COLLINS[1,2]
L.J.Collins@massey.ac.nz

PATRICK J BIGGS[1,2]
P.Biggs@massey.ac.nz

CLAUDIA VOELCKEL[1]
C.Voelckel@massey.ac.nz

SIMON JOLY[1,3]
S.Joly@massey.ac.nz

[1] *Allan Wilson Centre for Molecular Ecology and Evolution, Massey University, Palmerston North, New Zealand*
[2] *Institute of Molecular BioSciences, Massey University, Palmerston North, New Zealand*
[3] *Current address: Department of Biology, McGill University, Montreal, Quebec, Canada*

Transcriptome analysis using high-throughput short-read sequencing technology is straightforward when the sequenced genome is the same species or extremely similar to the reference genome. We present an analysis approach for when the sequenced organism does not have an already sequenced genome that can be used for a reference, as will be the case of many non-model organisms. As proof of concept, data from Solexa sequencing of the polyploid plant *Pachycladon enysii* was analysed using our approach with its nearest model reference genome being the diploid plant *Arabidopsis thaliana.* By using a combination of mapping and *de novo* assembly tools we could determine duplicate genes belonging to one or other of the genome copies. Our approach demonstrates that transcriptome analysis using high-throughput short-read sequencing need not be restricted to the genomes of model organisms.

Keywords: short-read sequencing; next generation sequencing *Pachycladon*; transcriptome analysis.

1. Introduction

High-throughput short-read sequencing is one of the latest sequencing technologies to be released to the genomics community. For example, on average a single run on the Illumina Genome Analyser can result in over 30 to 40 million single-end (~35 nt) sequences. However, the resulting output can easily overwhelm genomic analysis systems designed for the length of traditional Sanger sequencing, or even the smaller volumes of data resulting from 454 (Roche) sequencing technology.

Typically, the initial use of short-read sequencing was confined to matching data from genomes that were nearly identical to the reference genome. This enabled easy comparisons between genomes in order to investigate differences either in the genomic sequence itself (SNPs - single nucleotide polymorphisms, and other mutations), gene expression (transcriptomics), small RNAs, methylation or chromatin mapping (ChIP-sequencing) (examples [1; 2]). However, researchers are now pushing the boundaries of this technology to sequence more distantly related genomes. Our study presents an approach to transcriptome analysis of a non-model genome.

Transcriptome analysis on a global gene expression level is an ideal application of short-read sequencing. Traditionally such analysis involved complementary DNA (cDNA) library construction, Sanger sequencing of ESTs, and microarray analysis. Next generation sequencing has become a feasible method for increasing sequencing depth and coverage while reducing time and cost compared to the traditional Sanger method. A method for non-model organisms using 454 pyrosequencing data was recently published [3], highlighting how next-generation sequencing enables transcriptome analysis from any species. Short-read sequencing produces a far greater coverage even though the sequences produced are shorter than those produced by pyrosequencing. Genome projects are now looking not only to produce sequence counts of individual ESTs obtained using short-read sequencing, but to produce the EST sequences in the first place to investigate EST characteristics prior to counting. Our study introduces an approach enabling the latter, demonstrating its usefulness on data obtained from the *Pachycladon* transcriptome project.

The genus *Pachycladon* is an emerging non-model system in the study of plant speciation. The whole genus (2n=4x=20) is of allopolyploid origin from distant parents in the Brassicaceae family (S. Joly, P. Heenan and P. Lockhart, unpublished data), meaning that we expect (most) genes to be duplicated. Both genome copies present in *Pachycladon* diverged from the model species *Arabidopsis thaliana*, a functional diploid (2n=2x=10), relatively recently (ca. 7-10 Mya). The small number of species, its young age and its close relationship with *A. thaliana*, suit *Pachycladon* for evolutionary studies investigating the ecological drivers and the molecular basis of species diversification. Multiple approaches can be used to address these questions, including gene expression profiling, QTL mapping and candidate gene studies, all of which require molecular resources such as an EST library. These applications also require prior characterization of duplicate gene copies. Short-read sequences of amplified cDNA from roots and shoots of *Pachycladon enysii* obtained with the Illumina Genome Analyzer provided an opportunity to explore an efficient, inexpensive and reliable approach to EST sequencing that can be readily adopted by researchers studying non-model organisms. Our analysis resulted in the identification of duplicate gene candidates from *Pachycladon* ESTs, some of which could be matched to *A. thaliana* ESTs showing that analysis of short-read sequences is feasible when the reference genome is distantly related.

2. Approach Overview

Our overall approach to non-model organism transcriptome analysis (as shown in Figure 1) is to use high-throughput short-read sequences, optimize assembly and mapping parameters using partial data, then process the total data using these optimized mapping and *de novo* parameters. Assembled contigs are compared to themselves and also to the nominated reference genome using BLAST, leading to the extraction of candidate duplicate genes. Results are visualized at different stages for validation purposes.

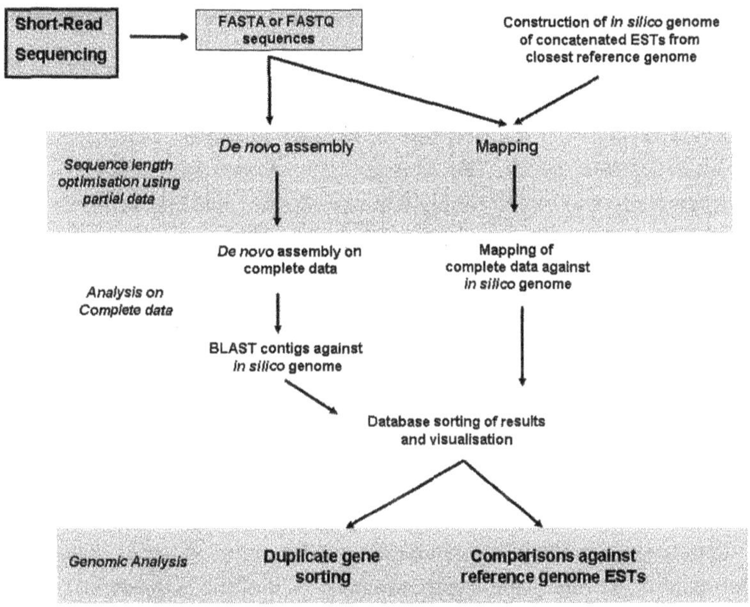

Figure 1. Overview of short-read based transcriptome analysis approach for non-model organisms. Mapping is an option only if a suitable genome is available, otherwise FASTA (or FASTQ) sequences can only proceed down the *de novo* track or into other project-specific analysis such as sequence counting (not shown). However mapping against even a distant genome can provide valuable information about genome conservation so should be done where possible.

Since the output is large, all data is managed and curated with a MySQL database from which genome areas of interest can be extracted. Reformatting and data extraction is handled through the use of Perl and MySQL scripts. Details about each stage of this approach are given below.

2.1 Dataset volumes and data management

Data output from short-read sequencing is large, consisting of millions of sequences and preliminary mapping output. To manage these data volumes as well as sequence and result curation we used the MySQL database system (version 5.0.45, running under Windows XP-Pro). This database was also used to store BLAST results, EST location and other relevant information. The MySQL database was also linked to the Gbrowse genome browser [4], to enable viewing of data subsets. We see no problems to other databases being used so long as they are robust enough to handle these data volumes, data types and genome viewer integration. We used data from the Illumina Genome Analyser (also known as Solexa Sequencing), but this approach is applicable to data produced from other platforms (such as the SOLiD platform from Applied Biosystems), so long as the sequence output has already been converted from any internal and/or proprietary forms (such as the SOLiD 'colour space') to the more standard FASTA or FASTQ format.

2.2 Data subset extraction and optimal parameter evaluation for mapping

At the end of the sequencing run the short-read sequences were converted to FASTA or FASTQ output and mapped against a nominated reference genome as part of the Illumina Genome Analyser Pipeline. However, this preliminary analysis can use parameters that may not produce optimal results. For example, the maximum number of mismatches allowed between reads and the reference sequence by the pipeline software (ELAND) is two, which could be a too restrictive value when one is using a more distant genome as a reference. A related parameter is the sequence length for ELAND mapping because longer reads mean more potential mismatches between the two genomes, and thus resulting in more non-mapped sequences.

One way to choose optimal parameters for analysis is by running simulations, but it is not possible to simulate data from a genome not already sequenced; this can only be done after the sequencing run. The primary parameter that required determination for this application was the sequence length used for both ELAND mapping and *de novo* assembly. Because of the large volumes of output from a single short-read run it is not efficient to determine experimental parameters on the entire dataset. Instead we use as standard the data from one lane (approximately 4 million sequences from a lower titration).

The Illumina pipeline software ELAND was run on this data subset with the sequence length parameter set for 17 initially, and then increased by one until the maximum of 32 was reached. This means the first 17 bases of the sequences are used for mapping to the reference genome. If the sequence length is set too short then we can expect to see a steep increase in the number of repeat-matches as the 'specificity' of the match lowers. However, if the sequence length is set too long then we run the risk of generating more non-matches as the number of differences between the sequenced genome and the reference genome will push the match beyond ELAND's limit of two mismatches. By rerunning a subset of data over a wide range of sequence lengths, an optimal length can thus be selected.

Another popular mapping software Maq (Release 0.5.0 [5]) was briefly compared. Maq offers the advantage of allowing a higher number of mismatches (three opposed to the two offered with ELAND) but is much slower when this is permitted. Maq uses FASTQ input incorporating quality information as well as sequence information. Users of Solexa produced FASTQ data should be aware that the scores are calculated differently from Sanger-type sequencing FASTQ and can include calibration from the initial mapping to a reference genome. When working with distantly related reference genomes, potential users of this software should specify 'uncalibrated quality scores' from a Solexa sequencing service. There are also some functions in Maq that have been specifically written for the SOLiD platform.

The third piece of software we compared was SOAP [6]. SOAP is similar to ELAND in that it uses hash look-up table algorithms to speed up analysis and runs comparably [6]. It also has a limit of two mismatches. Although we used ELAND for the proof of concept of our approach, any of these other software packages could theoretically be substituted for short-read sequences from any platform.

For analysis, an *in silico* reference genome must also be prepared from the many discontiguous sequences within EST sequence libraries. Mapping to each EST separately is possible so long as the conditions for running ELAND are met (ELAND documentation from Illumina). To construct the *in silico* EST 'genome' we concatenated the EST sequences leaving 50 'N's between each EST sequence. Co-ordinates for each sequence are retained during this process so that mappings against each EST can be determined separately.

2.3 De novo assembly

Out of the *de novo* assemblers available for handling short-read output (including Velvet [8], SSAKE [7] VCAKE [8]and SHARCGS [9]), we chose to primarily use Velvet (version 0.5) [10] as it was found to produce consistent and sizable contigs. Velvet was developed specifically for manipulating short-read sequences and uses de Bruijn graphs for sequence assembly. However, a downside is that it runs only in a 64 bit Linux environment. As with the mapping, the optimal 'k-mer' size (a Velvet parameter comparable to 'word' size used for BLAST searches) is determined using a subset of the entire data, although it is feasible for entire datasets to be assembled with a variety of k-mer lengths and the results compared. Assembled contigs are then BLASTed against themselves to find exact copies or against any other similar genomes using BLAST [11].

The results of the BLAST analyses are loaded into a MySQL database for further processing. The sequences of the contigs and coordinates of the hits to the *A. thaliana* EST genome were output so they could be viewed with a combination of Gbrowse and MySQL. The combination of reference genome mapping and BLASTing of contigs from *de novo* assembly then allows us to pull out regions corresponding to duplicate genes.

3. *Pachycladon* Transcriptome short-read analysis

The *Pachycladon enysii* Transcriptome project presented two genomic challenges. The first relates to the fact that the closest reference genome that could be used was the plant *A. thaliana*, a species that diverged 7-10 million years from both genomic copies present in *Pachycladon* (S. Joly, P. Heenan and P. Lockhart, unpublished data). Prior to this work, there were no published studies on whether a genome of a different species could be used as a reference genome in short-read sequencing. Thus our approach was used to both study this effect and to aid the construction of *Pachycladon* ESTs for further analyses. The second type of genomic challenge is that *P. enysii* is a polyploid organism with two genome copies whereas *A. thaliana* is a diploid organism with one genome copy. Polyploidy on any level creates issues for genome analysis. Our aim was to map *Pachycladon* orthologues to specific *A. thaliana* loci and find putative duplicate *Pachycladon* genes.

The *Pachycladon* RNA was extracted separately from the roots and leaves of one rosette-stage *P. enysii* specimen originating from Avalanche Peak, South Island, New Zealand using the Qiagen RNeasy kit (Biolab Ltd.). An equal amount of root and leaf RNA (12.5μg) was pooled and reverse-transcribed using the SuperScript™ Double-Stranded cDNA Synthesis Kit (Invitrogen) and oligo(dT) primers (Invitrogen). Double-stranded cDNA (3μl) was subsequently amplified using the Qiagen REPLI-g Mini Kit (Biolab Ltd.). Five μg of the REPLI-g-amplified *P. enysii* cDNA was then used as a template for Solexa Genomic DNA preparation. Solexa sequencing used the Genome DNA Sample Preparation kit (FC-102-1001, Illumina) over 36 cycles.

Solexa sequencing produced a total of 40 million single-end short-reads of 36 nucleotides (nt). Seven lanes were used and contained different numbers of sequences due to a titration of DNA concentrations being used to generate clusters on the flowcell. This data was analyzed using our approach and the results are described below.

3.1. *Mapping against Arabidopsis ESTs*

An *in silico* genome was constructed from *A. thaliana* ESTs by concatenating the TAIR7 EST dataset (TAIR7_cDNA_20070425) [12]. Each EST was separated by 50 'N's (to prevent short-read sequences mapping to more than one EST), and all coordinates recorded for later mapping. Because of the sequence distance between the *Pachycladon* genome and the *A. thaliana* EST reference genome, we recognized that using the full length of the sequence for the match may potentially exclude many sequences due to the mapping software (ELAND) only allowing up to two mismatches per sequence. However, even with a low percentage of unique matches expected, mapping to a nearby reference genome enables us to examine the conserved portion of the *Pachycladon* transcriptome.

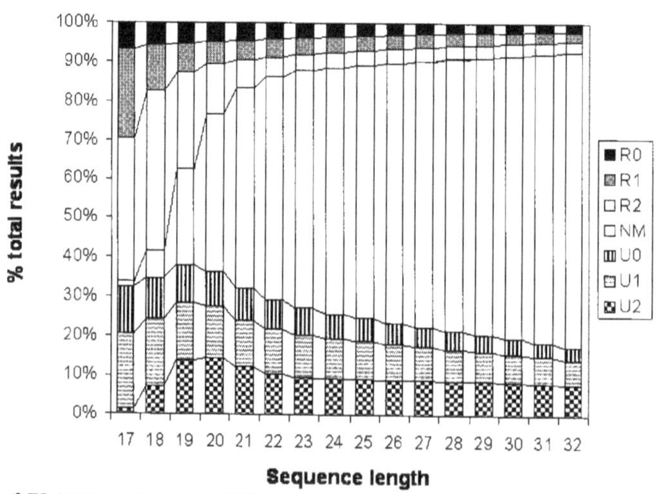

Figure 2. Graph of ELAND performed at different sequence lengths on one lane of sequences (4 million). Results are scored as a percentage of the total number of sequences. We get the greatest percentage of unique hits (37%) using a sequence length of 19.Key: U - Unique match U0 (no mismatches) U1 (1 mismatch) U2 (2 mismatches); R - Repeat match R0, R1 and R2 as for the unique hits; QC -Quality filter fail; NM - no match to reference genome. (QC results are omitted as they are too small to be seen on this graph.)

Table 1. Results from ELAND analysis of *Pachycladon* sequences against *A. thaliana* ESTs. Key: U - Unique match U0 (no mismatches) U1 (1 mismatch) U2 (2 mismatches); R - Repeat match R0, R1 and R2 as for the unique hits; QC -Quality filter fail; NM - no match to reference genome. The percentages obtained from a single lane of data were comparable with that from the entire 7 lanes.

ELAND result type	Length = 19 (1 lane)	% Total (1 lane)	Number (7 lanes)	Av % Total (7 lanes)	Std Dev (7 lanes)
U0	341804	8.99	3577678	8.86	1.2
U1	543248	14.29	5697913	14.23	0.34
U2	520442	13.69	5547612	13.85	0.29
R0	960119	25.26	1943961	24.84	0.93
R1	191895	5.05	2914570	4.88	0.22
R2	275036	7.24	10455270	7.27	0.16
QC	14	0.00	121	0.00	0.00
NM	968842	25.49	9916563	26.07	0.79
Total	**3801400**	100	**40053567**	100	-

ELAND was run using a wide range of input sequence lengths (17-32) and the number of matches, repeat matches and non-matched sequences noted. This was graphed (Figure 2) indicating that using a sequence length of 19 was optimal for further analysis. All data was then mapped using ELAND with a sequence length input of 19. The results of this mapping are summarized in Table 1.

Analysis of four duplicate genes (*NIA* (106 nt), *CHS* (1135 nt), *PRK* (476 nt), and *MS* (394 nt)) prior to the short-read sequencing (data not shown) gave an average distance (per nucleotide) between *A. thaliana* and *Pachycladon* of 0.064 ± 0.021 substitutions per site, and an average distance between the two *Pachycladon* copies of 0.058 ± 0.023 substitutions per site. Despite this distance there were a surprising number of unique matches to the *A. thaliana* EST library (37.0% U0, U1 and U2 combined for all ELAND-19 data). These mapped short-reads can be viewed to determine coverage but they are even more useful when assembled into longer contigs. We can then use these *A. thaliana*-mapped contigs to search for potential duplicate copies within the *Pachycladon* transcriptome. As expected, the number of repeat matches was higher in the shorter length mappings and we suspect that most of these mappings were later designated as non-matches as the mapping length increased. However, these repeat matches may be useful in future analysis of repeat regions in *Pachycladon*.

3.2. *de novo Assembly*

A single FASTA file of 40 million 35-mers was used as input for the assembly with Velvet [10]. Assemblies were performed independently with k-mers having the values 15, 17, 19, 21, 23 and 25 using default parameters and all assembled contigs being returned (Figure 3), thereby covering the range of the ELAND analyses (for computational reasons Velvet only allows odd numbered k-mers). Different k-mer lengths were tested as we were unsure as to how duplicate copies of the *Pachycladon* ESTs would affect *de novo* assembly.

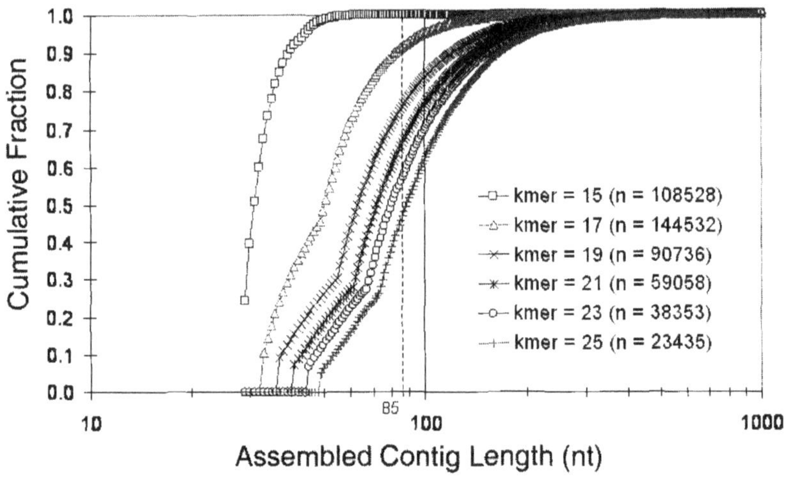

Figure 3 - Graph of *de novo* assembly results at different k-mer sizes. The length of contigs assembled under different k-mers is plotted as a cumulative fraction (the number of contigs generated is shown in the key). The dashed line at 85 nt shows the contig length cutoff that was used for further analyses (see text for details).

It can be seen from Figure 3 that the longer k-mer values resulted in longer assembled contigs but fewer of them, and that a k-mer of 15 gave very different results to all the other k-mers. Because of the previous ELAND results (Figure 2) a k-mer size of 19 was selected as optimal for the *de novo* assemblies. The resultant contigs were converted to a tab delimited form and loaded into a MySQL database in a way that kept the k-mer length as a searchable variable. The number of contigs generated and how they match to *A. thaliana* using BLAST is shown in Table 2. Contigs greater than 85 nt in length were then converted into a FASTA file. A cut-off of 85nt was chosen as it was a length where a reasonable fraction of all contigs made with a k-mer greater than 19 would be represented (34.4% of all contigs: Table 2) and analysis with all contigs becomes difficult to manage due to larger numbers of multiple lower scoring hits. From Table 2 it can be seen that the maximal number of contigs made is with a k-mer of 19, and that about 74% of all contigs have BLAST hits with a bit-score greater than 40.

Table 2 – *de novo* assembly results for contigs greater than 85 nt.

K-mer Size	Number of Contigs	BLAST hits	Number of Contigs with BLAST hits	% Contigs with BLAST hits	Number of genes hit		
					All alignments	Alignment >40 nt	Alignment >85 nt
15	2	3	1	50.00	3	2	0
17	12638	18844	8780	69.50	10579	9312	6786
19	22631	38535	16413	72.50	14906	13105	10304
21	20531	39554	15227	74.20	14594	12267	9504
23	16873	34486	12739	75.50	13209	10732	8222
25	12750	27360	9751	76.50	10871	8852	6595
Total	**85425**		**62911**	**73.65**			

To give an initial indication on the lengths of the resultant BLAST hits, the number of genes was calculated for all contigs irrespective of length, or where they were at least 40 or 85 nt long. Again the k-mer of 19 gives the highest number of genes hit (10304) with a contig length of at least 85 nt (Table 2). At this stage although we used *de novo* assembly, we were not attempting to completely assemble the *Pachycladon* EST transcriptome, but analyze sections of it for future experimentation.

3.3. *Gene Analysis*

The *Pachycladon* total contigs file as described in Table 2 (containing 85425 sequences), was BLASTed against the concatenated *A. thaliana* ESTs. The output was subsequently parsed with a filtering script to remove low bit-score values (less than 40) and to convert the remaining hits into a tab delimited format. To be conservative we used the set of *Pachycladon* contigs (assembled with a k-mer of 19) that mapped uniquely to a given *A. thaliana* EST, resulting in sequence alignment information for 4283 putative *Pachycladon* genes. The original *Pachycladon* contigs file (85425 sequences) was indexed using BLAST v2.16 [11] to convert it into a BLAST database and subsequently BLASTed against itself. These results were then used for duplicate gene analysis. Of the 4283 potential *Pachycladon* genes, 1155 showed evidence of overlap between contigs that mapped to a corresponding *A. thaliana* EST. The distribution of the amount of overlap for these *Pachycladon* genes is shown in Figure 4. We find 141 *Pachycladon* genes with a >100 nt overlap and 9 genes with a >300 nt overlap. These longer cases will be used in SNP and QTL analyses. One example is shown in Figure 5 where possible SNPs can be seen in the alignment. These SNPs are potentially useful for QTL analysis or estimates of genome divergence times.

We found another dataset of contigs that did not match to the *A. thaliana* ESTs but instead matched to other *Pachycladon* contigs. These contigs may represent gene copies which are different from *A. thaliana* and could be indicative of more recent duplications. Further analysis will be required to determine if this is the case. A sample of genes was viewed and analyzed in more detail for evidence that they are possibly duplicate genes from *Pachycladon*.

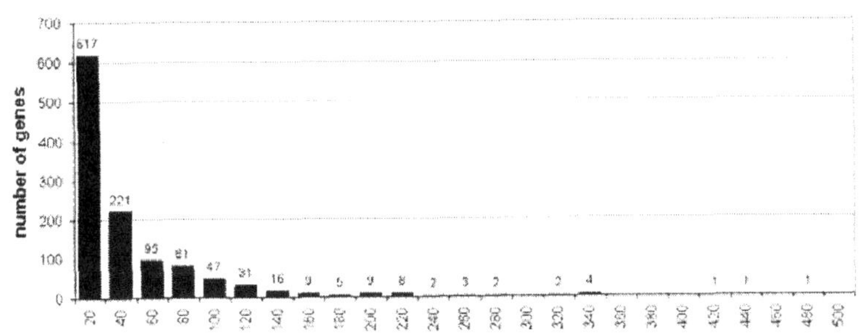

Figure 4 - Graph showing the number of *A. thaliana* ESTs to which *Pachycladon* contigs mapped with any overlap (bin size = 20 nt). For example, *Pachycladon* contigs were mapped 617 *A. thaliana* ESTs with an overlap of 1 to 20 nucleotides.

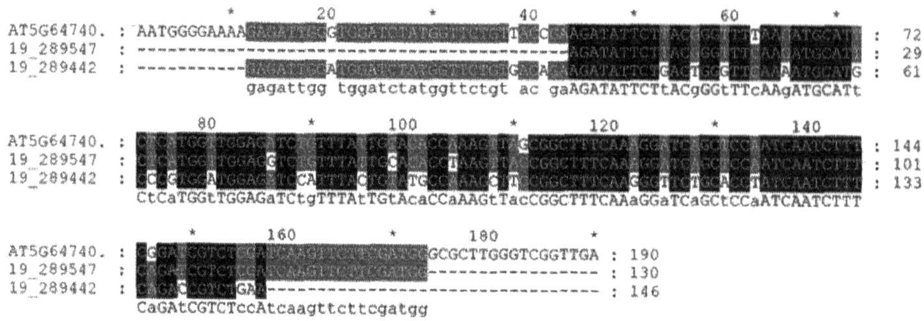

Figure 5 - Alignment of contigs from one of the duplicate *Pachycladon* genes against the equivalent *A. thaliana* gene (AT5G64740.1 also known as CESA6 or Cellulase Synthase 6). The darker the shading, the higher the conservation between sequences and the consensus sequence is given below the alignment (absolutely conserved positions in upper case and variable positions in lower case). SNPs between the *Pachycladon* copies as well as between *Pachycladon* and *A thaliana* can be seen.

To summarize; using an optimal ELAND mapping parameter of 19 nt, 37% of the *Pachycladon* short-reads mapped uniquely to the distant reference genome of 33122 *A. thaliana* ESTs. The 33122 *A. thaliana* ESTs correlate to 28152 gene loci, of which 24292 have only a single transcript (singletons). Our results uniquely mapped to 22438 of these singleton genes (92.4%). From a total of 40 million *Pachycladon* short-read sequences, 85425 contigs were assembled *de novo* under a variety of assembly conditions. 22631 *Pachycladon* contigs were assembled under the optimal assembly parameter k-mer = 19. BLAST results with the assembled contigs identified 4283 potential *Pachycladon* genes that matched *A. thaliana* ESTs. 1155 of these *Pachycladon* genes (27%) indicated some measure of *de novo* contig overlap which will enable future duplicate gene SNP and QTL analysis.

4. Discussion

Next-generation sequencing is an emerging technology that produces millions of short-read sequences and opens the way to rapid genome analysis of non-model organisms. Although the molecular biology and mechanics of this type of sequencing are well commercialized, the bioinformatics and especially practical downstream genomic approaches are not. Researchers receiving short-read output do have software tools available for mapping and *de novo* assembly but little guidance on how to apply them. Given the high data volumes of short-read sequencing, methods which in the past worked well for traditional Sanger sequencing, may fail especially for non-model genomes. Our approach was successful in showing that a distantly related reference genome could be used for mapping and for duplicate gene analysis. Other duplicate genes, not mapped to an *A. thaliana* equivalent were found after *de novo* assembly of contigs and comparison to the contig dataset. Although this was preliminary analysis of the *Pachycladon* short-read data, we gained valuable information leading to SNP detection between the duplicate copies (and *A. thaliana* where appropriate).

The *Pachycladon* transcriptome project posed problems not only due to the non-model nature of the genome, but results had the potential to be complicated by the polyploid nature of the genome. Having multiple copies of a gene in a genome (i.e. paralogy/gene families) is common, as polyploidy (having multiple copies of a genome) is extremely common in plants. Although our approach was ultimately targeted for the finding of near exact gene copies we cannot rule out that some copies may be from recent paralogous events. This again requires further research.

We found during the course of the *Pachycladon* analysis that the viewing of data was essential to understanding the genomic issues we faced. Using Gbrowse we were able to view the potentially duplicated genes as they mapped to *A. thaliana* and evaluate the consistency in nucleotide differences seen in each gene copy. It can also be used to connect data from other sources such as prior experiments. The use of longer reads from for example, the FLX-454 sequencing platform (Roche) can only enhance both the mapping and *de novo* aspects of our approach and this is planned for future work. Many *de novo* assemblers can now use a mixture of short and longer sequences.

A key part of our approach consists of testing parameters on a single lane of data prior to complete analysis. This is essential in situations where simulations cannot be done prior to the sequencing run. The basic idea of testing subsets of data to determine mapping and *de novo* assembly parameters can be applied to other applications using short-read output especially when even the analysis of a single lane of data takes an extraordinary length of time. Researchers at present only have a limited amount of software that can reliably handle large short-read datasets. As more software becomes available the same principle of testing parameters in these cases should of course apply.

In conclusion, we show that even though these are in fact early days in the use of high-throughput short-read sequencing technology, we can move beyond the analysis of the few model or well sequenced genomes and into the larger world of biological organisms and systems.

Acknowledgments

The authors would like to thank Peter Lockhart and the *Pachycladon* transcriptome Project team for the use of the Solexa-output, and the Genome Sequencing Facility at the Allan Wilson Centre, especially Lorraine Berry, Tim White and Maurice Collins. This work was funded by the Allan Wilson Centre and the New Zealand Marsden Fund. Claudia Voelckel holds a Feodor-Lynen Fellowship from the Alexander von Humboldt Foundation, and Simon Joly holds a post-doctoral fellowship from the National Sciences and Engineering Research Council of Canada. The authors would also like to thank Peter Lockhart and David Penny for valuable reading of this manuscript.

References

1. G. Robertson, M. Hirst, M. Bainbridge, M. Bilenky, Y. Zhao, T. Zeng, G. Euskirchen, B. Bernier, R. Varhol, A. Delaney, N. Thiessen, O.L. Griffith, A. He, M. Marra, M. Snyder, and S. Jones, Genome-wide profiles of STAT1 DNA association using chromatin immunoprecipitation and massively parallel sequencing. *Nat Methods*,4:651-7,2007.
2. N.L. Hiller, B. Janto, J.S. Hogg, R. Boissy, S. Yu, E. Powell, R. Keefe, N.E. Ehrlich, K. Shen, J. Hayes, K. Barbadora, W. Klimke, D. Dernovoy, T. Tatusova, J. Parkhill, S.D. Bentley, J.C. Post, G.D. Ehrlich, and F.Z. Hu, Comparative genomic analyses of seventeen Streptococcus pneumoniae strains: insights into the pneumococcal supragenome. *J Bacteriol*,189:8186-95,2007.
3. J.C. Vera, C.W. Wheat, H.W. Fescemyer, M.J. Frilander, D.L. Crawford, I. Hanski, and J.H. Marden, Rapid transcriptome characterization for a nonmodel organism using 454 pyrosequencing. *Mol Ecol*,17:1636-47,2008.
4. L.D. Stein, C. Mungall, S. Shu, M. Caudy, M. Mangone, A. Day, E. Nickerson, J.E. Stajich, T.W. Harris, A. Arva, and S. Lewis, The generic genome browser: a building block for a model organism system database. *Genome Res*,12:1599-610,2002.
5. http://maq.sourceforge.net
6. R. Li, Y. Li, K. Kristiansen, and J. Wang, SOAP: short oligonucleotide alignment program. *Bioinformatics*,24:713-4,2008.
7. R.L. Warren, G.G. Sutton, S.J. Jones, and R.A. Holt, Assembling millions of short DNA sequences using SSAKE. *Bioinformatics*,23:500-1,2007.
8. W.R. Jeck, J.A. Reinhardt, D.A. Baltrus, M.T. Hickenbotham, V. Magrini, E.R. Mardis, J.L. Dangl, and C.D. Jones, Extending assembly of short DNA sequences to handle error. *Bioinformatics*,2007.
9. J.C. Dohm, C. Lottaz, T. Borodina, and H. Himmelbauer, SHARCGS, a fast and highly accurate short-read assembly algorithm for de novo genomic sequencing. *Genome Res*,17:1697-706,2007.
10. D. Zerbino, and E. Birney, Velvet: Algorithms for De Novo Short Read Assembly Using De Bruijn Graphs. *Genome Res*,2008.
11. S.F. Altschul, T.L. Madden, A.A. Schaffer, J. Zhang, Z. Zhang, W. Miller, and D.J. Lipman, Gapped BLAST and PSI-BLAST: a new generation of protein database search programs. *Nucleic Acids Res*,25:3389-402,1997.
12. http://www.arabidopsis.org

Factoring local sequence composition in motif significance analysis

Patrick Ng, Uri Keich*

Department of Computer Science, Cornell University, Ithaca, NY, USA 14853

We recently introduced a biologically realistic and reliable significance analysis of the output of a popular class of motif finders [16]. In this paper we further improve our significance analysis by incorporating local base composition information. Relying on realistic biological data simulation, as well as on FDR analysis applied to real data, we show that our method is significantly better than the increasingly popular practice of using the normal approximation to estimate the significance of a finder's output. Finally we turn to leveraging our reliable significance analysis to improve the actual motif finding task. Specifically, endowing a variant of the Gibbs Sampler [18] with our improved significance analysis we demonstrate that de novo finders can perform better than has been perceived. Significantly, our new variant outperforms all the finders reviewed in a recently published comprehensive analysis [23] of the Harbison genome-wide binding location data [9]. Interestingly, many of these finders incorporate additional information such as nucleosome positioning and the significance of binding data.

Keywords: motif significance analysis; 3-Gamma approximation; local GC-content; Harbison dataset.

1. Introduction

Much of the recent progress in the area of motif finding can be attributed to leveraging additional pieces of data that are increasingly becoming available. These include quantitative binding assays (p-values) from ChIP-on-chip technology ([5], [32], [11], [8]), phylogenetic ([17] [33] [31] [21]), transcription factor structural class ([24], [22]), and nucleosome positioning information [23]. It has been convincingly demonstrated that finders incorporating such additional information can significantly outperform de novo finders† ([24], [23]). It is therefore somewhat surprising that we can report here on a de novo motif finding tool that outperforms all other finders reviewed in a recently published comprehensive analysis [23] of the Harbison genome-wide binding location data [9]. We stress that many of those finders incorporate additional data as described above suggesting that de novo finders can perform significantly better than has been perceived.

Local base composition has long been taken into consideration in sequence analysis. For example, isochores are taken into account in the GENSCAN gene finding

*to whom correspondence should be addressed

†A de novo motif finder is one that uses only the given sets and possibly a null reference set.

tool [4]. A considerable effort was made into incorporating sequence composition in pairwise local alignment significance analysis (e.g., [1]). Another example is the motif finder NestedMICA incorporating a "mosaic background" model. The latter is a mixture of several, differently parametrized, low order, Markov chains which allow one to factor in local composition [7]. Regardless of whether or not our finder incorporates such mixture models, we argue here that the local composition should be taken into account when analyzing the significance of its results. Intuitively, imagine a set of sequences containing stretches made only from A. In this case a motif such as AAAAAAAA should not be too surprising.

A reliable significance evaluation should be considered an essential component of any motif finder. Indeed, it is often the only information available to the users before they decide on whether to invest significant resources in further exploration or verification of the reported motifs. We recently introduced a reliable method to estimate "confidence" p-values from a small sample of the empirical null distribution of a motif finder's results [16]. In this paper, we naturally extend our confidence p-value approach to incorporate local base composition information. As the original confidence p-value estimate was rather robust and applicable to a wide range of finders and scoring schemes, we expect this extension to be fairly widely applicable as well. We demonstrate the ability of our local composition aware significance evaluation to reliably predict significant motifs in real biological setting.

Our confidence p-values are derived assuming the finder's null score follows a 3-parameter Gamma, or 3-Gamma, distribution‡ [16]. An often used alternative in this context is to derive the p-value using a point estimator assuming a normal distribution (e.g., [19], [9], [21], [23]). We provide multiple evidence that such an estimation tends to inflate the significance of the reported motif. In particular, using an FDR analysis [3] we show that our p-values are significantly better calibrated than the normal derived ones mentioned above.

Finally, we leverage our significance analysis to improve de novo motif finding. Specifically, we introduce GibbsMarkov [26], a new variant of the Gibbs Sampler [18], which relies on our p-values to choose between multiple suggested motifs of different widths. The result is a de novo finder that attains the surprising results mentioned above.

2. Factoring local base composition in motif significance analysis

2.1. *Background: 3-Gamma and the finder's null distribution*

In [25] we argue that the finder's null distribution is well suited for estimating the significance of a finder's output. This null distribution is defined as the distribution of the score of the finder on a randomly drawn set, generated for example by resam-

‡The distribution function of a 3-parameters Gamma with $\theta = (a, b, \mu)$ is a given by $F_\theta(s) = F_{\Gamma(a,b)}(s - \mu)$ where $F_{\Gamma(a,b)}$ is the Gamma distribution with it usual shape and scale parameters and μ is the location parameter [14].

pling a large genomic file. Note that this distribution varies not only with the null model that generates the dataset (including the set's dimensions), but also with the parameters of the finder (e.g., width). Since there are typically infinitely many combinations of these problem-parameters (finder and dataset) it is impossible to precompute this distribution.

For any specific set of problem-parameters we can approximate the finder's null distribution with an empirical null distribution. The latter is obtained by applying the motif finder to a sample of randomly drawn null sets. Increasing the sample size improves the quality of our approximation but at a significant cost: each new sample point essentially takes as much running time as the original run whose significance we are trying to estimate. Thus, using this non-parametric approach to reliably estimate small p-values, as we often need to when correcting for multiple hypotheses, is typically forbiddingly expensive (e.g. Harbison dataset has over 300 experiments [9]).

If, however, we know that the finder's null distribution can be well approximated by some parametric family then we only need to estimate these parameters. While the normal distribution is often used in this context ([19] [9] [21] [23]), we find that it consistently offers a relatively poor approximation to the finder's null distribution. In particular, using the normal approximation tends to inflate the significance of high scores which are the ones we are interested in (see Figure 2 below). Instead we find that the 3-parameter Gamma [14], or 3-Gamma for short, appears to fit very well the empirical null distribution for many combinations of motif finders and null models including the biologically realistic, genomic resampling (see Figure 2).

The parameters of the (Gumbel EVD) distribution of the optimal pairwise ungapped local alignment can be computed analytically [15] based on the theory of [6]. In our case the problem is complicated further by the dependence on the finder: our null distribution is of the *finder's* optimal score rather than the optimal alignment score [25]. Thus, it remains a challenging open problem whether a theory can be developed to estimate the parameters of the 3-Gamma from those of the problem. In the meantime we can resort to parametric statistical estimation. For example, suppose we want to estimate the p-value of the observed score s, denoted by $p(s)$. We can generate a *small* sample $X = (X_1, \ldots, X_n)$ from the finder's null distribution and find the 3-Gamma MLE (maximum likelihood estimator) $\hat{\theta} = \hat{\theta}(X)$. We can then find the MLE of $p(s)$, $\hat{p}(s) = \hat{p}(s, X)$, by using the popular plug-in method: $\hat{p}(s) = 1 - F_{\hat{\theta}}(s)$, where F_θ is the 3-Gamma CDF (cumulative distribution function).

As noted in [16] for a realistically small sample size such as $n = 20^§$, $\hat{p}(s)$ can grossly over-estimate the significance of the observed score s. This type of MLE estimation, albeit using the normal approximation, is used in ([19] [9] [21] [23]). We suspect that it further inflated the significance of the observed scores beyond that due to the selection of the normal approximation (see Figure 1 and Section 3.2

§A sample of size n increases the runtime by a factor of n.

for evidence).

Our conservative "confidence p-value", $\hat{p}_c(s, X)$, presented in [16] corrects the tendency of the point estimator $\hat{p}(s)$ to over-estimate the 3-Gamma p-value, $p(s)$. It does so by constructing a confidence interval for the estimated $p(s)$. In principle, the confidence p-value can be applied whenever the 3-Gamma distribution is expected to offer a reasonably good fit to the finder's null distribution.

Fig. 1: Comparing the estimators $\hat{p}_n$ and $\hat{p}_c(s, X)$ of p-value$= 10^{-3}$

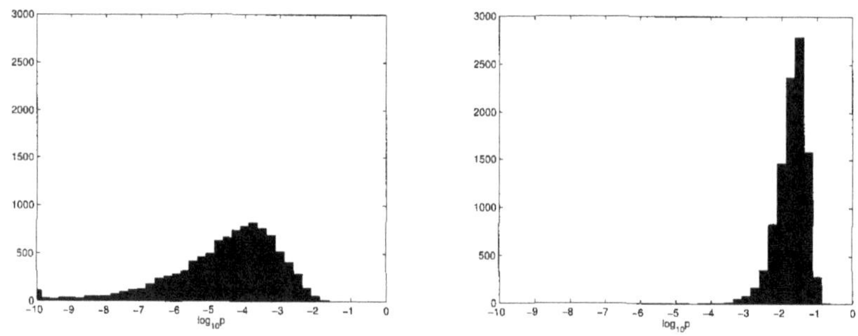

(a) $\hat{p}_n$ overestimates the significance (b) $\hat{p}_c(s, X)$ is mostly conservative

Histograms of 10^4 independent evaluations of the point estimator $\hat{p}_n(s)$and of the conservative $\hat{p}_c(s, X)$, where s was set to the empirical 0.999 quantile. $\hat{p}_n$ is the MLE plug-in estimator of the p-value assuming a normal approximation, and $\hat{p}_c(s, X)$ is our conservative "confidence p-value" assuming a 3-Gamma distribution. The quantile s was learned from the scores of GibbsMarkov on 10,000 resampled sets of 30 sequences each of length 1,000. The resampling was done from the human genomic file. This set of null scores was then used to create the 10,000 resamples X of size $n = 20$ drawn with repetitions. An ideal estimator of $p(s)$ should have all the mass concentrated on the point -3 because s was set to the 0.999 quantile. It is clear from the graphs that $\hat{p}_n$ has a considerably larger variance than $\hat{p}_c$ and that it can badly over-estimate the significance of the score s. GibbsMarkov was run in OOPS mode with the parameters `-l 23 -gibbsamp -best_ent -t 170 -L 100 -em 0 -markov 3 -p 0.10`. Statistical estimations were done in R [28].

2.2. *Incorporating local GC content in our confidence p-value*

We can factor local, or any other, composition information in our significance analysis in a rather straightforward manner. In principle, all we need to do is to condition our generated random sets on the relevant set of constraints. If the null distribution of the finder's score on these conditioned sets can be well approximated by the 3-Gamma distribution, then our confidence p-value method should be valid. Having no theory that could justify this approximation we resort to the empirical studies as we previously did. Indeed, we can simply think of our conditional generating model described below as just another null set generator. Figure 2 below compares the normal with the 3-Gamma approximation of such a conditional empirical null distribution.

Technically, our local GC-content adjusted resampling is done as follows. We first divide our genomic reference file into partially overlapping windows of a fixed size L (overlap size is $L/2$). We then place each window in one of K bins that uniformly cover the entire spectrum of GC-content. This preprocessing step need only be done once. Given an input set we generate local GC-content adjusted resampled images of it as follows. We first divide each sequence into non-overlapping windows of size L and determine their GC-content. We then replace each of the original windows

Fig. 2: Approximating a finder's null distribution conditioned on local GC-content

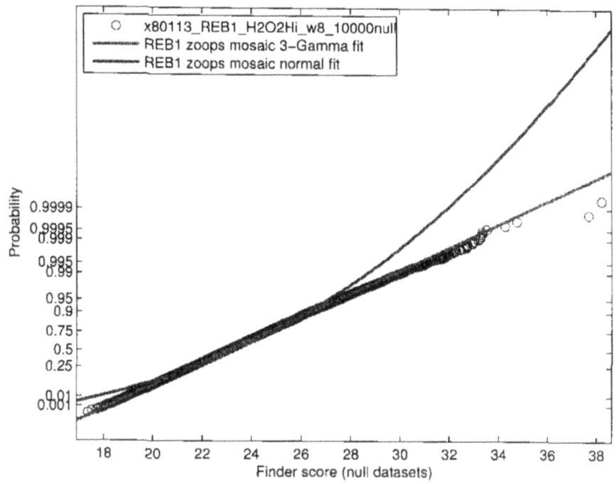

The figure demonstrates the difference between the quality of the normal and the 3-Gamma approximations to a finder's null distribution. In this example, GibbsMarkov was applied to 10,000 sets of GC-content adjusted resampled sequences ($L = 100, K = 20$). The sequences were resampled from the *S. cerevisiae* intergenic file. The mold, or input, set was the Harbison REB1_H2O2Hi dataset consisting of 48 sequences of average length 431bp [9]. The 3-Gamma seems to offer a reasonably good fit for this conditional null distribution while the normal does not. GibbsMarkov was run in ZOOPS mode with the parameters `-l 8 -gibbsamp -p 0.05 -best_ent -cput 300 -L 200 -em 0 -markov 5 -r 1 -ds -zoops 0.2`

with a randomly drawn genomic window from the appropriate bin. Note that within a set we draw windows without replacement as repetitive elements can wreak havoc on motif finding. For the same reason we exclude overlapping windows within a set. The same kind of exclusion applies to our "uniform" resampling strategy.

Does factoring local GC content make a difference in the significance analysis? We give two different types of evidence that it does. First, Figure 3 compares histograms of our GibbsMarkov run on null sets that were generated according to the two models we are comparing. One model was generating sets using uniform resampling of a *S. cerevisiae* intergenic file while the other was using the local GC content framework described above. Notice that the two histograms are distinctly different. For example, a score whose p-value, when factoring in local GC content, is 0.0002 has a p-value of only 0.001 when assuming the uniform model.

As we just saw, taking into account the local GC-content can considerably impact the significance of an observed score s. Our original construction of the confidence p-value [16] did not account even for the *global* base composition of the sample as outlined above. Indeed we followed the common procedure of resampling a relevant genomic file. To demonstrate the potential difference between such a naive approach and our local GC-content adjusted one we devised the following experiments. This experiment is realistic in the sense that it emulates a real problem we encountered when analyzing DNA replication origins in *Saccharomyces kluyveri*. We first generated 200 random datasets by resampling from our human genomic file (see Section 4). To make these sequences look closer to the *S. kluyveri* sequences we were analyzing, we accepted only sequences whose AT-content is above 65%. We then implanted in each sequence exactly one site generated from the *Saccharomyces*

Fig. 3: Comparing the uniform and the local composition aware null generators

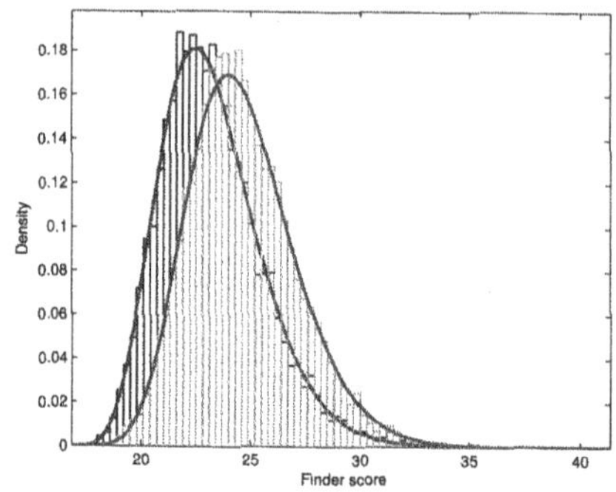

The data for "right" histogram was generated by applying GibbsMarkov to 10,000 sets that were resampled uniformly from the *S. cerevisiae* intergenic file. The "left" histogram was generated using the same local GC-content preserving scheme as described for Figure 2. To highlight the difference both histogram were ML-fitted with a 3-Gamma distribution. GibbsMarkov was run in ZOOPS mode with the parameters `-l 8 -gibbsamp -p 0.05 -best_ent -cput 300 -L 200 -em 0 -markov 5 -r 1 -ds -zoops 0.2`

cerevisiae AT-rich ACS profile (see Figure 4)¶. We next ran our GibbsMarkov in OOPS (one occurrence per sequence) mode on each of these 200 datasets, and noted the score, as well as whether or not the finder succeeded in uncovering the implanted ACS motif. Finally, we computed confidence p-values for each of these 200 scores in two different ways. The first was derived from our previous approach of uniform genomic resampling‖. The second was derived from the new local GC-content preserving resampling scheme. Table 1 summarizes the results. Notably, the latter identifies 50% more TPs. The FPs are under control in both cases as expected.

Fig. 4: The ACS motif

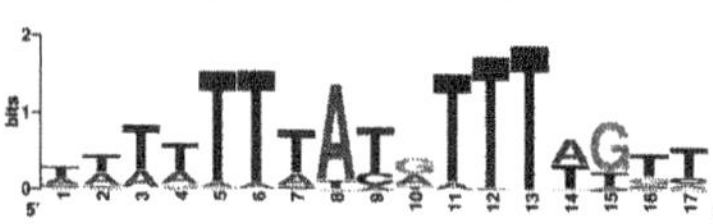

3. Results on the Harbison dataset

All the tests below refer to the Harbison dataset of 310 ChIP-chip, genome-wide location analysis, experiments of 203 yeast transcription factors [9]. By the "Narlikar test" we refer to the dataset consisting of the 156 sequence-sets from 80 TFs used in [23]. The literature consensus for each of these 80 TFs is published. We obtained these from [9], with the exception of DAL82, RTG1, and the modified CIN5 which we took from [21]. By the "MacIsaac test" we refer to the dataset consisting of 188 sequence-sets which include all 124 TFs whose matrices are reported in [21]. See

¶The ACS is a 17bp site to which the *S. cerevisiae* ORC (origin recognition complex) binds to initiate local chromosomal replication [29]. We expect its *S. kluyveri* analogue to be somewhat similar.

‖For technical reasons we used the same human genomic file which has roughly the same AT-level as that of *S. kluyveri*.

Table 1: The effect of base composition on significance analysis

p-value threshold	TP	TN	FP	FN
0.1	26/49	78/77	0/1	96/73
0.05	21/33	78/78	0/0	101/89

The first number in each entry is the number of sets (out of 200) for which the p-value derived from sets generated by a uniform genomic resampling (57% AT-content). The second number is for the locally adjusted p-value. Notably, the latter identifies 50% more TPs. The overall high number of FNs is partly due to the conservative nature of the confidence p-value and partly due to the fact that these sets were designed as twilight zone ones [25].
Each of the 200 implanted sets consists of 30 sequences of length 2500 resampled from the human genomic file conditional on having an AT-content $\geq$ 65%. Each sequence was implanted with exactly one site generated by drawing from the ACS matrix. This ACS matrix (Figure 4) was generated by us from a compiled list of confirmed ARSs on OriDB [27]. GibbsMarkov was run in OOPS mode with the parameters `-l 17 -gibbsamp -p 0.05 -best_ent -cput 300 -L 200 -em 0 -markov 3 -r 1 -s 123`. The confidence p-values were derived from sets resampled in two different ways. Both resampled from our human genomic file but one conditioned the resampling on the local GC-content observed in the input dataset. Note that each one of these 200 input sets had a different local GC-content pattern.

more details in Section 4. In the following analysis our confidence p-values factor in the local GC-content as described in Section 2.2.

3.1. *GibbsMarkov performance on the Narlikar test*

We compared our motif finder GibbsMarkov with results from the Supplementary of [23]. GibbsMarkov with fixed width $w = 8$ was run on the 156 sequence-sets. Using the same definition of success as defined in [23], GibbsMarkov successfully finds the correct motif in 71 of the 156 experiments. This is significantly better than all other *de novo* finders including PRIORITY-N [23] with 57 successes**. The full list which includes many more finders can be found in [23].

3.2. *How well calibrated are these p-values?*

If our p-values are well calibrated then the false discovery rate for any given threshold should be consistent with the rate guaranteed by the theory. To test that we applied the original FDR test [3] to find our p-value cutoff corresponding to an FDR of 5%. We applied this test separately to the p-values we assign to the 156 sets of the Narlikar test and then to the p-values we assign to the 188 sets of the MacIsaac test.

In order to get an accurate classification of predicted motifs, we disregarded motifs where (1) the consensus sequence of the predicted motif is AC-repeat or GT-repeat, and (2) the predicted motif does not match the literature motif but has a statisically significant match to a motif in the MacIsaac set of motifs [21] (see Section 4 for details). Type (1) motifs which we found in ACE2_YPD, AFT2_H2O2Hi, ARR1_YPD, and SWI5_YPD were disregarded because GT-repeats are possibly functional in yeast ([8], [12]). Type (2) motifs were disregarded because TFs often have co-factors that are DNA-binding. Such detected motifs should therefore not be considered false positives as they could still be biologically relevant.

** [23] reports that PRIORITY-N has 51 successes using a slightly different normalization. See Section 4.5.

At a 5% FDR threshold, for the MacIsaac test, our observed FDR comes at about 6.67%: 4/60, while for the Narlikar test it is about 7.41%: 4/54. At a 10% threshold, the observed FDR of the MacIsaac test and Narlikar test were 11.4% and 10.2%, respectively. Hence it is reasonable to conclude that our confidence p-values are well calibrated.

We also looked at the observed FDR of the results of [23] which are based on the normal MLE of the p-value. Their results were already disregarding the GT-repeats (type (1) from above), but we could not disregard possible type (2) motifs because we do not have access to their predicted motifs. At the 5% threshold their observed FDR on the Narlikar test is about 48%: 63/132, which is significantly higher than the expected 5%. For comparsion, we repeated the FDR analysis on our confidence p-value by disregarding only the GT-repeats so that the comparison was on equal footing. At that 5% threshold, our observed FDR comes to about 12%: 7/57.

3.3. *Using the p-values to improve our results*

GibbsMarkov was run with multiple widths on the 156 sets of the Narlikar test, and a single predicted motif among the multiple widths was selected based on our confidence p-values. In the Narlikar test, our results improved from 71 successes with $w = 8$ to 76 with multiple widths. This is better than all other finders although PRIORITY-DN [23] which uses nucleosome positioning information is a close second with 75 successes[††]. The improvement was more significant in the MacIsaac test: the multiple widths method correctly identified 114 motifs while GibbsMarkov using $w = 8$ found only 97 out of 188 sequence-sets.

To test our performance of using confidence p-values for multiple widths selection, we compare it against naively selecting widths according to average entropy. Thus instead of choosing a predicted motif among widths with the best confidence p-value, a prediction is chosen based on average entropy, which is simply the entropy score averaged over the width of a motif. In the MacIsaac test, width selection based on average entropy found 99 while selection based on confidence p-values found 114 as reported above.

We have yet to throughly explore our predictions but one interesting dimer of width 18 caught our eyes. It appears essentially the same in three different experiments: DIG1 Alpha, TEC1 Alpha, and STE12 Alpha (see Figure 5). In all three cases width 18 exhibits the most significant p-value at: 3.7e-15, 1.3e-04, and 7.2e-08 respectively. A closer inspection shows the dimer is made of a repetition of the known motif common to DIG1 and STE12 (see Figure 6). This dimer was recently independently reported in [12].

[††] [23] reports that PRIORITY-DN has 70 successes using a slightly different normalization. See Section 4.5.

Fig. 5: An interesting dimer picked up by GibbsMarkov

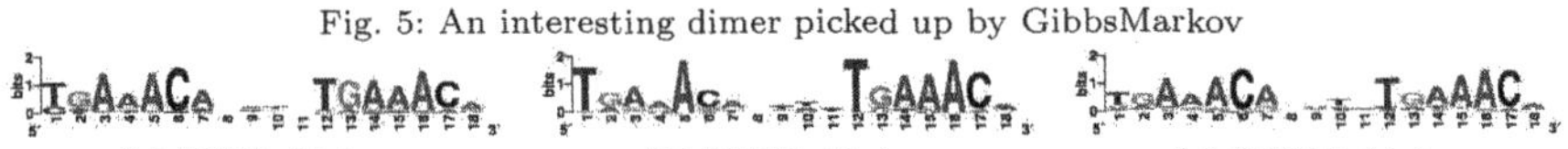

(a) DIG1 Alpha (b) TEC1 Alpha (c) STE12 Alpha

Fig. 6: Known motifs from [21]

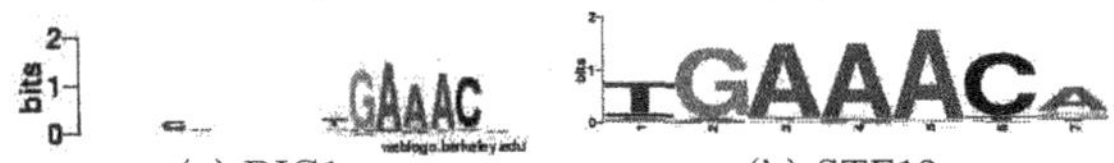

(a) DIG1 (b) STE12

4. Methods

4.1. *Confidence p-value*

All confidence p-values were computed in R [28] using functions described in [16]. The necessary samples were derived from resampled data generated as described in the text.

4.2. *GibbsMarkov*

By GibbsMarkov we refer here to our variant of a Gibbs Sampler finder [18]. Currently it handles an OOPS (one occurrence per sequence) or a ZOOPS (zero or one) model [2]. A detailed account of GibbsMarkov's sampling step and scoring function is described in [26].

4.3. *Genomic files*

For historical reasons we used two genomic files for resampling purposes. In both cases resampling was done by extracting contiguous sequences from a concatenated filtered genomic sequence. The "human genomic" contiguous sequence is from *Homo sapiens* chromosome 1 (HSA1). HSA1 was downloaded from the Ensembl Genome Browser v38 (NCBI build 36) [10]. RepeatMasker, TandemRepeatFinder, and DUST were applied to the data. The *S. cerevisiae* intergenic file was generated by removing from the *S. cerevisiae* genome downloaded from SGD [30] all protein and RNA coding sequences including tRNA, rRNA, snoRNA, snRNA, LTR, and other repetitive sequences.

4.4. *Is the predicted motif a known motif?*

Given a database of known motifs, we would like to determine whether a predicted motif has a statistically significant match to a known motif. For each predicted motif, we first obtained an empirical null distribution of maximal similarity scores (a higher score implies more similar motifs). Each score from this null is the maximal similarity score over all database PFMs against a random permutation of positions/columns of the predicted motif. Then the p-value for similarity is simply estimated from the null distribution described above and the similarity score between the predicted motif and its most similar motif within the database. Note that this technique accounts for evaluating statistical significance at the extreme value case of choosing the most similar motif within the database. In our FDR analysis, the empirical null of each predicted motifs was generated with 10,000 randomly permuted motifs as described above and ignored cases where the predicted motif does not match the literature but has a p-value < 0.05 for similarity.

4.5. *Harbison dataset*

All the consensus sequences were converted to PFM by the same method as [9]. For the MacIsaac tests, we used the same definition of success as defined in [9]. Likewise we used the definition of success defined in [23] for the Narlikar test with fixed width $w = 8$. However, in the last minute of this publication, we found out that the authors of both [9] and [23] made exactly the same typo in the published definition of the inter-motif distance. More precisely, both took the square root of the quadratic form but published it without the square root. It was too late to redo all our analysis which was based on the formula without the root. However to make sure the compared finders are on equal footing we re-evaluated the results in [23] using the previously *published* formula of inter-motif distance (without the square root).

For the Narlikar test with *multiple widths*, we slightly modified the average entropy constraint of inter-motif distance used in [23]. The average entropy of the predicted motif was taken over corresponding non-N positions of the literature consensus within an alignment, because predicted motifs such as GAL4 with literature consensus `CGGnnnnnnnnnnnCCG` should not be penalized for having degenerate positions at consensus positions with `n`.

GibbsMarkov was run with a fifth-order Markovian background estimated from the *S. cerevisiae* intergenic file. The strength of prior parameter in ZOOPS is $\alpha = 0.2$. The finder was allowed to run for 5 minutes with a plateau period of 200 iterations. All experiments were run under Red Hat Enterprise Linux 4 on a cluster with nodes that have AMD 248 2Ghz 64-bit processors with 2GB RAM and 1GB swap. The confidence p-values were computed from applying GibbsMarkov to 50 sequence-sets of local GC-content adjusted resampled sequences ($L = 100, K = 20$). For GibbsMarkov with multiple widths selection, GibbsMarkov parameterized with widths 8, 12, 15, and 18 were run separately on the input sequence-set, and then each were applied separately on the same 50 sequence-sets of local GC-content adjusted resampled sequences.

5. Conclusion & Future Research

We show that incorporating local base composition can improve the fidelity of our recently published confidence p-value method of estimating the significance of a finder's output [16]. We also demonstrate the practical advantage of this improvement over the previous method in identifying true motifs in a realistic experiment. We give evidence that the practice of using a normal approximation to estimate the significance of a finder's output is ill-advised on two counts. First, the normal distribution generally fits the finder's null distribution rather poorly. Second, the normal MLE point estimator of the p-value has a significant bias toward over-estimating the significance of the observed score. To drive home this point we show that the use of this p-value on a real biological dataset creates a FDR which is significantly higher than the stated one. In contrast, a FDR analysis based on our confidence

p-value is much closer to the declared rate. Our evaluation method is based on the validity of the 3-Gamma approximation of the finder's null distribution. As such, it is likely to be applicable to many more finders than the ones explored here.

We also develop GibbsMarkov, a variant of the Gibbs Sampler de novo motif finder. GibbsMarkov outperforms all the finders reviewed in a recent well designed study [23] of the Harbison genome-wide location analysis data [9]. Surprisingly, many of the finders that GibbsMarkov outperforms rely on additional information such as, the confidence of the binding, phylogenetic, and nucleosome positioning information [23]. Moreover, when we choose the best p-value among several Gibbs-Markov runs using different widths, we get a roughly 10% increase in our TP rate.

As far as future issues, we could benefit from a more sophisticated alternative to the window based method that we currently use to track the local GC-content. HMM models naturally fit in this context. Regardless, note that, in principle, our method can be extended to factor any local composition feature that the user might be interested in accounting for. Eventually it all boils down to two things: do we have sufficient data to generate random sets that satisfy the required local conditions and is the associated finder's null distribution well approximated by the 3-Gamma distribution.

Acknowledgement

It is our pleasure to acknowledge Anand Bhaskar for processing the list of MacIsaac matrices. This research uses computational resources funded by NIH grant 1S10RR020889 and is supported by the National Science Foundation Grant No. 0644136 to UK.

References

[1] SF Altschul, et al. Protein database searches using compositionally adjusted substi tution matrices. *FEBS J*, 272(20):5101–5109, Oct 2005.

[2] T Bailey and C Elkan. The value of prior knowledge in discovering motifs with meme. In *Proceedings of the Third ISMB*, pages 21–29, Menlo Park, California, 1995.

[3] Y Benjamini and Y Hochberg. Controlling the false discovery rate: A practical and powerful approach to multiple testing. *JRSS B (Methodological)*, 57(1):289–300, 1995.

[4] C Burge and S Karlin. Prediction of complete gene structures in human genomic DNA. *J Mol Biol*, 268(1):78–94, Apr 1997.

[5] H Bussemaker, H Li, and E Siggia. Regulatory element detection using correlation with expression. *Nat Genet*, 27(2):167–71, Feb 2001.

[6] A Dembo, S Karlin, and O Zeitouni. Limit distribution of maximal non-aligned twosequence segmental score, 1994.

[7] TA Down and TJP Hubbard. Nestedmica: sensitive inference of over-represented motifs in nucleic acid sequence. *Nucleic Acids Res*, 33(5):1445–1453, 2005.

[8] E Eden, D Lipson, S Yogev, and Z Yakhini. Discovering motifs in ranked lists of dna sequences. *PLoS Comput Biol*, 3(3):e39, Mar 2007.

[9] CT Harbison et al. Transcriptional regulatory code of a eukaryotic genome. *Nature*, 431(7004):99–104, 2004.

[10] E Birney et al. Ensembl 2006. *Nucleic Acids Res*, 34:D556–61, Jan 2006.

[11] BC Foat, AV Morozov, and HJ Bussemaker. Statistical mechanical modeling of genome-wide transcription factor occupancy data by matrixreduce. *Bioinformatics*, 22(14):e141–e149, Jul 2006.
[12] N Habib, T Kaplan, H Margalit, and N Friedman. A novel bayesian dna motif comparison method for clustering and retrieval. *PLoS Comput Biol*, 4(2), Feb 2008.
[13] S Jensen, X Liu, Q Zhou, and J Liu. Computational discovery of gene regulatory binding motifs: a bayesian perspective. *Statistical Science*, 19(1):188–204, 2004.
[14] NL Johnson, S Kotz, and N Balakrishnan. *Continuous Univariate Distributions, 2nd edition.* Wiley Series in Probability and Statistics, 1994.
[15] S Karlin and S Altschul. Methods for assessing the statistical significance of molecular sequence features by using general scoring schemes. *PNAS*, 87(6):2264–8, Mar 1990.
[16] U Keich and P Ng. A conservative parametric approach to motif significance analysis. In *The 18th International Conference on Genome Informatics*, Singapore, 2007.
[17] M Kellis et al. Sequencing and comparison of yeast species to identify genes and regulatory elements. *Nature*, 423(6937):241–254, 2003.
[18] C Lawrence, et al. Detecting subtle sequence signals: a Gibbs sampling strategy for multiple alignment. *Science*, 262(5131):208–14, Oct 1993.
[19] X Liu, DL Brutlag, and JS Liu. BioProspector: discovering conserved DNA motifs in upstream regulatory regions of co-expressed genes *PSB*, 127–38, 2001.
[20] JS Liu, A Neuwald, and C Lawrence. Bayesian models for multiple local sequence alignment and gibbs sampling strategies. *J. Amer. Stat. Assoc.*, 90:1156–1169, 1995.
[21] KD Macisaac, et al. An improved map of conserved regulatory sites for saccharomyces cerevisiae. *BMC Bioinformatics*, 7(1), March 2006.
[22] AV Morozov and ED Siggia. Connecting protein structure with predictions of regulatory sites. *Proc Natl Acad Sci U S A*, 104(17):7068–7073, Apr 2007.
[23] L Narlikar, R Gordân, and AJ Hartemink. Nucleosome occupancy information improves *de novo* motif discovery. In *RECOMB*, pages 107–121, 2007.
[24] L Narlikar, R Gordân, U Ohler, and AJ Hartemink. Informative priors based on transcription factor structural class improve *de novo* motif discovery. In *ISMB (Supplement of Bioinformatics)*, pages 384–392, 2006.
[25] P Ng, N Nagarajan, N Jones, and U Keich. Apples to apples: improving the performance of motif finders and their significance analysis in the Twilight Zone. *Bioinformatics*, 22(14):e393–401, Jul 2006.
[26] P Ng and U Keich. GIMSAN: A Gibbs motif finder with significance analysis. *Bioinformatics. In press.*, 2008.
[27] CA Nieduszynski, et al. Oridb: a dna replication origin database. *Nucleic Acids Res*, 35:D40–D46, Jan 2007.
[28] R Development Core Team. *R: A Language and Environment for Statistical Computing.* R Foundation for Statistical Computing, Vienna, Austria, 2006.
[29] RA Sclafani and TM Holzen. Cell cycle regulation of dna replication. *Annu Rev Genet*, 41:237–280, 2007.
[30] SGD project. Saccharomyces genome database. http://www.yeastgenome.org/.
[31] R Siddharthan, ED Siggia, and EV Nimwegen. PhyloGibbs: a Gibbs sampling motif finder that incorporates phylogeny. *PLoS Comput Biol*, 1(7):e67, Dec 2005.
[32] A Tanay. Extensive low-affinity transcriptional interactions in the yeast genome. *Genome Res*, 16(8):962–72, Aug 2006.
[33] T Wang and GD Stormo. Combining phylogenetic data with co-regulated genes to identify regulatory motifs. *Bioinformatics*, 19(18):2369–80, Dec 2003.

A New Model of Multi-Marker Correlation for Genome-Wide Tag SNP Selection

WEI-BUNG WANG[1] TAO JIANG[1]
weiw@cs.ucr.edu jiang@cs.ucr.edu

[1] *Department of Computer Science, University of California - Riverside*

Tag SNP selection is an important problem in computational biology and genetics because a small set of tag SNP markers may help reduce the cost of genotyping and thus genome-wide association studies. Several methods for selecting a smallest possible set of tag SNPs based on different formulations of tag SNP selection (block-based or genome-wide) and mathematical models of marker correlation have been investigated in the literature. In this paper, we propose a new model of multi-marker correlation for genome-wide tag SNP selection, and a simple greedy algorithm to select a smallest possible set of tag SNPs according to the model. Our experimental results on several real datasets from the HapMap project demonstrate that the new model yields more succinct tag SNP sets than the previous methods.

1. Introduction

Single nucleotide polymorphisms (SNPs) represent the most frequent form of genetic variations in the human genome. They play an important role in genome-wide association studies that intend to help us understand the correlation between genetic variations and human diseases. Assaying (or genotyping) all SNP markers in the involved genomes would be desirable, but it is expensive and unnecessary. Since SNPs are often not independent, a subset of SNPs may be sufficiently informative and allow us to infer all the other SNPs. The *tag SNP selection* problem is thus to find a smallest possible set of tag SNPs that would enable us to infer all the other SNPs with a certain level of confidence [9]. Clearly, the smaller the tag SNP set, the more genotyping cost it could help save.

Two frameworks for tag SNP selection have been studied in the literature: block-based and genome-wide. The block-based tag SNP selection framework focuses on *haplotype* patterns in a population.[a] The approach assumes that the chromosomes can be partitioned into blocks separated by recombination hotspots, so that there are few recombinations within a block. Then it attempts to identify a smallest possible set of tag SNPs for each block so that all the possible halpotype patterns formed by the SNPs in the block can be fully represented by the haplotype patterns formed by the tag SNPs [14]. The genome-wide framework does not partition a

[a]Recall that humans are diploids and our chromosomes form pairs, each of which consists of a paternal chromosome and a maternal chromosome. A haplotype refers to the set of SNPs from a single chromosome. A pair of corresponding paternal and maternal haplotypes form a genotype.

chromosome into blocks. Instead, it considers the correlation between SNP markers across the entire genome [1]. Typically, a SNP marker has two states in a population. The state with a higher frequency is called the *major allele* and the other is called the *minor allele*. In the other words, the SNP markers are usually *bi-allelic*. It is a common practice to consider only SNPs whose *minor allele frequency* (MAF) is at least 5%. Genome-wide tag SNP methods generally follow two approaches. Halldórsson *et al.* [3] define "informativeness" of SNPs and attempt to find the most informative set of SNPs. The other approach, such as the one adopted by Carlson *et al.* [1], usually evaluates the *linkage disequilibrium* (LD) between the states of two SNP markers using the correlation coefficient r^2, which indicates the dependency between the two markers, and aims at finding a smallest set of tag SNPs such that all the other SNPs are strongly linked to the selected tag SNPs in terms of the LD coefficient r^2 (more precisely, each of them is linked to some tag SNPs with an r^2 coefficient above a certain threshold). The tag SNPs selected by this approach are shown to be effective in disease association mapping studies, since the coefficient r^2 is directly related to the statistical power of association mapping. Genome-wide tag SNP selection based on the r^2 LD statistics has gained popularity among researchers in the SNP community [1, 2, 8, 12, 15, 18], because it has a comparable performance at a lower computational cost than many other methods [17, 18]. In this paper, we will be focused on genome-wide tag SNP selection using the r^2 LD statistics.

Most of the existing tag SNP selection methods in this framework consider the r^2 coefficient between a pair of SNP markers [1, 11, 12, 15]. Hence, each of the SNPs is guaranteed to be tagged by a single tag SNP selected. Recently, Hao *et al.* [4, 5] extended the r^2 statistics to describe the statistical correlation between a group of (*e.g.* two or three) markers and another marker. We will simply refer to this as the *multi-marker correlation model*. In this model, a SNP is tagged by a group of tag SNPs if it is correlated to the group with an r^2 coefficient above a certain threshold. Hao *et al.* [4, 5] presented a greedy algorithm for selecting tag SNPs to cover a certain (large) fraction of a given set of SNPs and showed that the multi-marker correlation model is more effective than the traditional pairwise correlation model in terms of reducing the number of required tag SNPs.

In this paper, we generalize the multi-marker correlation model in [4, 5] to further improve its effectiveness. Comparing with the model in [4, 5], our model is more natural and supports more succinct tag SNP sets. We will also present a simple greedy algorithm to select a smallest possible set of tag SNPs according to this multi-marker model, and compare its performance with those of the previous methods on real HapMap data.

Genome-wide tag SNP selection methods can also be classified as *haplotype-based* or *haplotype-independent*, depending on how the r^2 statistics is obtained. For genotype data, the r^2 statistics is usually estimated using a maximum likelihood approach [6, 10], which could be time consuming on a large set of SNPs. However, when phased haplotypes are available, the r^2 coefficients can be calculated very easily and efficiently. The haplotype-based methods require phased haplotype data while the haplotype-independent methods do not. In this work, we will consider both types of data.

The rest of the paper is organized as follows. In Section 2, we introduce a new multi-marker correlation model and discuss how to calculate the r^2 LD statistics under the model for both haplotype and genotype data. Section 3 presents the simple greedy algorithm for selecting tag SNPs. In Section 4, we discuss the implementation of the algorithm and test its performance on some real HapMap datasets. We also compare the performance of our algorithm with those of the most recent algorithms on genome-wide tag SNP selection given in [4, 5, 11]. Section 5 concludes the paper with a few remarks. For the ease of reading, we defer some illustrative figures and a detailed mathematical proof required in the calculation of the multi-marker correlation coefficient r^2 to Appendix A.

2. The New Multi-Marker Correlation Model

In this section, we propose a new multi-marker correlation model that generalizes the model introduced in [4, 5]. We also discuss how to calculate the r^2 statistics under the new model for both haplotype and genotype data.

2.1. *Multi-Marker Correlation on Haplotype Data*

The statistical correlation between a group of k markers and another marker will be referred to as k-marker correlation. For simplicity, we define below the 2-marker correlation model. The generalization of the model to 3 or more markers is straightforward. Consider three bi-allelic SNPs A, B and C. Each of them has possible alleles A/a, B/b and C/c, respectively. Here, the uppercase letters represent both the SNPs as well their major alleles and the lowercase ones represent the minor alleles. Given the states (*i.e.* alleles) of SNPs A and B, it might be possible for us to infer the state of SNP C, if SNP C is correlated with both SNPs A and B. Clearly, if $Pr(C \mid AB) > 0.5$, we would opt to predict the major allele C instead of the minor allele c when the haplotype AB is observed.

For a fixed population of haplotype data and any haplotype h, let n_h denote the number of times that the haplotype h is observed in the population. Consider three SNPs A, B and C again. For each haplotype $h \in \{AB, Ab, aB, ab\}$, if $n_{hC} > n_{hc}$, then we would opt to predict allele C when observing haplotype h (assuming that the SNP C is unassayed). We put all the haplotypes $h \in \{AB, Ab, aB, ab\}$ such that $n_{hC} > n_{hc}$ into a *major bucket* and the others into a *minor bucket*. For example, if $n_{ABC} > n_{ABc}, n_{abC} > n_{abc}$ and $n_{AbC} < n_{AbC}, n_{aBC} < n_{aBc}$, then the major bucket will contain haplotypes $\{AB, ab\}$ while the minor bucket contains haplotypes $\{Ab, aB\}$. This would suggest a prediction of the allele C when any of the haplotypes $\{AB, ab\}$ in the major bucket is observed.

To define the r^2 correlation coefficient, we introduce a new bi-allelic (compound) marker M that combines the SNPs A and B. The major and minor alleles of M are M/m. We say that the marker M is in state (allele) M if any of the haplotypes in the major bucket is observed, or otherwise it is in state m. Hence, the numbers of observations of alleles M and m are defined as $n_M = n_{AB} + n_{ab}$ and $n_m = n_{Ab} + n_{aB}$. We can define the r^2 statistics between the two markers {A,B} and the marker C as the usual r^2 statistics between the new marker M and the marker C.

Occasionally, we may have a tie between haplotype counts in the population, such as $n_{hC} = n_{hc}$. In this case, we would have to decide whether to put the haplotype h in the major bucket or the minor bucket. The following claim shows that it is usually advantageous to put the haplotype in the minor bucket.

Claim 2.1. *Consider three SNP markers with alleles A/a, B/b, and C/c, and the correlation coefficient r^2 between the markers {A, B} and the marker C. If h is an observed haplotype on the markers A and B, and the numbers of observations satisfy $n_{hC} = n_{hc}$, then putting h in the minor bucket leads to a higher r^2 value most of the time.*

Proof. See Appendix A. □

Since there are 4 possible haplotypes on markers A and B, there are $2^4 = 16$ ways to fill the major bucket. After eliminating symmetric ways and the empty set, there are $2^4/2 - 1 = 7$ different ways to separate the 4 possible haplotypes into two buckets. Note that, a split of the four haplotypes like $\{AB, Ab\}/\{aB, ab\}$ really represents the single-marker correlation between markers A and C. Therefore, the seven different separations correspond to two single-marker and five 2-marker correlations.

In [4, 5], Hao *et al.* proposed a very similar 2-marker correlation model to define the correlation between markers {A,B} and marker C. However, they require that one of the buckets must contain exactly one haplotype (unless the split actually represents a single-marker correlation). For example, a split like $\{AB\}/\{Ab, aB, ab\}$ would be allowed but the split $\{AB, ab\}/\{Ab, aB\}$ is not. Therefore, the 2-marker correlation model in [4, 5] allows a total of $2 + 4 = 6$ different splits, two of which correspond to single-marker correlations. Clearly, our new model is more flexible and gives us the opportunity to cover more SNPs with the same set of tag SNPs. Therefore, it may help reduce the number of tag SNPs required. This flexibility is even more obvious when we consider the correlation between a group of three markers and another marker. To infer a fourth SNP D from three SNPs A, B and C, our model allows $2^{2^3}/2 - 1 = 127$ possible splits of the 8 haplotypes on the SNPs A, B, and C into the major and minor buckets (modulo symmetry). However, because the model of Hao *et al.* in [4, 5] requires that one of the buckets must contain exactly one haplotype, it only allows $3 + 3 \cdot 4 + 8 = 23$ different splits, including 3 splits corresponding to single-marker correlations and another 12 corresponding to two-marker correlations.

2.2. *Calculating r^2 Values on Genotype Data*

Obtaining r^2 values from haplotype data is trivial. However, if the SNP data is in the form of unphased genotypes, we cannot obtain r^2 values directly since the above definition is based on haplotype data. There are two ways to deal with genotype data. One is to use some haplotype inference program such as PHASE [13, 16] to convert the genotype data into a haplotype data. The other way is to estimate k-marker haplotype frequencies directly from the population without phasing. The former method is trivial. So, here we discuss the latter method.

Hill [6] proposed in 1974 a maximum likelihood method to estimate the degree of LD between two loci (*i.e.* markers) given the frequencies of diploid genotypes in a random-mating population. Then he generalized the method to estimate haplotype frequencies at several loci in 1975 [7]. This method has been used to estimate LD r^2 statistics for more than 30 years. For example, it was used in [10] to estimate the LD among multi-allelic markers.

Hill's method works as follows. For simplicity, let us only consider estimating the frequency of 3-marker haplotypes. Consider a sample of population data from N random-mating individuals. Let n_g be the number of times that genotype g is observed in the sample. Denote as f_h the frequency of haplotype h. Let $\hat{f}_h$ be the maximum likelihood estimation of f_h. For three SNPs A, B and C, the frequency of haplotype ABC satisfies the following equation (due to Hardy-Weinberg equilibrium):

$$\begin{aligned}\hat{f}_{ABC} = \frac{1}{2N}\Bigg(& 2n_{AABBCC} + n_{AABBCc} + n_{AABbCC} + n_{AaBBCC} \\ & + n_{AABbCc}\frac{\hat{f}_{ABC}\hat{f}_{Abc}}{\hat{f}_{ABC}\hat{f}_{Abc} + \hat{f}_{ABc}\hat{f}_{AbC}} \\ & + n_{AaBBCc}\frac{\hat{f}_{ABC}\hat{f}_{aBc}}{\hat{f}_{ABC}\hat{f}_{aBc} + \hat{f}_{ABc}\hat{f}_{aBC}} \\ & + n_{AaBbCC}\frac{\hat{f}_{ABC}\hat{f}_{abC}}{\hat{f}_{ABC}\hat{f}_{abC} + \hat{f}_{AbC}\hat{f}_{aBC}} \\ & + n_{AaBbCc}\frac{\hat{f}_{ABC}\hat{f}_{abc}}{\hat{f}_{ABC}\hat{f}_{abc} + \hat{f}_{ABc}\hat{f}_{abC} + \hat{f}_{AbC}\hat{f}_{aBc} + \hat{f}_{Abc}\hat{f}_{aBC}}\Bigg).\end{aligned} \tag{1}$$

We can set up equations for the frequencies of the other seven haplotypes on SNPs A, B, and C similarly. Solving these equations can be done by a standard *expectation-maximization* (EM) algorithm [6, 10] The EM algorithm is iterative. It begins with a random guess of the frequencies. The frequencies obtained at the left hand side in Equation (1) will be repeatedly inserted into the right hand side to improve the estimation. When the improvement is sufficiently small (*e.g.* smaller than a pre-determined threshold, typically 10^{-15}), the algorithm terminates and starts a new round with another random guess. After a sufficient number of rounds, it outputs all feasible solutions. We merge the solutions with distances smaller than a threshold (*e.g.* $\epsilon = 10^{-4}$), and obtain the r^2 value using these estimated 3-marker haplotype frequencies.

There are two things that we have to be careful with when applying Hill's method. The first is that the method assumes the population was produced from random mating and Hardy-Weinberg equilibrium holds. Therefore, datasets consisting of related individuals (such as the CEU dataset in HapMap) would not be suitable. The CEU data consists of family trios, not random-mating individuals. The second is that errors caused by the EM algorithm may lead to wrong assignment of haplotypes into the major and minor buckets. For example, Claim 2.1 says that when $n_{hC} = n_{hc}$, it is advantageous to assign the haplotype h to the minor bucket instead of the major bucket. However, if $f_{hC} = f_{hc}$ but $\hat{f}_{hC}$ happens to be slightly higher than $\hat{f}_{hc}$ due to some error in the EM computation, we will assign h to the major bucket without caution. This could lead to a reduced r^2 value. To avoid this, we assign h to the minor bucket as long as $\hat{f}_{hC} < \hat{f}_{hc} + \epsilon$ for some small $\epsilon > 0$.

3. The Greedy Algorithm for Selecting Tag SNPs

In this section, we first define some notations that will be useful in the algorithm, and then describe the algorithm. For simplicity, we present the algorithm for the 2-marker correlation model first, and then generalize it to work for the multi-marker model. At the end of the section, we analyze the time complexity of the algorithm.

3.1. *Some Notations*

In the rest of the paper, we call a group of three SNPs, which includes two potential *tagging* SNPs s_i, s_j and one SNP s_k to be tagged, a *triplet* and denote it as $(s_i, s_j \triangleright s_k)$. Similarly, a *quartet* is a group of four SNPs including three potential tagging SNPs and SNP to be tagged. The triplets are used in the 2-marker correlation model and the quartets in the 3-marker correlation model. Each such triplet or quartet has a correlation coefficient r^2 value. We will only be interested in triplets and quartets whose correlation coefficient values r^2 are above a certain threshold. It is convenient to think of the triplets or quartets as edges in a hypergraph. Let us regard SNPs as vertices in the hypergraph. The tagging SNPs in a triplet or a quartet have an *outgoing edge* to the SNP to be tagged. This edge can be also thought of as an *incoming edge* of the tagged SNP from the tagging SNPs. Figure A1 shows an example hypergraph with five triplets.

During a tag SNP selection process, a SNP has three possible states: *uncovered*, *covered* and *picked*. A SNP is picked if it has been selected as a tag SNP. A SNP s is covered if either s is picked or there is a triplet $(s_i, s_j \triangleright s)$ where s_i, s_j are picked. In this case we say that SNPs s_i, s_j *cover* s. A SNP is uncovered if it is not picked nor covered. Sometimes, we may use the term *partially covered*. A SNP s is partially covered if it is uncovered and there is a triplet $(s_i, s_j \triangleright s)$ such that either s_i or s_j is picked but not both.

3.2. *The Algorithm for the 2-Marker Correlation Model*

An outline of our algorithm is shown in Figure A2. To avoid considering SNPs that cannot possibly be linked, we set a window size of W bps (in terms of the physical distance on a chromosome). For every triplet of SNPs within the window size, we compute its r^2 value as previously described. Then we run an iterative greedy-based algorithm to select a set of tag SNPs as follows. We first initialize all SNPs as uncovered. In each iteration, we pick an appropriate SNP, put it in the tag SNP set, and then check if any uncovered SNPs are now covered due to the newly selected SNP. We repeat this process until all SNPs are covered.

So the main issue is how to pick an appropriate SNP in each iteration. Our first preference is an uncovered SNP that has no incoming edges. A SNP without incoming edges cannot be tagged by any other SNPs and has to be picked as a tag SNP sooner or later. Therefore, we always check if there is such a SNP. If all SNPs have incoming edges, we pick a SNP (covered or uncovered) that can cover the largest number of uncovered SNPs. If there is a tie, the SNP that partially covers the most uncovered SNPs is preferred. Note that, a covered SNP may also be picked in the above if it covers many other SNPs.

After picking each SNP, we need update and remove some triplets that are no longer useful. A triplet $t = (s_i, s_j \rhd s_k)$ should be removed if any one of the following conditions holds:

(1) s_k is covered, and therefore t is useless.
(2) s_i and s_j are both picked. In this case, s_i and s_j together tag s_k. After changing the state of s_k to covered, t is no longer useful.
(3) There is another triplet $t' = (s_i, s'_j \rhd s_k)$ where s'_j is picked. In this case, the triplet t is superseded by the triplet t' and thus redundant.

Note that, although the condition 3 seems optional and unnecessary, it is actually important since keeping useless triplets in the algorithm may actually affect the final result when useless triplets are involved in the partial coverage of SNPs (and ties have to be broken in the algorithm).

Algorithm 3.1 MMTAGGER(for 2-Marker Model)

Require: set of triplets
1: **while** there are SNPs uncovered **do**
2:     **if** there is a SNP s with no incoming edges **then**
3:         $s^* \leftarrow s$
4:     **else**
5:         $s^* \leftarrow$ a SNP that covers the most uncovered SNPs
6:     Put s^* in the tag SNP set /* s^* is picked */
7:     **for each** triplets t of form $(s_., s_. \rhd s^*)$ **do**
8:         remove t and its corresponding edges
9:     **for each** triplets t of form $(s^*, s_i \rhd s_j)$ or $(s_i, s^* \rhd s_j)$ **do**
10:         **if** s_i is picked **then**
11:             put s_j into covered SNP set
12:             remove all triplets of form $(s_., s_. \rhd s_j)$ or $(s_., s_. \rhd s_j)$
13:         **else**
14:             remove all triplets of form $(s_i, s_. \rhd s_j)$ or $(s_., s_i \rhd s_j)$

Algorithm 3.1 illustrates the pseudocode of the algorithm. In the algorithm, lines 2–5 pick the next SNP. The subsequent lines update the states of the SNPs and remove useless/redundant triplets.

3.3. Extension to the 3-Marker Correlation Model

The extension is straightforward. The outline in Figure A2 still works except that we need now calculate r^2 values for quartets. The above greedy algorithm can also be kept the same, although we should modify the removal of useless/redundant quartets slightly. The third condition should be changed to: if there is another quartet $q' = (s_i, s'_j, s'_k \rhd s_l)$ where s'_j, s'_k are picked, then we remove the quartet q.

It is also straightforward to extend the algorithm to the k-marker correlation model, although calculating r^2 values for groups of k SNPs from haplotype data could be very demanding when k is larger than 4, not to mention doing the calculation for genotype data.

3.4. Time Complexity

Suppose that there are m SNPs $s_1, s_2, \ldots, s_m$ on a chromosome sorted by their positions. For simplicity, we assume that there are at most w SNPs within each window of W bps. We need compute the r^2 values of all possible triplets involving three SNPs from the same windows. If the first SNP with the smallest index is among $s_1, s_2, \ldots, s_{m-w}$, there will be $\binom{w-1}{2}$ combinations for the second and the third SNPs. If the first SNP is among $s_{m-w+1}, \ldots, s_m$, then there are totally $\binom{w}{3}$ combinations for all three SNPs. The time complexity of computing the r^2 values is therefore $(m-w)\binom{w-1}{2}+\binom{w}{3} = O(mw^2)$. Similarly, the time complexity to compute r^2 values of all possible quartets is $O(mw^3)$.

Assume that there are T triplets with sufficiently high r^2 values. During the selection of tag SNPs, we maintain a data structure where each SNP has two linked-lists to the triplets containing the SNP. One list contains all the triplets corresponding to the outgoing edges and the other contains all the triplets corresponding to the incoming edges. For each SNP, we also keep track of the number of triplets containing the SNP, and various other statistics on these triplets. Therefore, in each iteration of the selection algorithm, we need only scan all the SNPs and use these numbers to pick an appropriate one. To keep the data structure up-to-date, we need update a triplet $t = (s_i, s_j, \rhd s_k)$ when

(1) s_i or s_j is picked;
(2) s_k is covered and t needs to be removed; or
(3) t is superseded by another triplet and needs to be removed.

If it takes $O(1)$ time to retrieve each triplet that we need update, then the time complexity will be reasonably low. In cases 1 and 2, we can access each of the involved triplets in $O(1)$ time given the data structure. To achieve $O(1)$ access time in case 3, we sort all the triplets in each linked list corresponding to outgoing edges in preprocessing. As a result, if s_i is picked as a tag SNP, then $(s_i, s_j \rhd s_k)$ will supersede all triplets of the form $(s_h, s_j \rhd s_k)$ for some h. These triplets $(s_h, s_j \rhd s_k)$ must be neighbors of $(s_i, s_j \rhd s_k)$ on s_j's outgoing linked list. Therefore, we can access to each of these triplets in $O(1)$ time. Since a triplet may be updated at most 3 times, the time to select tag SNPs is $O(T)$. The preprocessing may take $O(T \log T)$ time.

In practice, the algorithm spends most of its time on evaluating r^2 values. Therefore, we say that the time complexity of the algorithm is $O(mw^2)$ (or $O(mw^3)$) for the 2-marker correlation (or 3-marker correlation) models, respectively.

4. Experimental Result

We have implemented the above algorithm as a C program, simply called MMTagger. In this section, we compare MMTagger with the program LRTag in [11] and the program MultiTag in [4] on real datasets from the HapMap project. The following is a brief summary of the features of the three programs to be compared.

- LRTag [11] uses the traditional single-marker correlation model and works for a single population as well as multiple populations. The algorithm is based on a powerful combinatorial optimization technique called *Lagrangian relaxation.* According to the extensive tests in [11], LRTag outperforms other state-of-the-art single-marker programs such as FESTA [15] and LD-Select [1] in terms of the number of selected tag SNPs. It requires the pairwise r^2 statistics as the input.
- MultiTag [4] uses a multi-marker correlation model which is more restricted than our model. It is a greedy algorithm. The input to MultiTag must be a population haplotype data.
- MMTagger is a greedy algorithm using a more general multi-marker correlation model. Its input is a population data, either in the form of haplotypes or genotypes.

In order to compare these three programs, we need phased haplotype data. We downloaded the CEU ENCODE region data from the HapMap project[b] and use the first 5 of the 10 sample datasets. For LRTag, we need a preprocessing step to calculate the pairwise r^2 values. For both MMTagger and MultiTag, we use a window size W of 100K bps so that SNPs farther than W bps apart are not considered as correlated. To make it fair, we also apply this restriction when calculating r^2 values for LRTag.

Table 1 shows the numbers of the tag SNPs selected by LRTag, MultiTag and MMTagger using different parameters. The reduction of tag SNPs by using the multi-marker correlation models is obvious. However, the running time of the programs based on the multi-marker correlation models (MultiTag and MMTagger) is much longer. LRTag requires only pairwise r^2 values, but MultiTag and MMTagger need r^2 values for each group of three or four SNPs. In general, MMTagger selected fewer tag SNPs than MultiTag. In fact, the improvement is quite significant when the threshold for r^2 is 0.9 or larger.

When comparing the performance of MultiTag and MMTagger, we should also take into account the running time and memory usage. We thus downloaded the entire chromosomal data of the Japanese and Chinese populations from HapMap[c] and used chromosomes 19, 21 and 22 as our test data.

---

[b] http://www.hapmap.org/downloads/phasing/2005-03_phaseI/ENCODE/
[c] http://www.hapmap.org/downloads/phasing/2006-07_phaseII/phased/

Table 1. Numbers of tag SNPs selected in CEU ENCODE region

Region	ENm010	ENm013	ENm014	ENr112	ENr113
# SNP	459	731	874	868	1035
$r^2 \geq 0.8$					
LRTag	119	88	134	148	133
2-marker MultiTag	75	57	80	87	75
2-marker MMTagger	72	52	78	85	73
3-marker MultiTag	68	53	75	78	64
3-marker MMTagger	62	48	75	68	59
$r^2 \geq 0.9$					
LRTag	148	121	172	204	190
2-marker MultiTag	100	76	111	118	122
2-marker MMTagger	92	73	100	109	115
3-marker MultiTag	91	66	102	101	100
3-marker MMTagger	79	58	85	81	81
$r^2 \geq 0.95$					
LRTag	192	148	196	268	247
2-marker MultiTag	127	96	131	157	156
2-marker MMTagger	117	92	122	141	149
3-marker MultiTag	120	83	119	138	145
3-marker MMTagger	97	66	102	107	112

Hao [4] mentioned two different methods to implement his greedy algorithm and handle a large number of input SNPs: (1) Preprocess and compute all r^2 values, and (2) Calculate r^2 values on the fly while selecting tag SNPs. The former method would lead to heavy memory load and/or file I/O load. The latter method may lead to redundant r^2 value computation. MultiTag employs the latter method. In our implementation of MMTagger, we choose the former method to speed up the computation.

Table 2. MMTagger vs. MultiTag

Chromosome	# SNP	mode	r^2	program	# SNPs Selected	Time (hours)	Memory (M bytes)
JPT+CHB chr19	28931	2-marker	0.9	MultiTag	9600	26hrs	30–35
				MMTagger	9145	2mins	125
		3-marker	0.95	MultiTag	N/A	>700hrs	30–35
				MMTagger	10032	<1hr	657
JPT+CHB chr21	28914	2-marker	0.9	MultiTag	7115	42hrs	30–35
				MMTagger	6766	2mins	187
		3-marker	0.95	MultiTag	N/A	>700hrs	30–35
				MMTagger	7404	<1hr	1210
JPT+CHB chr22	26595	2-marker	0.9	MultiTag	7557	93hrs	30–35
				MMTagger	7221	2mins	183
		3-marker	0.95	MultiTag	N/A	>700hrs	30–35
				MMTagger	7788	3hrs	1216

Note: Both programs were run on a desktop PC with dual AMD Athlon(tm) processors of 2.1 GHz.

Table 2 illustrates a head-to-head comparison between MultiTag and MMTagger. Note that, for the memory usage, we were able to insert some code into MMTagger to obtain the precise maximum memory used by the program. However, we were not able to get the precise memory usage numbers for MultiTag and could only provide a rough estimate. The following gives a detailed comparison between the two programs.

- MMTagger is able to achieve a smaller tag SNP set than MultiTag mostly because our multi-marker correlation model is more general and flexible.
- MMTagger's heuristic to always pick uncovered SNPs with no incoming edges first may also be a factor in its improved performance. This heuristic can be easily incorporated into MultiTag.
- MMTagger may pick a SNP that has been covered if it covers many other SNPs. However, MultiTag always picks an uncovered SNP. Modifying MultiTag to allow covered SNPs to be picked would cost its more time since it calculates r^2 values on the fly. However, this does not impact the running time of MMTagger much because it pre-calculates all r^2 values.
- MMTagger is much faster than MultiTag. Its running time mostly depends on the window size W, since it spends most time on calculating the r^2 values. The running time of MultiTag depends on both the window size W and the number of tag SNPs selected. Hence, it requires more time for higher r^2 thresholds since more tag SNPs would be required. Hao [4] reported that the program took about 300 hours to process the human chromosome 2 data on a typical workstation (Intel Xeon 2.80 GHz CPU and 512 MB memory).
- MMTagger requires much more memory. Its memory usage grows when the r^2 threshold decreases, as more triplets/quartets would be qualified. To run the program on a large chromosome such as human chromosome 2, it require about 4 GB of memory for the 3-marker correlation model when the r^2 threshold is 0.9. However, MultiTag's memory usage is pretty reasonable even for large chromosomes and low r^2 thresholds.
- MMTagger and MultiTag use the window size W in slightly different ways. MMTagger requires that all SNPs in a triplet/quartet should be in the same window, while MultiTag requires that a covered SNP and each of its tagging SNPs should not be farther than W. Therefore, the distance of the two tagging SNPs of a triplet may actually be as far as $2W$ in MultiTag.

As observed before, the 2-marker correlation model improves on the single-marker correlation model significantly. A similar significant improvement from the 2-marker model to the 3-marker model is also shown in Table 2. Although it is likely that the 4-marker model will show further improvements, we are not able to extend the results to the 4-marker model because MMTagger would require too much time and memory on any realistic datasets. For the same reason, MultiTag was only implemented for the 2-marker and 3-marker models in [4, 5]

5. Conclusion

We have introduced a new multi-marker correlation model that generalizes a previous result in the literature. A greedy algorithm is designed to select tag SNPs based on the model. Our experimental results on real datasets from the HapMap project demonstrate that the algorithm produces the most succinct tag SNP sets compared with the previous algorithms.

Acknowledgements

The research is supported in part by NSF grant IIS-0711129 and NIH grant LM008991.

References

[1] Carlson, C., *et al.* Selecting a maximally informative set of single-nucleotide polymorphisms for association analyses using linkage disequilibrium, *The American Journal of Human Genetics*, 74(1):106–120, 2004.
[2] De Bakker, P., *et al.* Transferability of tag SNPs in genetic association studies in multiple populations, *Nature Genetics*, 38(11):1298–1303, 2006.
[3] Halldórsson, B. V., *et al.* Optimal haplotype block-free selection of tagging SNPs for genome-wide association studies, *Genome Research*, 14:1633–1640, 2004.
[4] Hao, K., Genome-wide selection of tag SNPs using multiple-marker correlation, *Bioinformatics*, 23(23):3178–3184, 2007.
[5] Hao, K., Di, X., and Cawley, S., LdCompare: rapid computation of single- and multiple-marker r^2 and genetic coverage, *Bioinformatics*, 23(2):252–254, 2007.
[6] Hill, W., Estimation of linkage disequilibrium in randomly mating populations, *Heredity*, 33(2):229–239, 1974.
[7] Hill, W., Tests for association of gene frequencies at several loci in random mating diploid populations, *Biometrics*, 31(4):881–888, 1975.
[8] Hinds, D., *et al.* Whole-genome patterns of common DNA variation in three human populations, *Science*, 307(5712):1072–1079, 2005.
[9] Johnson, G., *et al.* Haplotype tagging for the identification of common disease genes, *Nature Genetics*, 29:233–237, 2001.
[10] Kalinowski, S. and Hedrick, P., Estimation of linkage disequilibrium for loci with multiple alleles: basic approach and an application using data from bighorn sheep, *Heredity*, 87:698–708, 2001.
[11] Liu, L., Wu, Y., Lonardi, S., and Jiang, T., Effcient algorithms for genome-wide tagSNP selection across populations via linkage disequilibrium criterion, *Proc. 6th Annual International Conference on Computational Systems Bioinformatics*, 67–78, 2007.
[12] Mägi, R., Kaplinski, L., and Remm, M., The whole genome tagSNP selection and transferability among HapMap populations, *Pacific Symposium on Biocomputing*, 11:535–543, 2006.
[13] Marchini, J., *et al.* A comparison of phasing algorithms for trios and unrelated individuals, *The American Journal of Human Genetics*, 78:437–450, 2006.
[14] Patil, N., *et al.* Blocks of limited haplotype diversity revealed by high-resolution scanning of human chromosome 21, *Science*, 294(5547):1719–1723, 2001.
[15] Qin, Z., Gopalakrishnan, S., and Abecasis, G., An effient comprehensive search algorithm for tagSNP selection using linkage disequilibrium criteria, *Bioinformatics*, 22(2):220–225, 2006.
[16] Stephens, M., Smith, N., and Donnelly, P., A new statistical method for haplotype reconstruction from population data, *The American Journal of Human Genetics*, 68:978–989, 2001.
[17] Stram, D., *et al.* Choosing haplotype tagging SNPs based on unphased genotype data using a preliminary sample of unrelated subjects with an example from the multiethnic cohort study, *Human Heredity* 55(1):27–36, 2003.
[18] Zhang, Kun and Jin, Li, HaploBlockFinder: Haplotype block analyses, *Bioinformatics*, 19(10):1300–1301, 2003.

Appendix A. The Missing Proof and Figures

Proof of Claim 2.1: Let us consider the frequency table as shown in Table A1, where A is a SNP to be covered/tagged and M is a compound marker representing several (*e.g.* two or three) SNPs. Let n_{AM} denote the number of times that the haplotype AM is observed in the population, $n_A = n_{AM} + n_{Am}$, and n the total number of haplotypes.

Table A1. Number of observations of each haplotype

	A	a	
M	n_{AM}	n_{aM}	n_M
m	n_{Am}	n_{am}	n_m
	n_A	n_a	n

For any haplotype h on M, if $n_{Ah} > n_{ah}$, we would put h in the major bucket, otherwise we put it in the minor bucket. However, when $n_{Ah} = n_{ah}$, it seems that we could put h in either the major bucket or the minor bucket. We show in the following that putting h in the minor bucket leads to a bigger r^2 value between M and A. By definition of the r^2 statistics,

$$\begin{aligned} r^2 &= \frac{(p_{AM} - p_A p_M)^2}{p_A p_a p_M p_m} \\ &= \frac{(n_{AM} \cdot n - n_A n_M)^2}{n_A n_a n_M n_m} \\ &= \frac{(n_{AM} n_{am} - n_{Am} n_{aM})^2}{(n_{AM} + n_{Am})(n_{aM} + n_{am})(n_{AM} + n_{aM})(n_{Am} + n_{am})} \end{aligned}$$

We take the partial derivative of r^2 with respect to n_{AM} and obtain

$$\frac{\partial r^2}{\partial n_{AM}} = \frac{(n_{AM} n_{am} - n_{Am} n_{aM})}{n_A n_a n_M n_m} \cdot \left(2n_{am} - \frac{(n_{AM} n_{am} \quad n_{Am} n_{aM})(2n_{AM} + n_{Am} + n_{aM})}{(n_{AM} + n_{Am})(n_{AM} + n_{aM})} \right)$$

By simplifying the equation, we get

$$\begin{aligned} \frac{\partial r^2}{\partial n_{AM}} &= c\left(2n_{am} - \frac{X(n_A + n_M)}{n_A n_M}\right) \\ \frac{\partial r^2}{\partial n_{Am}} &= c\left(-2n_{aM} - \frac{X(n_A + n_m)}{n_A n_m}\right) \\ \frac{\partial r^2}{\partial n_{aM}} &= c\left(-2n_{Am} - \frac{X(n_a + n_M)}{n_a n_M}\right) \\ \frac{\partial r^2}{\partial n_{am}} &= c\left(2n_{AM} - \frac{X(n_a + n_m)}{n_a n_m}\right) \end{aligned}$$

where $c = \frac{(n_{AM} n_{am} - n_{Am} n_{aM})}{n_A n_a n_M n_m}$, $X = (n_{AM} n_{am} - n_{Am} n_{aM})$.

Suppose that $n_{Ah} = n_{ah}$. If we put haplotype h in the major bucket, then the r^2 value would change by approximately $n_{Ah} \cdot \frac{\partial r^2}{\partial n_{AM}} + n_{ah} \cdot \frac{\partial r^2}{\partial n_{aM}}$. If we put h in the minor bucket, then the r^2 value would change by approximately $n_{Ah} \cdot \frac{\partial r^2}{\partial n_{Am}} + n_{ah} \cdot \frac{\partial r^2}{\partial n_{am}}$. Let

$$\begin{aligned}
\Delta_M &= \frac{\partial r^2}{\partial n_{AM}} + \frac{\partial r^2}{\partial n_{aM}} \\
&= c\left(2n_{am} - 2n_{Am} - X\left(\frac{1}{n_A} + \frac{1}{n_a} + \frac{2}{n_M}\right)\right) \\
\Delta_m &= \frac{\partial r^2}{\partial n_{Am}} + \frac{\partial r^2}{\partial n_{am}} \\
&= c\left(2n_{AM} - 2n_{aM} - X\left(\frac{1}{n_A} + \frac{1}{n_a} + \frac{2}{n_m}\right)\right)
\end{aligned}$$

We have

$$\begin{aligned}
\Delta_m - \Delta_M &= 2c(n_{AM} - n_{aM} + n_{Am} - n_{am}) + cX\left(\frac{2}{n_M} - \frac{2}{n_m}\right) \\
&= 2c(n_A - n_a) + 2cX\left(\frac{1}{n_M} - \frac{1}{n_m}\right)
\end{aligned}$$

We need check if $\Delta_m - \Delta_M \geq 0$ holds. By multiplying both side with $\frac{n_M n_m}{2c}$ we get

$$\begin{aligned}
&\frac{1}{2c} n_M n_m (\Delta_m - \Delta_M) \\
&= (n_A - n_a) n_M n_m + (n_{AM} n_{am} - n_{Am} n_{aM})(n_m - n_M) \\
&= (n_{AM} + n_{Am} - n_{aM} - n_{am})(n_{AM} + n_{aM})(n_{Am} + n_{am}) \\
&\qquad + (n_{AM} n_{am} - n_{Am} n_{aM})(n_{Am} + n_{am} - n_{AM} - n_{aM}) \\
&= n_{AM}(n_{AM} + n_{aM}) n_{Am} + n_{Am} n_{AM}(n_{Am} + n_{am}) \\
&\qquad - n_{aM}(n_{AM} + n_{aM}) n_{am} - n_{am} n_{aM}(n_{Am} + n_{am}) \\
&= n_{AM} n_{Am} \cdot n - n_{aM} n_{am} \\
&= n(n_{AM} n_{Am} - n_{aM} n_{am})
\end{aligned}$$

where $n = n_{AM} + n_{Am} + n_{aM} + n_{am}$. Therefore, $\Delta_m \geq \Delta_M$ if and only if $n_{AM} n_{Am} \geq n_{aM} n_{am}$. When the latter inequality holds, putting the haplotype h in the minor bucket will result in a higher r^2 value.

Since $n_{AM} + n_{Am} = n_A > n_a = n_{aM} + n_{am}$, $n_{AM} n_{Am}$ tends to be greater than $n_{aM} n_{am}$ in practice. Moveover, even when $n_{AM} n_{Am} < n_{aM} n_{am}$, putting the haplotype h in the minor bucket would increase n_{Am} and n_{am} at the same time, and hence result in a greater increase in $n_{AM} n_{Am}$ than in $n_{aM} n_{am}$ since n_{AM} is usually larger than n_{aM}. This could help improve the r^2 value in the long run. Therefore, putting h in the minor bucket may still be better in this case. For example, suppose $n_{AM} = 100$, $n_{Am} = 0$, $n_{aM} = 5$, and $n_{am} = 20$ before haplotype h is considered. If $n_{Ah} = n_{ah} = 1$, then putting h in the major (or minor) bucket results in $r^2 = 0.7261$ (or $r^2 = 0.7235$, respectively). However, if $n_{Ah} = n_{ah} = 3$, then putting h in the major (or minor) bucket leads to $r^2 = 0.6628$ (or $r^2 = 0.6631$, respectively).

Note that, the tag SNP selection program MultiTag in [4, 5] considers all the possible splits of the haplotypes in question and picks the one that results in the highest r^2 value. So, ties between haplotype counts are not an issue. However, we cannot afford doing this in our tag SNP selection program MMTagger (to be introduced in Section 4) because our multi-marker correlation model allows for many more possible splits. Trying all such splits would be very inefficient. Since the above analysis shows that putting haplotype h in the

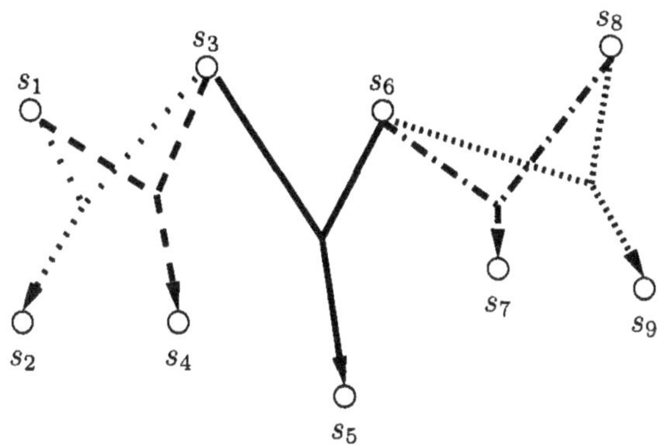

Fig. A1. An example with five triplets: $(s_1, s_3 \rhd s_2)$, $(s_1, s_3 \rhd s_4)$, $(s_3, s_6 \rhd s_5)$, $(s_6, s_8 \rhd s_7)$ and $(s_6, s_8 \rhd s_9)$.

minor bucket is generally better when we have a tie $n_{Ah} = n_{ah}$, MMTagger always puts h in the minor bucket when such a tie arises. □

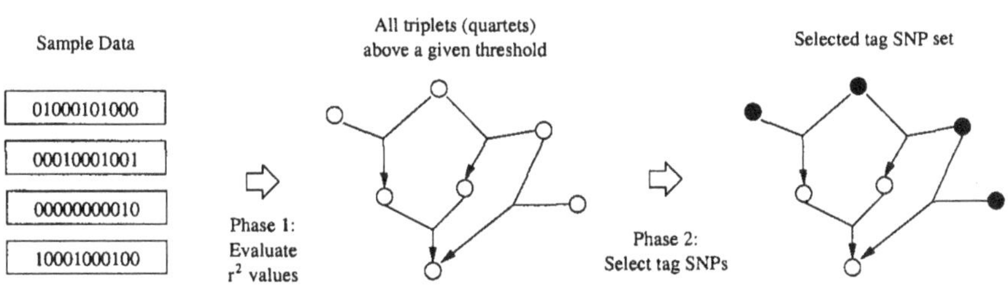

Fig. A2. An outline of our algorithm.

PHENOTYPE PROFILING OF SINGLE GENE DELETION MUTANTS OF *E. COLI* USING BIOLOG TECHNOLOGY

YUKAKO TOHSATO[1]
yukako@sk.ritsumei.ac.jp

HIROTADA MORI[2,3]
hmori@gtc.naist.jp

[1] *Department of Bioscience and Bioinformatics, Ritsumeikan University, 1-1-1 Nojihigashi, Kusatsu, Shiga, 525-8577, Japan*
[2] *Graduate School of Biological Sciences, Nara Institute of Science and Technology, 8916-5 Takayama, Ikoma, Nara 630-0101, Japan*
[3] *Institute for Advanced Biosciences, Keio University, Tsuruoka, Yamagata 997-0017, Japan*

Phenotype MicroArray (PM) technology is high-throughput phenotyping system [1] and is directly applicable to assay the effects of genetic changes in cells. In this study, we performed comprehensive PM analysis using single gene deletion mutants of central metabolic pathway and related genes. To elucidate the structure of central metabolic networks in *Escherichia coli* K-12, we focused 288 different PM conditions of carbon and nitrogen sources and performed bioinformatic analysis. For data processing, we employed noise reduction procedures. The distance between each of the mutants was defined by Manhattan distance and agglomerative Ward's hierarchical method was applied for clustering analysis. As a result, five clusters were revealed which represented to activate or repress cellular respiratory activities. Furthermore, the results might suggest that Glyceraldehyde-3P plays a key role as a molecular switch of central metabolic network.

Keywords: Phenotype MicroArray, phenotype, clustering, metabolic pathway

1. Introduction

The definition and testing of phenotypes has had a key role in genetics and this is also true in present systems biology. For a long way to complete understanding metabolic network in a cell, even though numerous accumulation of knowledge of enzymes genetically and biochemically, still it is too short to understand the whole system of this network. Since genome sequencing project, especially in 1990s, new comprehensive technology, such as DNA microarray for transcription and yeast two hybrid or pull-down assay for protein-protein interaction by Mass spectrometry, have been developed. And combinatorial analysis has had big contribution not only basic scientific knowledge but seeking potential pharmacological targets etc. The central metabolic pathway is one of the well-studied cellular enzymatic networks, however, the whole regulatory mechanism of this pathway including transcription, translation and enzymatic activity is still remain to be analyzed. "Robustness" is one of the most important features of cellular organisms and this is also the case in the central metabolic pathway of *Escherichia coli. E. coli* cell, even such small bacterial cell, accepts single gene deletion of most of the steps of central metabolic pathway easily. Ishii and his colleagues proposed compensatory mechanism of such gene deletion by alteration of transcription, enzyme copy number and their activities to maintain cellular homeostasis [2]. This is clear "Robustness" phenotype plausibly by activation of alternative enzymes or bypass pathways, etc. In this study, analysis using Phenotype MicroArray (PM) data [1] was performed to discover new alternative pathways and identify functions of genes for which the functions have yet to be determined.

PM technology was originally developed by Bochner to open up opportunity for finding the unique traits of individual organisms and for recognizing traits common to group of organisms, such as species [3] and expanding as a high-throughput tool for global analysis of cellular phenotypes in post-genomic era [1]. This system allows monitoring of cellular respiration during cell growth on 96-well microtiter plates under a maximum of 1920 different medium conditions by colorimetrically detection of generation of purple colored Formazane from Tetrazolium dye corresponding to the intracellular reducing state by NADH simultaneously.

Several studies using PM have been reported [4, 5, 6], but most of those used the absolute values generated by PM. However, experimental data, especially by such comprehensive high-throughput analyses system, generally includes a great deal of noises. In this study, to reduce noises and make analysis more reliable, relative ratio and vector data from reference wild type and mutant cells were used. We report here the results obtained by applying the proposed method to PM data from wild-type cell and 45 single gene deletion strains.

2. Materials and Methods

2.1. *Phenotype MicroArray Data and E. coli Strains*

Selected 45 single gene deletion mutants of glycolysis, TCA cycle and pentose phosphate pathway from Keio collection [7] were used and listed in Table 1. The wild-type host strain of Keio collection (BW25113 [8]) was used as a reference strain.

Fig. 1 shows examples of ten times repeats Biolog test of wild type BW25113 with time (hrs., X-axis) and NADH production level (Y-axis). Figs 1a and 1b show the results with α-D-Glucose and Glycerol medium conditions respectively.

96 time points at every 15 min for 24 hours under 288 different conditions (Biolog Assay Plate No. 1 to 3) of carbon and nitrogen sources were collected. These 288 screening conditions were listed in Appendix. Experiments were repeated twice for each mutant strains, and ten times for the wild-type strain under the same conditions.

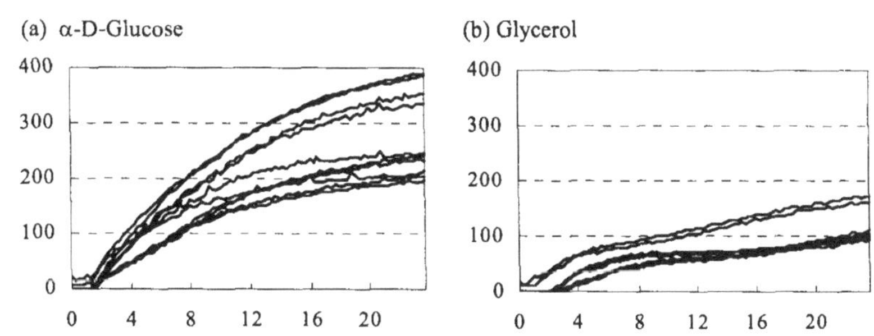

Figure 1. Actual example of PM data of wild-type

Table 1. List of 45 single-gene-knockout mutants used in this analysis. The genes deleted were assigned to metabolic maps according to the KEGG database [9]. Glycolysis (G), TCA cycle (T) and Pentose phosphate pathway (P) in Map column. All the assigned pathways are listed.

Gene detected	Function	Map
aceF	pyruvate dehydrogenase, dihydrolipoyltransacetylase component E2	G
acs	acetyl-CoA synthetase	G
adhC	alcohol dehydrogenase class III	G
adhE	CoA-linked acetaldehyde dehydrogenase, iron-dependent alcohol dehydrogenase	G
adhP	alcohol dehydrogenase	G
agp	glucose-1-phosphatase	G
ascF	PTS family enzyme IIBC component,cellobiose/salicin/arbutin-specific	G
crr	PTS family enzyme IIA component	G
eda	2-keto-4-hydroxyglutarate aldolase, oxaloacetate decarboxylase	P
edd	6-phosphogluconate dehydratase	P
fbaA	fructose-bisphosphate aldolase, class II	G, P
fbaB	fructose-bisphosphate aldolase class I	G, P
fbp	fructose-1,6-bisphosphatase	G, P
frdA	fumarate reductase, anaerobic, catalytic and NAD/flavoprotein subunit	T
frdB	fumarate reductase, anaerobic, Fe-S subunit	T
frdC	fumarate reductase, anaerobic, membrane anchor polypeptide	T
frdD	fumarate reductase, anaerobic, membrane anchor polypeptide	T
fruA	PTS family enzyme IIB'BC, fructose-specific	G
galM	galactose-1-epimerase (mutarotase)	G
glk	glucokinase	G
glpX	fructose 1,6-bisphosphatase II, in glycerol metabolism	G
gltA	citrate synthase	T
gndC	gluconate-6-phosphate dehydrogenase, decarboxylating	P
icdA	e14 prophage; isocitrate dehydrogenase, specific for NADP+	T
malX	PTS family enzyme IIBC component, maltose/glucose-specific	G
pck	phosphoenolpyruvate carboxykinase	T
pfkA	6-phosphofructokinase I	G, P
pfkB	6-phosphofructokinase II	G, P
pgi	glucosephosphate isomerase	G, P
pgm	phosphoglucomutase	G
ptsG	PTS family enzyme IIBC component, glucose-specific	G
pykA	pyruvate kinase II	G
pykF	pyruvate kinase I	G
rpe	D-ribulose-5-phosphate 3-epimerase	P
rpi	ribosephosphate isomerase, constitutive	P
rpiB	ribose 5-phosphate isomerase B	P
sucC	succinyl-CoA synthetase, beta subunit	T
tktA	transketolase 1, thiamin-binding	P
tktB	transketolase 2, thiamin-binding	P
tpiA	triosephosphate isomerase	G
ybhE	putative isomerase	P
ybiC	putative dehydrogenase	T
yccX	predicted acylphosphatase	G
yibO	phosphoglycerate mutase III, cofactor-independent	G
zwf	glucose-6-phosphate dehydrogenase	P

2.2. Vectorization of Data

First, "zero-substitution" procedure was performed as follows; the original raw data from each strain under 288 medium conditions less than a certain threshold were substituted with zero. The distribution of the observed data frequency for the wild-type strain were used to determine the threshold value.

In the PM data, the observation time is expressed as $i=1,\ldots,m$, and medium condition is expressed as $j=1,\ldots,n$. The observation strength is x_{ij} when observation time is i and medium condition is j. The moving average is calculated by first obtaining the moving average a_{ij} between time t_i and t_{i+k}.

$$a_{ij} = \frac{1}{k+1}\sum_{i'=i}^{i+k} x_{i'j} \tag{1}$$

Here, original data were smoothed by taking an average of consecutive five observation points (k=5).

Regression analysis was performed using Eq. (2). Here, S_{ta} indicates covariance of t and a, S_t and S_a are standard deviation for t and a, respectively. Respiratory activity of medium condition j and time t_i is a_{ij}. Eq. (3) was used to calculate the slope b_{ij}.

$$u_{ijk} = \frac{S_{ta}}{S_{tt}} = \frac{\sum_{g=i}^{i+f}(t_g - \bar{t})(a_{gj} - \overline{a_j})}{\sum_{g=i}^{i+f}(t_g - \bar{t})^2} \tag{2}$$

$$b_{jk} = Max\{u_{1jk}, u_{2jk}, \ldots, u_{(m-f-k)jk}\} \tag{3}$$

where $i=1,\ldots,(m\text{-}f\text{-}k)$, $j=1,\ldots,n$, $k=1,\ldots,288$ and f=9. This will allow each well to be expressed with its maximum slope, and therefore PM data for each strain can be considered as n-dimensional vector data $b_k=(b_{k1},b_{k2},\ldots,b_{kn})$.

The ratio of each respiration rate for vector data of gene deletion strain $b_k=(b_{k1},b_{k2},\ldots,b_{kn})$ and of the wild-type strain $b_w=(b_{w1},b_{w2},\ldots,b_{wn})$ was calculated,

$$\frac{b_{ki}}{b_{wi}} \ (1 \le i \le n) \tag{4}$$

and the data were substituted with +1 for values of 1.2 or higher, with −1 for those less than 0.8, and with 0 for all other values.

$$v_k = (v_{k1}, v_{k2}, \ldots, v_{kn}) \tag{5}$$

Here, v_{ki} = 0 or 1 or −1 ($1 \le i \le n$). v_{ki} = 1 indicates that the gene deletion activate the respiratory activity, and v_{ki} = −1 indicates that the gene deletion repress the respiratory activity.

In this study, we calculated "the reference data" from the averages of ten repeated experiments of the wild-type strain. Then, we calculated relative ratio of the array data v

from mutants to the reference data. For each mutant, two array data are reconverted to one array data by setting zero to different bits. Thereafter, this array data is simply called "vector data."

2.3. *Hierarchical Clustering*

The degrees of dissimilarity $d(v_x, v_y)$ of the vector data $v_x=(v_{x1},v_{x2},...,v_{xn})$ of strain x and the vector data $v_y=(v_{y1},v_{y2},...,v_{yn})$ of strain y data are calculated using the Manhattan distance as follows.

$$d(v_x, v_y) = \sum_{k=1}^{n} \left| v_{xk} - v_{yk} \right| \tag{6}$$

The degree of similarity using the Manhattan distance tends to become larger for pairs of vector data that are less similar, and outlying data are slightly emphasized [10].

After obtaining all the distances between two strains, the strains were classified according to the Ward method, which is a type of hierarchical clustering. In the Ward method, the fluctuation within a cluster created by joining two clusters becomes larger than the sum of fluctuations of the clusters before joining them, and the amount of increase in the fluctuation is set as the distance between the clusters [10]. This method is considered to show good results as compared to other hierarchical clustering methods.

2.4. *Assignment of Conditions and P-values to Clusters*

We calculate a P-value for each experimental condition using the following formula [11].

$$P_{\pm 1} = 1 - \sum_{i=0}^{k-1} \frac{\binom{C}{i}\binom{G-C}{n-i}}{\binom{G}{n}} \tag{7}$$

where G is the number of all strain data, C is the number of the selected group of strains, n is the number of strains with a value of +1 (or −1), k is the number of strains with a value of +1 (or −1) within the selected strain group.

2.5. *Metabolic Pathway Data and Extraction of Path from Graph*

The metabolic pathway information is extracted from KEGG ver. 43 [9]. The step between the two compounds in the same metabolic map can be extracted using shortest paths algorithms (e.g., Dijkstra's algorithm [12]). However, pathway reconstruction using a shortest paths algorithm has major problems caused by traversing irrelevant shortcuts through highly connected nodes, such as H_2O and ATP etc [12]). Therefore, in this study, we used "reaction main" dataset in KEGG to avoid this problem. The major path data is represented one adjacency matrix of a directed graph. We calculated a length between any pair of compounds in the adjacency matrix using Dijkstra's algorithm.

3. Results and Discussion

3.1. Selection of Threshold Value for Zero-Substitution

When looking at the respiration rate, medium conditions that result in overall low observation strength may lead to unstable experimental measurement. Therefore, we attempted to neutralize the observed values that may have a negative influence on the analysis by substituting them with zero.

Maximum values of each medium condition by the wild-type strain were collected and the frequency was shown in Fig. 2. Based on these results, the value of 100 was set as the threshold for zero-substitution step. Zero-substitution procedure effects reduction of the noise.

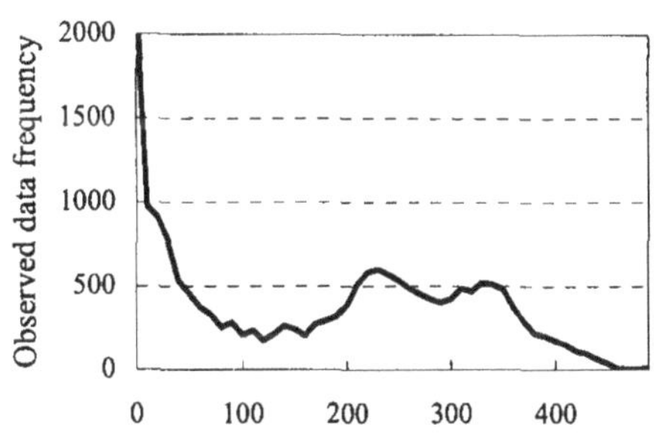

Figure 2. Observed data frequency of each observation strength for wild-type strain data.

3.2. Clustering Results

The result of clustering analysis is shown in Fig. 3. Three major clusters, C_1 to C_3 , were obtained. Cluster C_2 and C_3 are divided into sub-clusters, C_{21} and C_{22}, C_{31} and C_{32}, respectively. Their Map position on central metabolic pathway was shown in Fig. 4. Phenotype profiles of these clusters are summarized in Table 2.

Clustering analysis revealed that the seven mutants of cluster C_1, Δ*pfkA*, Δ*glpX*, Δ*pgi*, Δ*pfkB*, Δ*fbaB*, Δ*fbaA* and Δ*fbp*, are located at the early stage of the glycolysis. Genes in cluster C_2 and C3 represent up-and down-regulation in their respiratory activity by their deletion, respectively. Four mutants, Δ*aceF*, Δ*gltA*, Δ*icd*, Δ*pykF* in five steps in cluster C_{31} (blue) are closely related to the TCA cycle but the Δ*crr* mutant is located in the pentose-phosphate pathway.

Based on the different profiles between steps in glycolysis before and after Glyceraldehyde-3P, it might be plausible that this compound had a switching mechanism of metabolic systems. In addition, in the enzyme reaction from D-glucose to α-glucose-6P, PtsG and Crr form enzyme II complex and transport sugar as PTS (phosphotransferase) system. The results shown in Fig. 5, however, revealed that deletions of *ptsG* and *crr* genes affect opposite direction in phenotype profiles. This observation might be consistent with the previous knowledge that Crr might function as switching for further steps after transportation of Glucose [13].

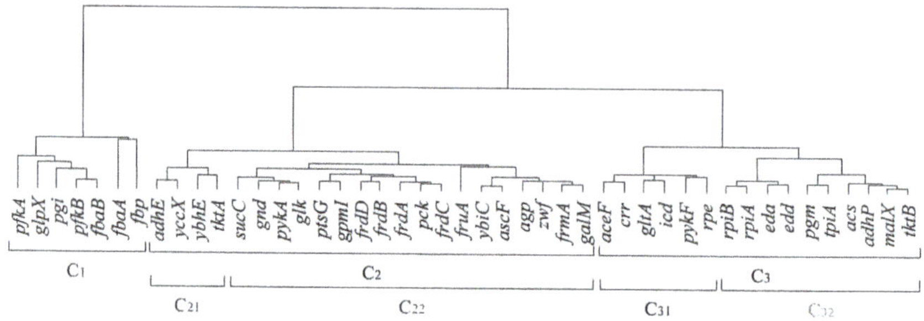

Figure 3. Clustering result of 45 single-gene-knockout mutants in central metabolism under 288 conditions.

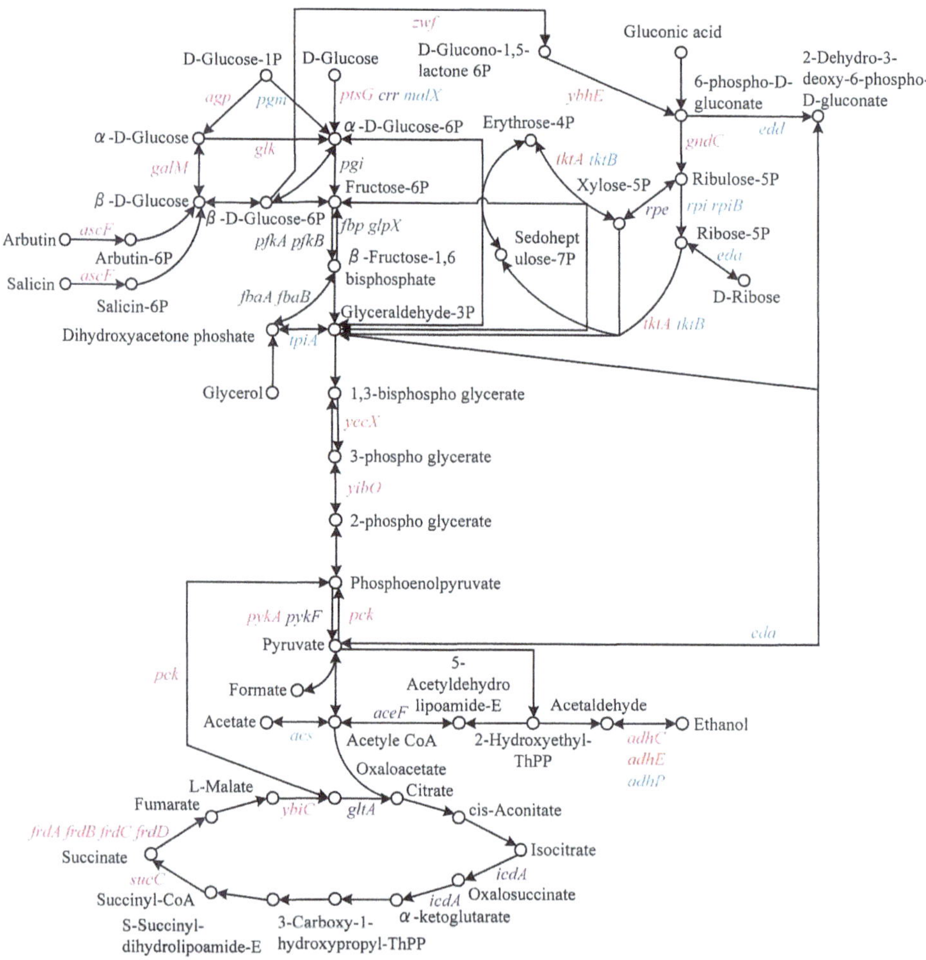

Figure 4. Distribution of gene-knockout affects in central metabolism. Compounds are considered as the nodes, and the arrows indicate direction of the reactions. The compounds' names are shown beside their compounds. The abbreviations (italic) (e.g., *ptsG*) represent the *E. coil*'s gene names corresponding to names shown in Table 1. Color code: ***green***, cluster C_1; ***red***, cluster C_{21}; ***pink***, cluster C_{22}; ***blue***, cluster C_{31}; ***light blue***, cluster C_{32}.

Table 2. Distribution of *P*-values for medium conditions in different phenotype categories. The *P*-values were calculated by using the Eq. (7) (see Methods), measuring whether a gene subset was activated or restricted cellular respiratory activity (only conditions that P-values less than 0.1 are shown).

	C_1	C_{21}	C_{22}	C_{31}	C_{32}
1-A03			+0.05	−0.1	
1-A04				−0.05	
1-A05				−0.05	
1-A06		+0.1	+0.05		
1-A07				−0.05	
1-A08		+0.05			
1-A09		+0.05		−0.05	
1-A10				−0.05	
1-A11	+0.1		+0.05	−0.05	
1-A12		+0.05		−0.1	
1-B01		+0.1			
1-B02			+0.05		
1-B03				−0.05	
1-B04				−0.05	
1-B06	+0.05				
1-B07		+0.05		−0.05	
1-B08			+0.05	−0.05	
1-B09		+0.1		−0.05	
1-B10		+0.05			
1-B11			+0.05	−0.05	
1-C01		+0.05			
1-C02				−0.05	
1-C03		+0.05		−0.05	
1-C04			+0.05	−0.05	
1-C07			+0.05	−0.05	
1-C08				−0.05	
1-C09			+0.05	−0.05	
1-C10		+0.05		−0.05	
1-C11		+0.1			
1-C12				−0.05	
1-D01		+0.05		−0.05	

	C_1	C_{21}	C_{22}	C_{31}	C_{32}
1-D06				−0.1	
1-D08		+0.05		−0.05	
1-D12		+0.1		−0.05	
1-E01				−0.05	−0.1
1-E02		+0.1		−0.05	
1-E05		+0.05			
1-E07		+0.1		−0.05	
1-E08		+0.05		−0.05	
1-E10			+0.1		
1-E11		+0.1		−0.05	
1-E12				−0.05	
1-F01				−0.1	
1-F05		+0.05		−0.05	
1-F06			+0.05	−0.05	
1-F07				−0.05	
1-F08	+0.1	+0.05		−0.05	
1-F09				−0.05	
1-F10				−0.05	−0.1
1-F12		+0.05		−0.05	
1-G01		+0.05			
1-G03		+0.05		−0.05	
1-G04		+0.05			
1-G05		+0.1		−0.05	−0.1
1-G06				−0.05	
1-G07		+0.1			
1-G08	+0.05			−0.05	
1-G09		+0.05		−0.1	
1-G10	+0.1	+0.05		−0.05	
1-G11		+0.1		−0.05	
1-G12		+0.1		−0.05	
1-H01		+0.05			

	C_1	C_{21}	C_{22}	C_{31}	C_{32}
1-H05		+0.1		−0.05	+0.05
1-H07					−0.1
1-H08	+0.1		+0.1	−0.05	
1-H09					
1-H10	+0.05	+0.05		−0.1	
2-A06		+0.1		−0.05	
2-A12		+0.05	−0.1		
2-B02				−0.05	
2-B03			+0.05	−0.05	
2-E05				−0.05	
2-E12		+0.05		−0.05	
2-F01			−0.05		
2-F03		+0.05			
2-G02		+0.05			
2-H09			+0.05	−0.1	
3-A02				−0.05	
3-A07			+0.05	−0.05	
3-A08				−0.05	
3-A09		+0.05		−0.05	
3-A10			+0.05	−0.05	
3-A11				−0.05	
3-A12				+0.1	
3-B01			+0.1	−0.05	
3-B02			+0.05	−0.05	
3-B06			+0.05	−0.05	−0.1
3-B07			+0.1	−0.1	−0.05
3-B08			+0.05	−0.05	−0.05
3-B09			+0.05		
3-B10	+0.1	+0.05		−0.05	
3-B11			+0.1	−0.05	−0.1
3-B12				−0.05	

	C_1	C_{21}	C_{22}	C_{31}	C_{32}
3-C03			+0.05	−0.05	
3-C04			+0.05		
3-C08			+0.05	−0.05	
3-C12		+0.05		−0.05	
3-D11	+0.05			−0.05	
3-E06			+0.05	−0.05	
3-E08			+0.05	−0.05	
3-E11			+0.05	−0.1	
3-F01			+0.05		
3-F02			+0.05	−0.05	
3-F03			+0.05	−0.05	
3-F04			+0.1	−0.05	
3-F05			+0.1	−0.05	
3-G01				−0.05	
3-G02			+0.05	−0.05	
3-G03			+0.05	−0.1	−0.1
3-G04					−0.1
3-G07			+0.1	−0.05	−0.05
3-G08			+0.05		
3-G11		+0.1	+0.05	−0.05	
3-H01				−0.1	−0.1
3-H02			+0.05		
3-H03			+0.05	−0.1	
3-H04				−0.05	
3-H05				−0.05	
3-H06				−0.05	
3-H07			+0.1	−0.05	
3-H09			+0.05		
3-H10		+0.05		−0.05	
3-H11			+0.05	−0.05	
3-H12				−0.05	

(+0.05) P_{+1} <= 0.05 (+0.1) P_{+1} <= 0.1 (−0.05) P_{-1} <= 0.05 (−0.1) P_{-1} <= 0.1

3.3. *Phenotypic and Metabolic Pathway Relationship*

The minimal pathway distances for all strain pairs whose knockout genes are involved in central metabolism were calculated (Fig. 5). We defined the distance between two genes as the number of the compounds between given genes (refer to the chapter of Methods for details). For example, the *ptsG* and *pgi* genes have a pathway distance of 1. For the established pairs, phenotypic similarity was determined. This result shows no correlation between phenotypic similarity and pathway distance.

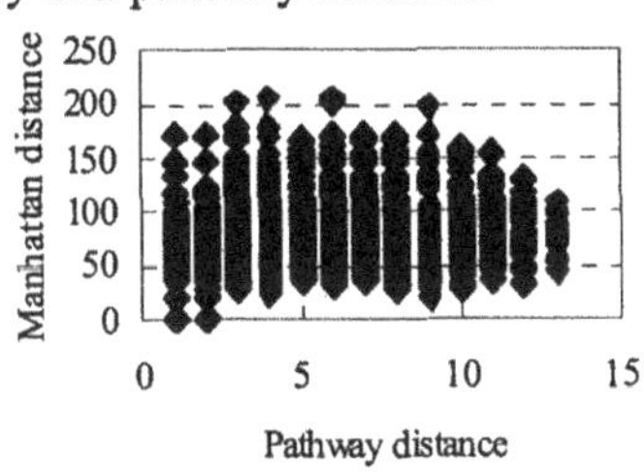

Figure 5. Pathway distance and phenotypic similarity. Phenotypic similarities were calculated by using the Eq. (6). At each pathway distance (X-axis), the phenotypic distances of mutant pairs are plotted.

4. Conclusions

This study was performed to analyze further insight into central metabolic pathway network by applying various statistical analyses to Phenotype MicroArray data. These results suggested the possibility of metabolism steps with unknown bypass routes, as well as metabolic steps that could be the key steps in redox reactions. In addition, medium conditions that activate or repress cellular respiratory activities for different strain groups were identified. However, our results suggest that proposal methods have insufficient sensitivity to continue to identify functions of genes of uncertain function or to analysis for further large-scale data. The most likely causes are robustness and unknown alternative passes within metabolic pathways. Therefore, we plan to propose a computational method for prediction about bond strength among known reactions, realize double gene knockout experiments, and combine PM data and another high-throughput data in future studies.

Appendix

Table A.1: List of medium conditions.

#Cond.	Medium Condition	#Cond.	Medium Condition	#Cond.	Medium Condition
1-A01	Negative-Control	1-D03	D-Glucosaminic-Acid	1-G05	L-Alanine
1-A02	L-Arabinose	1-D04	1,2-Propanediol	1-G06	L-Alanyl-Glycine
1-A03	N-Acetyl-D-Glucosamine	1-D05	Tween-40	1-G07	Acetoacetic-Acid
1-A04	D-Saccharic-Acid	1-D06	α-Keto-Glutaric-Acid	1-G08	N-Acetyl-ß-D-Mannosamine
1-A05	Succinic-Acid	1-D07	α-Keto-Butyric-Acid	1-G09	Mono-Methyl-Succinate
1-A06	D-Galactose	1-D08	ß-Methyl-D-Galactoside	1-G10	Methyl-Pyruvate
1-A07	L-Aspartic-Acid	1-D09	α-D-Lactose	1-G11	D-Malic-Acid
1-A08	L-Proline	1-D10	Lactulose	1-G12	L-Malic-Acid
1-A09	D-Alanine	1-D11	Sucrose	1-H01	Glycyl-L-Proline
1-A10	D-Trehalose	1-D12	Uridine	1-H02	p-Hydroxy-Phenyl-Acetic-Acid
1-A11	D-Mannose	1-E01	L-Glutamine	1-H03	m-Hydroxy-Phenyl-Acetic-Acid
1-A12	Dulcitol	1-E02	M-Tartaric-Acid	1-H04	Tyramine
1-B01	D-Serine	1-E03	D-Glucose-1-Phosphate	1-H05	D-Psicose
1-B02	D-Sorbitol	1-E04	D-Fructose-6-Phosphate	1-H06	L-Lyxose
1-B03	Glycerol	1-E05	Tween-80	1-H07	Glucuronamide
1-B04	L-Fucose	1-E06	α-Hydroxy-Glutaric-Acid-G-	1-H08	Pyruvic-Acid
1-B05	D-Glucuronic-Acid	1-E07	α-Hydroxy-Butyric-Acid	1-H09	L-Galactonic-Acid-G-Lactone
1-B06	D-Gluconic-Acid	1-E08	ß-Methyl-D-Glucoside	1-H10	D-Galacturonic-Acid
1-B07	D,L-A-Glycerol-Phosphate	1-E09	Adonitol	1-H11	Phenylethylamine
1-B08	D-Xylose	1-E10	Maltotriose	1-H12	2-Aminoethanol
1-B09	L-Lactic-Acid	1-E11	2-Deoxy-Adenosine	2-A01	Negative-Control
1-B10	Formic-Acid	1-E12	Adenosine	2-A02	Chondroitin-Sulfate-C
1-B11	D-Mannitol	1-F01	Glycyl-L-Aspartic-Acid	2-A03	α-Cyclodextrin
1-B12	L-Glutamic-Acid	1-F02	Citric-Acid	2-A04	ß-Cyclodextrin
1-C01	D-Glucose-6-Phosphate	1-F03	M-Inositol	2-A05	γ-Cyclodextrin
1-C02	D-Galactonic-Acid-G-Lactone	1-F04	D-Threonine	2-A06	Dextrin
1-C03	D,L-Malic-Acid	1-F05	Fumaric-Acid	2-A07	Gelatin
1-C04	D-Ribose	1-F06	Bromo-Succinic-Acid	2-A08	Glycogen
1-C05	Tween-20	1-F07	Propionic-Acid	2-A09	Inulin
1-C06	L-Rhamnose	1-F08	Mucic-Acid	2-A10	Laminarin
1-C07	D-Fructose	1-F09	Glycolic-Acid	2-A11	Mannan
1-C08	Acetic-Acid	1-F10	Glyoxylic-Acid	2-A12	Pectin
1-C09	α-D-Glucose	1-F11	D-Cellobiose	2-B01	N-Acetyl-D-Galactosamine
1-C10	Maltose	1-F12	Inosine	2-B02	N-Acetyl-Neuraminic-Acid
1-C11	D-Melibiose	1-G01	Glycyl-L-Glutamic-Acid	2-B03	ß-D-Allose
1-C12	Thymidine	1-G02	Tricarballylic-Acid	2-B04	Amygdalin
1-D01	L-Asparagine	1-G03	L-Serine	2-B05	D-Arabinose
1-D02	D-Aspartic-Acid	1-G04	L-Threonine	2-B06	D-Arabitol

#Cond.	Medium Condition	#Cond.	Medium Condition	#Cond.	Medium Condition
2-B07	L-Arabitol	2-G05	Glycine	3-D03	L-Pyroglutamic-Acid
2-B08	Arbutin	2-G06	L-Histidine	3-D04	Hyroxylamine
2-B09	2-Deoxy-D-Ribose	2-G07	L-Homoserine	3-D05	Methylamine
2-B10	I-Erythritol	2-G08	Hydroxy-L-Proline	3-D06	N-Amylamine
2-B11	D-Fucose	2-G09	L-Isoleucine	3-D07	N-Butylamine
2-B12	3-0-β-D-Galactopyranosyl	2-G10	L-Leucine	3-D08	Ethylamine
2-C01	Gentiobiose	2-G11	L-Lysine	3-D09	Ethanolamine
2-C02	L-Glucose	2-G12	L-Methionine	3-D10	Ethylenediamine
2-C03	Lactitol	2-H01	L-Ornithine	3-D11	Putrescine
2-C04	D-Melezitose	2-H02	L-Phenylalanine	3-D12	Agmatine
2-C05	Maltitol	2-H03	L-Pyroglutamic-Acid	3-E01	Histamine
2-C06	α-Methyl-D-Glucoside	2-H04	L-Valine	3-E02	B-Phenylethylamine
2-C07	β-Methyl-D-Galactoside	2-H05	D,L-Carnitine	3-E03	Tyramine
2-C08	3-Methyl-Glucose	2-H06	Sec-Butylamine	3-E04	Acetamide
2-C09	β-Methyl-D-Glucuronic-Acid	2-H07	D.L-Octopamine	3-E05	Formamide
2-C10	α -Methyl-D-Mannoside	2-H08	Putrescine	3-E06	Glucuronamide
2-C11	β -Methyl-D-Xyloside	2-H09	Dihydroxy-Acetone	3-E07	D,L-Lactamide
2-C12	Palatinose	2-H10	2,3-Butanediol	3-E08	D-Glucosamine
2-D01	D-Raffinose	2-H11	2,3-Butanone	3-E09	D-Galactosamine
2-D02	Salicin	2-H12	3-Hydroxy 2- Butanone	3-E10	D-Mannosamine
2-D03	Sedoheptulosan	3-A01	Negative-Control	3-E11	N-Acetyl-D-Glucosamine
2-D04	L-Sorbose	3-A02	Ammonia	3-E12	N-Acetyl-D-Galactosamine
2-D05	Stachyose	3-A03	Nitrite	3-F01	N-Acetyl-D-Mannosamine
2-D06	D-Tagatose	3-A04	Nitrate	3-F02	Adenine
2-D07	Turanose	3-A05	Urea	3-F03	Adenosine
2-D08	Xylitol	3-A06	Biuret	3-F04	Cytidine
2-D09	N-Acetyl-D-Glucosaminitol	3-A07	L-Alanine	3-F05	Cytosine
2-D10	δ-Amino-Butyric-Acid	3-A08	L-Arginine	3-F06	Guanine
2-D11	δ-Amino-Valeric-Acid	3-A09	L-Asparagine	3-F07	Guanosine
2-D12	Butyric-Acid	3-A10	L-Aspartic-Acid	3-F08	Thymine
2-E01	Capric-Acid	3-A11	L-Cysteine	3-F09	Thymidine
2-E02	Caproic-Acid	3-A12	L-Glutamic-Acid	3-F10	Uracil
2-E03	Citraconic-Acid	3-B01	L-Glutamine	3-F11	Uridine
2-E04	Citramalic-Acid	3-B02	Glycine	3-F12	Inosine
2-E05	D-Glucosamine	3-B03	L-Histidine	3-G01	Xanthine
2-E06	2-Hydroxy-Benzoic-Acid	3-B04	L-Isoleucine	3-G02	Xanthosine
2-E07	4-Hydroxy-Benzoic-Acid	3-B05	L-Leucine	3-G03	Uric-Acid
2-E08	B-Hydroxy-Butyric-Acid	3-B06	L-Lysine	3-G04	Alloxan
2-E09	G-Hydroxy-Butyric-Acid	3-B07	L-Methionine	3-G05	Allantoin
2-E10	α-Keto-Valeric-Acid	3-B08	L-Phenylalanine	3-G06	Parabanic-Acid
2-E11	Itaconic-Acid	3-B09	L-Proline	3-G07	D,L-A-Amino-N-Butyric-Acid
2-E12	5-Keto-D-Gluconic-Acid	3-B10	L-Serine	3-G08	γ-Amino-N-Butyric-Acid
2-F01	D-Lactic-Acid-Methyl-Ester	3-B11	L-Threonine	3-G09	ε-Amino-N-Caproic-Acid
2-F02	Malonic Acid	3-B12	L-Tryptophan	3-G10	D,L-A-Amino- Caprylic-Acid
2-F03	Melibionic-Acid	3-C01	L-Tyrosine	3-G11	δ -Amino-N-Valeric-Acid
2-F04	Oxalic-Acid	3-C02	L-Valine	3-G12	α-Amino-N-Valeric-Acid
2-F05	Oxalomalic-Acid	3-C03	D-Alanine	3-H01	Ala-Asp
2-F06	Quinic-Acid	3-C04	D-Asparagine	3-H02	Ala-Gln
2-F07	D-Ribono-1,4- Lactone	3-C05	D-Aspartic-Acid	3-H03	Ala-Glu
2-F08	Sebacic-Acid	3-C06	D-Glutamic-Acid	3-H04	Ala-Gly
2-F09	Sorbic-Acid	3-C07	D-Lysine	3-H05	Ala-His
2-F10	Succinamic-Acid	3-C08	D-Serine	3-H06	Ala-Leu
2-F11	D-Tartaric-Acid	3-C09	D-Valine	3-H07	Ala-Thr
2-F12	L-Tartaric-Acid	3-C10	L-Citrulline	3-H08	Gly-Asn
2-G01	Acetamide	3-C11	L-Homoserine	3-H09	Gly-Gln
2-G02	L-Alaninamide	3-C12	L-Ornithine	3-H10	Gly-Glu
2-G03	N-Acetyl-L-Glutamic-Acid	3-D01	N-Acetyl-D,L-Glutamic-Acid	3-H11	Gly-Met
2-G04	L-Arginine	3-D02	N-Phthaloyl-L-Glutamic-Acid	3-H12	Met-Ala

References

[1] Biochner, B.R., Gadzinski, P., and Panomitros, E., Phenotype microarrays for high-throughput phenotypic testing and assay of gene function, *Genome Res.*, 11(7):1246-1255, 2001.
[2] Ishii, N., Nakahigashi, K., Baba, T., Robert, M., Soga, T., Kanai A., Hirasawa T., Naba M., Hirai K., Hoque A., Ho P.Y., Kakazu Y., Sugawara K., Igarashi S., Harada S., Masuda T., Sugiyama N., Togashi T., Hasegawa M., Takai Y., Yugi K., Arakawa K., Iwata N., Toya Y., Nakayama Y., Nishioka T., Shimizu K., Mori H., and Tomita M., Multiple high-throughput analyses monitor the response of E. coli to perturbations, *Science*, 316(5824):593-597, 2007.
[3] Bochner, B.R., Sleuthing out bacterial identities, *Nature*, 339(6220):157-158, 1989.
[4] Koo, B.M., Yoon, M.J., Lee, C.R., Nam, T.W., Choe, Y.J., Jaffe, H., Peterkofsky, A., and Seok, Y.J., A novel fermentation/respiration switch protein regulated by enzyme IIAGlc in Escherichia coli., *J Biol Chem.*, 279(30):31613-31621, 2004.
[5] Sauer, J.D., Bachman, M.A., and Swanson, M.S., The phagosomal transporter A couples threonine acquisition to differentiation and replication of Legionella pneumophila in macrophages, *Proc. Natl. Acad. Sci. U.S.A*, 102(28):9924-9929, 2005.
[6] Ito, M., Baba, T., and Mori, H., Functional analysis of 1440 Escherichia coli genes using the combination of knock-out library and phenotype microarrays, *Metabolic Engineering*, 7(4):318-327, 2005.
[7] Baba, T., Ara, T., Hasegawa, M., Takai, Y., Okumura, Y., Baba, M., Datsenko, K.A., Tomita, M., Wanner, B.L., and Mori, H., Construction of Escherichia coli K-12 in-frame, single-gene knockout mutants: the Keio collection, *Mol Syst Biol.*, 2:2006 0008, 2006.
[8] Datsenko, K.A. and Wanner, B.L., One-step inactivation of chromosomal genes in Escherichia coli K-12 using PCR products, *Proc. Natl. Acad. Sci. U.S.A*, 97(12): 6640-6645, 2000.
[9] Kanehisa, M., Goto, S., Kawashima, S., Okuno, Y., and Hattori, M., The KEGG resource for deciphering the genome, *Nucleic Acids Res.*, 32:D277-280, 2004.
[10] Everitt, B. S., Landau, S., and Leese, M., *Cluster Analysis*, 4th edition. Arnold Publishers, 2001.
[11] Tavazoie, S., Hughes, J.D., Campbell, M.J., Cho, R.J., and Church, G.M., Systematic determination of genetic network architecture, *Nat Genet.* 22(3):281-285, 1999.
[12] Arita, M., The metabolic world of Escherichia coli is not small, *Proc. Natl. Acad. Sc. U.S.A*, 101(6):1543-1547, 2004.
[13] Inada, T., Kimata, K., and Aiba, H., Mechanism responsible for glucose-lactose diauxie in Escherichia coli: challenge to the cAMP model, *Genes to Cells*, 1(3):293-301, 1996.

IMPROVED ALGORITHMS FOR ENUMERATING TREE-LIKE CHEMICAL GRAPHS WITH GIVEN PATH FREQUENCY

YUSUKE ISHIDA[1]
yusukei@amp.i.kyoto-u.ac.jp

LIANG ZHAO[1]
liang@amp.i.kyoto-u.ac.jp

HIROSHI NAGAMOCHI[1]
nag@amp.i.kyoto-u.ac.jp

TATSUYA AKUTSU[2]
takutsu@kuicr.kyoto-u.ac.jp

[1] *Department of Applied Mathematics and Physics, Graduate School of Informatics, Kyoto University, Yoshida, Kyoto 606-8501, Japan*
[2] *Bioinformatics Center, Institute for Chemical Research, Kyoto University, Gokasho, Uji, Kyoto 611-0011, Japan*

This paper considers the problem of enumerating all non-isomorphic tree-like chemical graphs with given path frequency, where "tree-like" means that the graph can be viewed as a tree if multiple edges (i.e., edges with the same end points) and a benzene ring are treated as one edge and one vertex, respectively, and "path frequency" is a vector of the numbers of specified vertex-labeled paths that must be realized in every output. This and related problems have several potential applications such as classification of chemical compounds, structure determination using mass-spectrum and/or NMR and design of novel chemical compounds.

For this problem, several studies have been done. Recently, Fujiwara *et al.* (2008) showed two formulations and for each of them, they gave a branch-and-bound algorithm, which combined efficient enumeration of non-isomorphic trees with bounding operations based on the path frequency and the atom-atom bonds to avoid the generation of invalid trees. In this paper, based on their work and a result of Nagamochi (2006), we introduce two new bounding operations, the detachment-cut and the H-cut, to further reduce the size of the search space. We performed computational experiments to compare our proposed algorithms with those of Fujiwara *et al.* (2008) using some chemical compound data obtained from the KEGG LIGAND database (`http://www.genome.jp/kegg/ligand.html`). The results show that our proposed algorithms are much faster than their algorithms.

Keywords: chemical graph enumeration; chemical tree enumeration; path frequency; feature vector; detachment.

1. Introduction

Enumerating chemical graphs is one of the fundamental issues in chemoinformatics and bioinformatics which can go back to the 19th century (Cayley [6]). Its applications include structure determination using mass-spectrum and/or NMR-spectrum [5, 11], virtual exploration of chemical universe [9, 15], reconstruction of molecular structures from their signatures [8, 12], and classification of chemical compounds [7]. In these applications, enumeration of chemical graphs satisfying given constraints is important [2].

This paper considers to enumerate chemical compounds with given path frequency, i.e., the numbers of specified vertex-labeled paths that must be realized in every output. The problem was motivated from the *pre-image problem* in machine learning [4]. In the pre-image problem, given a *feature vector*, a structure consistent with the feature vector is computed. The pre-image problem for chemical graphs has a potential application to design of novel chemical compounds [2, 4], which is an important target of bioinformatics. Suppose that we have some potential function in a feature space, which reflects the pharmacological activity of chemical compounds and may be learned from training data. Then, a desired object is computed as a point in the feature space using the potential function and some optimization technique. Finally, a pre-image of the point is computed as a candidate of a novel chemical compound. Though this approach has not yet been shown to be better than existing approaches, there is a possibility that chemical structures, which have better pharmacological activity than training data, are obtained. Since feature vectors based on frequency of labeled paths were successfully applied to classification of chemical compounds [13, 14], we consider the graph pre-image problem with given path frequency.

Akutsu and Fukagawa [1] first studied the computational complexity of the pre-image problem. They proved that the problem is NP-hard even if chemical graphs are restricted to trees. Since the problem is NP-hard and it is quite difficult to handle all chemical graphs, they developed a branch-and-bound algorithm for tree-like chemical graphs [2]. Recently, Fujiwara *et al.* [10] proposed a much more efficient branch-and-bound algorithm, which combined the tree enumeration algorithm of Nakano and Uno [17, 18] to generate non-isomorphic trees with bounding operations based on the path frequency and the atom-atom bonds to avoid the generation of invalid trees. To improve the efficiency, they also gave an alternative formulation of the problem by removing all hydrogens and replacing each group of multiple edges with a new virtual atom and two new single edges. Experimental results show that some instances up to 61 atoms could be enumerated within 30 minutes using a normal PC. Their algorithms can also be applied to a classical problem of enumeration of alkanes (C_nH_{2n+2}), which was considered by Cayley [6], and the latter was shown to be at least as fast as the fastest existing algorithm [3].

In order to further improve the efficiency, we introduce two new bounding operations in this paper. The first, the *detachment-cut* is based on a result of Nagamochi [16]. Another, the *H-cut* can only be applied to the second formulation, which uses the information of the removed hydrogens. We show that they can effectively reduce the size of the search space and thereby reduce the running time. Applying to the same instances, we show that our algorithms are much faster than those of [10]. The proposed algorithms are faster than those of [10] in all the examined cases and are dozens of times faster in many cases. As in [10], our algorithms can be extended for treating benzene rings too.

The rest of the paper is organized as follows. Section 2 gives preliminaries and the first formulation. Section 3 shows the branch-and-bound framework following

[10] and the new detachment-cut bounding operation. Section 4 gives the second formulation and the new algorithm that employs both the detachment-cut and the H-cut. We report in Section 5 some experimental results and conclude in Section 6.

2. Preliminaries and problem formulation

A graph is called a *multigraph* if multiple edges are allowed; otherwise it is *simple*. A *multitree* is a multigraph with no cycle nor self-loop. A *path* P is a sequence $v_0, e_1, v_1, e_2, v_2, \ldots, e_k, v_k$ of distinct vertices v_i and edges e_j which join v_{j-1} and v_j, $j = 1, \ldots, k$. Without confusion we may write $P = (v_0, v_1, \ldots, v_k)$. The length $|P|$ of path P is defined by k, i.e., the number of edges.

We are given a set $\Sigma = \{\ell_1, \ell_2, \ldots, \ell_s\}$ of s labels, which correspond chemical elements. Let each label ℓ be associated with a valence $val(\ell) \in \mathbb{Z}_+$, where $\mathbb{Z}_+$ denotes the set of non-negative integers. A multigraph G is said Σ-*labeled* if each vertex v has a label $\ell(v) \in \Sigma$, and is called (Σ, val)-*labeled* if, in addition, the degree of each vertex v is $val(\ell(v))$, i.e., the valence of the element $\ell(v)$. Chemical compounds that can be viewed as (Σ, val)-labeled, self-loopless and connected multigraphs, where vertices and labels represent atoms and elements, respectively. For a path $P = (v_0, v_1, \ldots, v_k)$, we call $\ell(P) = \ell(v_0), \ell(v_1), \ldots, \ell(v_k)$ the *label sequence* of P. Given a label sequence t, let $\#t$ denote the number of paths P with $\ell(P) = t$ in the graph, where multiple edges are treated as a single edge and paths are considered "directed." The *feature vector* $f_K(G)$ *of level* $K \in \mathbb{Z}_+$ of G is defined as the $p(K, s)$-dimensional vector whose entry $f_K(G)[t]$ ($|t| \leq K$) represents $\#t$, where $p(K, s) = (s^{K+2} - s)/(s - 1)$ for $s > 1$ and $p(K, 1) = K + 1$. Fig. 1 illustrates an example.

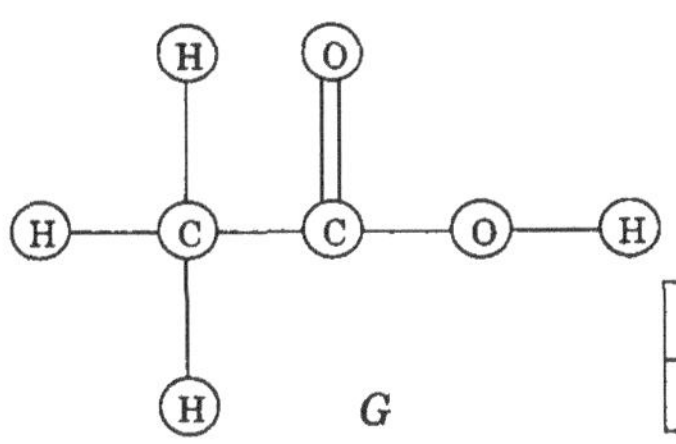

$\Sigma = \{C, O, H\}$

$val(C) = 4$, $val(O) = 2$, $val(H) = 1$

feature vector of level 1

H	O	C	HH	HO	HC	OH	OO	OC	CH	CO	CC
4	2	2	0	1	3	1	0	2	3	2	2

Fig. 1. An illustration of a (Σ, val)-labeled multitree G and $f_1(G)$, where multiple edges are treated as one edge and paths are considered "directed."

Let $deg(v; G)$ denote the degree of a vertex v in a graph G. The problem can be formulated as follows (an alternative formulation will be given in Section 4).

Problem 1. *Given a set Σ of s labels, a valence function $val : \Sigma \to \mathbb{Z}_+$ and a feature vector g of level K, find all (Σ, val)-labeled multitrees T such that $f_K(T) = g$ and $deg(v; T) = val(\ell(v))$ for all vertices $v \in T$.*

For a given feature vector g, the entry $g(t)$ specifies $\#t$ in an output graph. In particular, the number n of vertices is decided by $\sum_{\ell \in \Sigma} g(\ell)$. To solve the problem, we start with an empty graph, and repeatedly extend the current tree T by appending a new vertex with each label $\ell \in \Sigma$ to obtain a *valid* tree (a tree that has not violated any constraints on output trees) by one vertex until we get n vertices. In order to avoid duplicate outputs, we follow the branch-and-bound framework of [10] which first defines a canonical representation for isomorphic trees, then lists them using the algorithm of [17, 18] (the branching operation) and discards invalid trees using some bounding operations.

3. The enumeration algorithm

Given a simple tree with n vertices, the valence constraint uniquely determines the multiplicities of all edges. Thus we consider listing all non-isomorphic Σ-labeled *simple* trees and get/check the corresponding multitrees by the valence constraint. The framework of the enumeration algorithm is shown in Fig. 2.

```
Main:
  FOR all labels ℓ ∈ Σ DO
    Let T be the tree consisting of one (root) vertex labeled by ℓ
    Gen(T)
  DONE

Gen(T):
  IF T has n vertices THEN
    Check if T is valid. If it is, output it.
  ELSE
    Extend T to T_1, T_2, ..., T_p for some finite p by appending a new leaf vertex
    FOR all such trees T_i DO
      Check if T_i is valid. If it is, call Gen(T_i) (do nothing otherwise).
    DONE
  ENDIF
```

Fig. 2. The framework of the enumeration algorithm.

Section 3.1 reviews the way of extending trees (the branching operation), which is exactly the same as [10]. Section 3.2 describes how the validity is checked by several bounding operations, three from [10] and a new *detachment-cut*.

3.1. *Canonical representation of trees and the branching operation*

First of all, we need a representation for the output that must be unique for isomorphic trees. For this purpose, we use the idea of *centroid-rooted left-heavy tree* [10], where centroid is defined from the next theorem (see also [3]).

Theorem 3.1 (Jordan 1869). *For any tree with n vertices, either there exists a unique vertex v^* such that each subtree obtained by removing v^* contains at most*

$\lfloor \frac{n-1}{2} \rfloor$ vertices, or there exists a unique edge e^* such that both of the subtrees obtained by removing e^* contain exactly $\frac{n}{2}$ vertices. □

Such a vertex v^* (resp., edge e^*) is called the *unicentroid* (resp., *bicentroid*) of the tree. For example, the tree in Fig. 1 has a bicentroid (the C-C edge). To introduce"left-heaviness," we need an *ordering* among rooted trees.

Let T be a tree of n vertices rooted at a vertex v_0 (which is not necessarily its centroid). Suppose that it is embedded in the plane, where v_0 is the top. Let $v_0, v_1, \ldots, v_{n-1}$ be indexed by the depth-first search (DFS) that starts from v_0 and visits vertices from the left to the right. The *depth* $d(v)$ of a vertex v is defined as the length of the path from v_0 to v in T. The *depth-label sequence* of T is defined as

$$DL(T) = (d(v_0), \ell(v_0), d(v_1), \ell(v_1), \ldots, d(v_{n-1}), \ell(v_{n-1})).$$

We say that T is rooted at an edge (v_0, v_1) if v_0 and v_1 are the two tops, where we define $d(v)$ by the minimum of the lengths of the v_0, v-path and the v_1, v-path. Then $DL(T)$ can be defined as before. Now we have a one-to-one mapping between plane-embedded trees and label sequences. See Fig. 3 for an illustration.

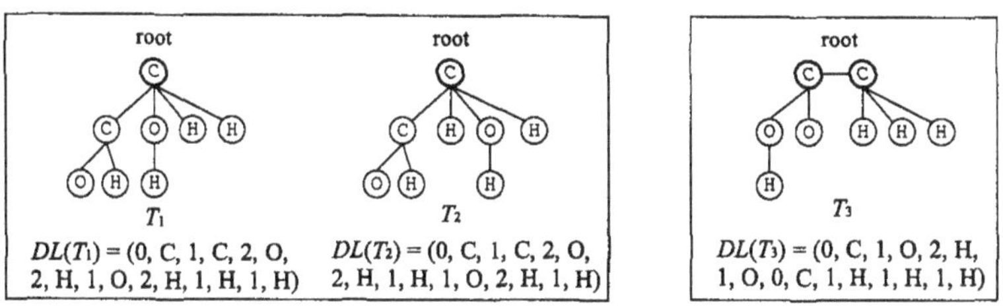

Fig. 3. Rooted trees and their depth-label sequences. Notice that T_1 and T_2 are isomorphic.

Given an (arbitrary) order of labels, define the order of depth-label sequences as follows. For any T_1 and T_2, we say $DL(T_1) > DL(T_2)$ if $DL(T_1)$ is *lexicographically* larger than $DL(T_2)$. Similarly we can define $DL(T_1) \geq DL(T_2)$ straightforward. In Fig. 3, we have $DL(T_1) > DL(T_2) > DL(T_3)$, supposing C > O > H. The *canonical representation* of a rooted tree is defined by the *largest* depth-label sequence among all its plane embeddings. This is equivalent to the *left-heavy* plane embedding (see [17, 18]); i.e., any two siblings (vertices having the same parent or the two vertices of the edge root) v_i and v_j with $i < j$ satisfy $DL(\mathbf{T}(v_i)) \geq DL(\mathbf{T}(v_j))$, where $\mathbf{T}(v)$ denotes the subtree consisting of v and all its descendants. For example, T_1 and T_3 in Fig. 3 are left-heavy whereas T_2 is not.

Thus our branching task is to list all centroid-rooted left-heavy trees with n vertices and m or less labels. Following the scheme of [17, 18], we define a *parent-child* relation between two left-heavy trees. The *parent* $P(T)$ of a left-heavy tree T is obtained from T by removing its *rightmost* leaf. If T is rooted at a vertex or an edge (v_0, v_1) but v_1 is not the rightmost leaf, then the root of $P(T)$ remains unchanged. Otherwise we change the root to *vertex* v_0 since v_1 is removed. Clearly $P(T)$ is still left-heavy. In this way we can define a *family tree* $\mathcal{F}(n, m)$ of left-heavy trees whose leaves are exactly what we want, i.e., the centroid-rooted left-heavy trees with n vertices and m or less labels. Notice that, in general, a non-leaf node in the family tree may not be rooted at its own centroid.

Therefore we only need to enumerate the (leaf) nodes of $\mathcal{F}(n, m)$. This can be done by starting from the empty tree (the root node of $\mathcal{F}(n, m)$) and repeatedly appending a new leaf to some appropriate place on the rightmost path. For that purpose, our branching operation employs the algorithm due to [17, 18], which extends the current tree T (i.e., finds a child of T) in *constant* time. See [10] for detail.

3.2. *Bounding operations*

Next we explain how to check the validity of a tree T generated during the branching operation. If we can conclude that T and all its descendants are not valid, then we can discard T, i.e., skip the task of appending leaves to T. Our branching operation discards T if at least one of the following criteria is violated.

(C1) The root of T remains the centroid of an output (the centroid constraint);
(C2) $f_K(T) \leq g$ (the feature vector constraint);
(C3) $deg(v; T) \leq val(\ell(v))$ for all $v \in T$ (the valence constraint);
(C4) T can be extended to a connected and loopless tree with n vertices (the detachment constraint).

The first three are the same as [10], and not difficult to check (see [10]). In the following, we explain how to check the last one. We need some definitions.

Let $G = (V, E)$ be a multigraph which may have self-loops. Given a function $r : V \to \mathbb{Z}_+$, an *r-detachment* of G is a multigraph H obtained from G by splitting each vertex $v \in V$ into a set of $r(v)$ copies of v, denoted by $W_v = \{v^1, v^2, \ldots, v^{r(v)}\}$, so that each edge (u, v) in G is mapped to a distinct edge (u^i, v^j) in H for some $u^i \in W_u$ and $v^j \in W_v$, where a self-loop (u, u) in G may be mapped to a self-loop (u^i, u^i) or a non-loop edge (u^i, u^j) in H. Notice that, for all vertex pairs $\{u, v\} \subseteq V$, the number of edges in H between W_u and W_v is equal to that in G between u and v. (We note that an r-detachment may not be unique in general.)

An *r-degree specification* is a set ρ of vectors $\rho(v) = (\rho_1^v, \rho_2^v, \ldots, \rho_{r(v)}^v)$ such that $\sum_{1 \leq i \leq r(v)} \rho_i^v = deg(v; G)$ for all $v \in V$. An r-detachment H is called a *ρ-detachment* if $deg(v^i; H) = \rho_i^v$, for all $v \in V$, and $i = 1, 2, \ldots, r(v)$. See Fig. 4 for an illustration.

Theorem 3.2 (Nagamochi [16]). *Given $G = (V, E)$, $r : V \to \mathbb{Z}_+$ and an r-degree specification ρ, G has a connected and loopless ρ-detachment if and only if*

$$r(X) + c(G - X) - d(X, V; G) \leq 1, \qquad \emptyset \neq X \subseteq V,$$
$$1 \leq \rho_i^v \leq d(v; G) + d(\{v\}, \{v\}; G), \quad v \in V, \ i = 1, 2, \ldots, r(v),$$

where $r(X) = \Sigma_{v \in X} r(v)$, $G - X$ denotes the graph obtained from a graph G by removing the vertices in X together with all edges incident to them, $c(G-X)$ denotes

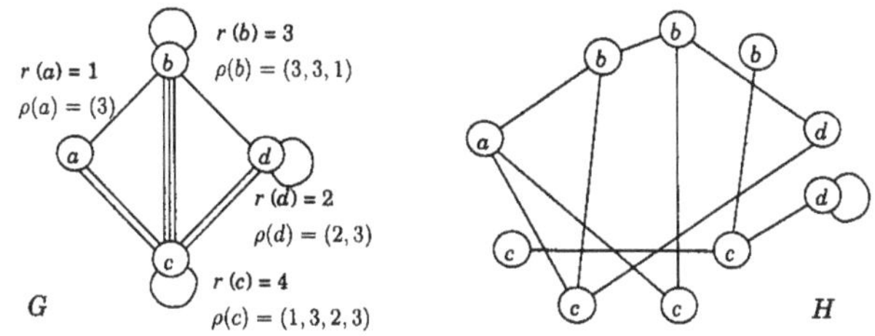

Fig. 4. An illustration of a multigraph G and a ρ-detachment H of G

the number of connected components of graph $G - X$, and $d(A, B; G)$ denotes the number of edges $(u, v) \in E$ with $u \in A$ and $v \in B$. □

Using this theorem, we can check if a partial multitree T violates **(C4)**. Let $RP(T) = (r_0, r_1, \ldots, r_k)$ be the rightmost path of T, and let $r_0, \ldots, r_h$ $(h \le k)$ be the vertices to which a new leaf can be attached without violating the left-heavy property (see [10] for how to do this). Recall $\ell_1, \ell_2, \ldots, \ell_s$ and g are the given labels and the feature vector, respectively. Let n_i^R $(1 \le i \le s)$ be the number of vertices r_j $(0 \le j \le h)$ with $\ell(r_j) = \ell_i$. Introducing a new label ℓ_{s+1} of valence $h+1$, we define a new feature vector g' of level 1 by

$$g'(\ell_i) = \begin{cases} g(\ell_i) - \#\ell_i + n_i^R & 1 \le i \le s \\ 1 & i = s+1, \end{cases}$$

$$g'(\ell_i \ell_j) = \begin{cases} g(\ell_i \ell_j) - \#\ell_i \ell_j & 1 \le i,\ j \le s \\ n_i^R & 1 \le i \le s,\ j = s+1. \end{cases}$$

(Recall $\#t$ denotes the number of paths in T of label sequence t.) We construct an auxiliary graph $G = (V, E)$ by $V = \{\ell_1, \ldots, \ell_s, \ell_{s+1}\}$ and $E = \{e_{ij} | e_{ij} = (\ell_i, \ell_j), d(\{\ell_i\}, \{\ell_j\}; G) = g'(\ell_i \ell_j), 1 \le i, j \le s+1\}$ where $d(\{\ell_i\}, \{\ell_j\}; G)$ means the multiplicity of edge e_{ij}. The function r and the degree specification ρ are defined as follows (see Fig. 5 for an illustration of G).

$$r(v) = g'(\ell_i), \qquad \ell(v) = \ell_i, 1 \le i \le s+1,$$

$$\rho_i^v = \begin{cases} val(\ell(v_i)) & v_i \notin \{r_0, \ldots, r_h\},\ 1 \le i \le r(v) \\ val(\ell(v_i)) - deg(v_i; T) + 1 & v_i \in \{r_0, \ldots, r_h\},\ 1 \le i \le r(v). \end{cases}$$

If G has no ρ-detachment, then T cannot be extended to a connected and loopless tree with n vertices. The new label (label A in Fig. 5) is introduced in order to ensure the existence of the edges (r_i, r_{i+1}), $i = 0, 1, \ldots, h-1$. By Theorem 3.2, we only need to check if one or more of the next two conditions is violated.

(a) $\Sigma_{1 \le i \le r(v)}\ \rho_i^v \ge deg(v; G)$, $\forall v \in V$.
(b) $r(X) + c(G - X) - d(X, V; G) \le 1$, $\emptyset \ne X \subseteq V$.

Notice that condition (a) is not equality because the feature vector counts multiple edges as one edge. Our detachment-cut discards T if any of (a) and (b) is violated.

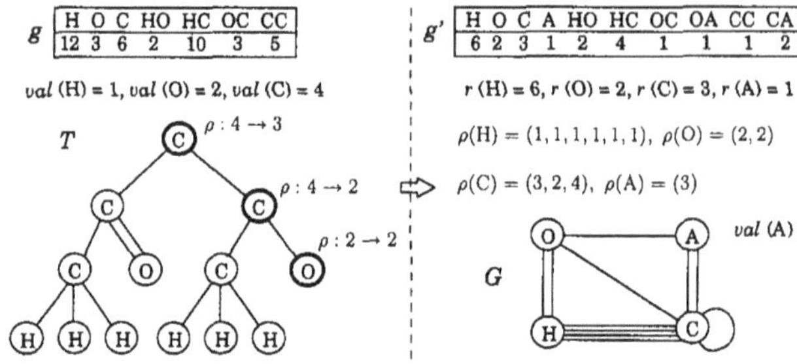

Fig. 5. An illustration of how to construct a graph G for checking the validity of T using the detachment-cut, where we omit symmetric and zero entries in the feature vectors.

We remark that condition (b) has $2^{s+1} - 1$ inequalities, but usually it is small because s is very small. E.g., s is 2 for alkanes and 5 in our experiments.

4. Alternative problem formulation

We also follows the second problem formulation in [10], which use two kinds of graph transformation. First the *H-removal transformation* reduces the size of compounds by removing hydrogens. Then the *single-bond transformation* replaces multiple edges with a new virtual atom and two new simple edges joining the same end points. Fig. 6 illustrates these two transformations.

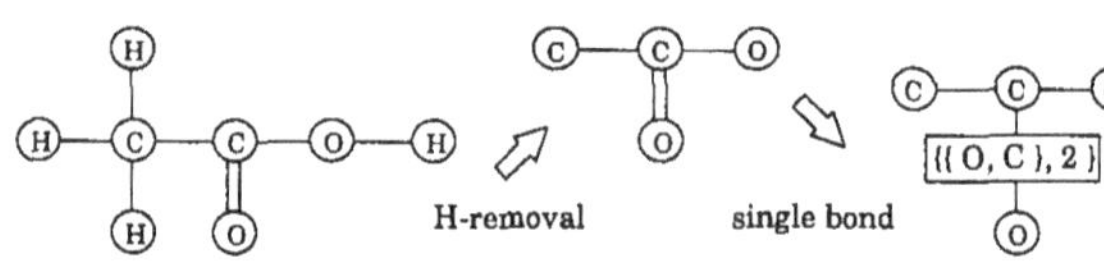

Fig. 6. An illustration of the H-removal and single-bond transformations.

When the single-bond transformation replaces multiple edges (u, v) by a new vertex w and two new simple edges (u, w) and (w, v), we define the *bond label* $\ell(w)$ of w by $\ell(w) = (\{\ell(u), \ell(v)\})$, and define the *bond valence* of $\ell(w)$ by the multiplicity of (u, v). Let C_Σ be the set of all such bond labels and $\Sigma^* = \Sigma \cup C_\Sigma$. For each vertex $v \in \Sigma^*$, its *bond degree*, y $\widetilde{deg}(v; T)$ is defined as the number of vertices adjacent to v. We consider the next formulation.

Problem 2. *Given a set of labels Σ^*, a feature vector g of level K, and a valence function $val : \Sigma \to \mathbb{Z}_+$, find all Σ^*-labeled simple trees $T^* = (V^*, E^*)$ that satisfy $f_K(T^*) = g$ and $\widetilde{deg}(v; T^*) \leq val(\ell(v))$ for all $v \in V^*$.*

To solve this, we follow the aforementioned framework with the same branching operation. The bounding operations are somewhat different, however. In fact, we can still employ bounding operations based on the four criteria **(C1)-(C4)** as stated in Section 3.2 (notice that Problem 2 considers only simple trees). Moreover, we introduce a new *H-cut* bounding operation, which discards the partial tree T being checked if the number of hydrogens that must be appended to T and any of its descendants in order to restore the compound exceeds a pre-calculated limit.

Formally, we first calculate the numbers $h^*(\ell)$, $\ell \in \Sigma$, of hydrogens that must be appended to vertices labeled ℓ. It is easy to see that this can be done from the input feature vector of level 1 and the valence function. The H-cut checks if (a lower bound on) the number of hydrogens that must be appended to the ℓ-labeled vertices in T exceeds $h^*(\ell)$ for each $\ell \in \Sigma$. We use the next lower bound

$$h(\ell; T) = \Sigma\{(val(\ell(v)) - deg(v; T)) \mid v \in T \setminus RP(T), \ell(v) = \ell\}.$$

(Recall T and all descendants of T in the family tree share the common structure of $T \setminus RP(T)$.) See an illustration in Fig. 7.

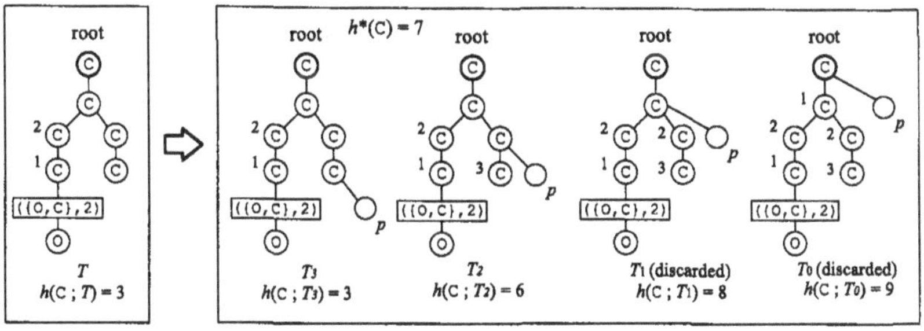

Fig. 7. An illustration of the H-cut procedure, where only label C is being considered, in which numbers $val(\ell(v)) - deg(v; T)$ are shown near each carbons not on the rightmost path.

5. Experimental Results

We conducted computational experiments to compare the running time of our algorithms with [10] using the same instances, which were obtained by randomly picking up some tree-like compounds from the KEGG LIGAND database (`http://www.genome.jp/kegg/ligand.html`) and replacing each benzene ring by a new virtual element of valence 6. Feature vectors were calculated for levels $1, 2, \ldots, 7$. For Problem 2, we preprocessed the instances with the H-removal and single-bond transformations. The experimental results were performed on a PC with a Pentium4 3.00GHz CPU. Tables 1 and 2 show the experimental results for Problems 1 and 2, respectively. We observe that the new algorithms run considerably faster than [10].

6. Conclusion

In this paper, we showed two branch-and-bound algorithms for enumerating tree-like chemical graphs from given path frequency, which are based on the framework of [10] and improved their results. In particular, we have proposed two bounding operations, the detachment-cut and the H-cut. As a future work, we are considering

Table 1. Experimental Results of Problem 1.

Entry Formula	n_1	K	Fujiwara *et al.*'s algorithm [10] time	nnt	fs	Our algorithm (this paper) time	nnt	fs
		1	T.O.	1,334,417,908	N.F.	158.23	25,149,700	570,773
		2	3.11	830,298	9	0.48	46,311	9
C03343		3	3.25	614,413	2	0.30	28,106	2
	37	4	3.06	428,440	1	0.27	21,688	1
$C_{16}H_{22}O_4$		5	3.42	391,046	1	0.26	18,616	1
		6	2.33	210,246	1	0.21	12,129	1
		7	1.85	146,605	1	0.19	10,551	1
		1	T.O.	1,407,334,896	N.F.	109.27	7,966,323	73,711
		2	50.55	16,339,119	55	1.40	95,639	55
C07530		3	16.78	3,265,086	1	0.61	35,025	1
	43	4	7.14	994,926	1	0.34	15,734	1
$C_{17}H_{28}N_2O$		5	3.28	366,628	1	0.18	7,929	1
		6	3.37	299,518	1	0.16	6,862	1
		7	3.88	299,518	1	0.18	6,862	1
		1	T.O.	1,237,087,310	N.F.	500.78	31,003,703	70,170
		2	51.72	15,827,372	16	3.51	158,597	16
C07178		3	4.26	915,962	2	0.32	15,427	2
	46	4	0.94	146,789	1	0.16	6,677	1
$C_{21}H_{28}N_2O_5$		5	1.02	123,251	1	0.15	5,485	1
		6	1.13	118,295	1	0.16	5,450	1
		7	1.00	93,947	1	0.15	5,036	1
		1	T.O.	1,428,804,364	N.F.	T.O.	456,703,633	N.F.
		2	T.O.	499,544,612	N.F.	318.68	32,927,230	1,198
C03690		3	T.O.	338,357,072	N.F.	188.13	16,574,164	8
	61	4	T.O.	254,834,091	N.F.	44.07	3,469,929	4
$C_{24}H_{38}O_4$		5	T.O.	198,785,929	N.F.	36.54	2,385,611	2
		6	T.O.	129,353,817	1	16.02	854,956	1
		7	1287.30	77,002,582	1	10.27	477,305	1

Note: (1) C03343, C07530, C07178, and C03690 are the entries of 2-Ethylhexyl phthalate, Etidocaine, Trimethobenzamide, and Bis (2-ethylhexyl) phthalate in the KEGG LIGAND database, respectively; (2) n_1 is the number of atoms in an instance preprocessed by replacing each benzene ring with a new atom with valence 6; (3) K is the level of a given feature vector; (4) "time" is the CPU time in seconds; (5) "T.O." means "time over" (the time limit was set to 1800 seconds); (6) "nnt" is the number of nodes in the family trees that are checked; (7) "fs" is the number of feasible solutions found in the time limit; and (8) "N.F." means "not found".

to enumerate more general graph classes, e.g., outerplanar graphs which are known to cover most of the chemical graphs. A preliminary work can be found in [19].

We note that the depth label sequences defined in this paper only represent the graphical structures of compounds in the viewpoint of planarity but may lose information of stereochemistry, especially for stereoisomers. Thus, designing better representations is another interesting topic for future research.

Acknowledgments

This work was supported in part by Grant-in-Aid #19200022 from the Ministry of Education, Culture, Sports, Science and Technology (MEXT) of Japan. We thank Hiroki Fujiwara and Jiexun Wang for their helpful discussions.

Table 2. Experimental Results of Problem 2.

Entry Formula	n_2	K	Fujiwara *et al.*'s algorithm [10] time	nnt	fs	Our algorithm (this paper) time	nnt	fs
		1	66.06	28,683,656	570,773	13.31	5,865,685	570,773
		2	0.03	5,157	9	0.01	3,091	9
C03343		3	0.03	4,607	2	0.02	2,780	2
	17	4	0.04	4,086	1	0.02	2,453	1
$C_{16}H_{22}O_4$		5	0.04	3,470	1	0.02	2,098	1
		6	0.04	2,909	1	0.02	1,739	1
		7	0.04	2,675	1	0.02	1,596	1
		1	10.26	4,029,246	73,711	1.00	424,121	73,711
		2	0.16	43,513	55	0.06	14,900	55
C07530		3	0.09	16,090	1	0.04	6,385	1
	16	4	0.06	8,006	1	0.02	3,736	1
$C_{17}H_{28}N_2O$		5	0.04	5,624	1	0.02	2,522	1
		6	0.04	4,642	1	0.02	2,245	1
		7	0.04	4,642	1	0.02	2,245	1
		1	222.29	96,006,467	70,170	9.03	3,909,283	70,170
		2	0.11	21,460	16	0.02	4,321	16
C07178		3	0.09	11,950	2	0.02	2,984	2
	19	4	0.03	3,152	1	0.01	1,062	1
$C_{21}H_{28}N_2O_5$		5	0.02	2,143	1	0.01	819	1
		6	0.02	2,088	1	0.01	794	1
		7	0.02	2,088	1	0.01	794	1
		1	T.O.	664,265,016	5,305,243	T.O.	708,264,977	60,257,365
		2	23.36	2,984,162	1,198	8.10	1,113,024	1,198
C03690		3	15.87	1,464,436	8	5.66	570,616	8
	25	4	7.12	509,870	4	2.46	197,027	4
$C_{24}H_{38}O_4$		5	4.97	283,418	2	1.90	120,718	2
		6	2.66	132,434	1	1.12	60,310	1
		7	2.10	101,097	1	0.88	46,319	1
		1	T.O.	734,327,164	2,653,617	T.O.	759,794,526	11,587,705
		2	T.O.	228,786,134	161	1543.37	300,524,875	2,520
C04036		3	184.54	14,517,014	1	45.36	4,745,395	1
	29	4	11.86	638,457	1	3.60	262,162	1
$C_{19}H_{39}O_7P$		5	5.95	225,966	1	2.27	107,378	1
		6	4.34	127,250	1	1.57	60,557	1
		7	3.38	81,532	1	1.26	40,403	1
		1	T.O.	667,687,809	3,959	T.O.	639,689,202	96,245
		2	T.O.	168,054,487	77	T.O.	239,538,772	1,736
C03630		3	T.O.	115,797,466	11	438.19	37,803,253	13
	33	4	118.48	5,104,899	11	25.65	1,519,286	11
$C_{21}H_{39}O_7P$		5	50.63	1,554,928	9	12.24	515,752	9
		6	27.83	673,426	7	6.44	225,620	7
		7	11.97	244,166	5	3.14	92,431	5

Note: (1) C03343, C07530, C07178, C03690, C04036, and C03630 are the entries of 2-Ethylhexyl phthalate, Etidocaine, Trimethobenzamide, Bis (2-ethylhexyl) phthalate, 1-Palmitoylglycerol 3-phosphate, and Oleoylglycerone phosphate in the KEGG LIGAND database, respectively; (2) n_2 is the number of vertices preprocessed by replacing benzene rings with new atoms of valence 6 and by the H-removal and single-bond transformations; (3) K is the level of a given feature vector; (4) "time" is the CPU time in seconds; (5) "T.O." means "time over" (the time limit was set to 1800 seconds); (6) "nnt" is the number of nodes in the family trees that are checked; (7) "fs" is the number of feasible solutions found within the time limit.

References

[1] Akutsu, T., Fukagawa, D., Inferring A Graph from Path Frequency, *LNCS*, 3537, 371-382, 2005.

[2] Akutsu, T., Fukagawa, D., Inferring a Chemical Structure from a Feature Vector Based on Frequency of Labeled Pathsand Small Fragments, *Series on Advances in Bioinformatics and Computational Biology*, Proc. 5th Asia-Pacific Bioinformatics Conf. Sankoff, D., Wang, L., Chin, F., Eds.; Imperial College Press, 165–174, 2007.

[3] Aringhieri, R., Hansen, P., Malucelli, F., Chemical Trees Enumeration Algorithms, *4OR*, 1, 67–83, 2003.

[4] Bakır, G. H., Zien, A., Tsuda, K., Learning to Find Graph Pre-Images, *LNCS*, 3175, 253–261, 2004.

[5] Buchanan, B. G., Feigenbaum, E. A., DENDRAL and Meta-DENDRAL - Their Applications Dimension, *Artif. Intell.*, 1, 5–24, 1978.

[6] Cayley, A., On the Analytic Forms Called Trees, with Applications to the Theory of Chemical Combinations, *Reports British Assoc. Adv. Sci.*, 45, 257-305, 1875.

[7] Deshpande, M., Kuramochi, M., Wale, N., Karypis, G., Frequent Substructure-Based Approaches for Classifying Chemical Compounds, *IEEE Transactions on Knowledge and Data Engineering*, 17, 1036–1050, 2005.

[8] Faulon, J. L., Churchwell, C. J., Visco, Jr., D.P., The Signature Molecular Descriptor. 2. Enumerating Molecules from Their Extended Valence Sequences, *J. Chem. Inf. Comp. Sci.*, 43, 721–734, 2003.

[9] Fink, T., Reymond, J. L., Virtual Exploration of the Chemical Universe up to 11 Atoms of C, N, O, F: Assembly of 26.4 Million Structures (110.9 Million Stereoisomers) and Analysis for New Ring Systems, Stereochemistry, Physicochemical Properties, Compound Classes, and Drug Discovery, *J. Chem. Inf. Comp. Sci.*, 47, 342–353, 2007.

[10] Fujiwara, H., Wang, J., Zhao, L., Nagamochi, H., Akutsu, T., Enumerating Tree-like Chemical Graphs with Given Path Frequency, *J. Chem. Inf. Model.*, 2008 (to appear).

[11] Funatsu, K., Sasaki, S., Recent Advances in the Automated Structure Elucidation System, CHEMICS. Utilization of Two-Dimensional NMR Spectral Information and Development of Peripheral Functions for Examination of Candidates, *J. Chem. Inf. Comp. Sci.*, 36, 190–204, 1996.

[12] Hall, L. H., Dailey, E. S., Design of Molecules from Quantitative Structure-Activity Relationship Models. 3. Role of Higher Order Path Counts: Path 3, *J. Chem. Inf. Comp. Sci.*, 33, 598–603, 1993.

[13] Kashima, H., Tsuda, K., Inokuchi, A., Marginalized Kernels between Labeled Graphs, Proc. 20th International Conference on Machine Learning, Fawcett, T., Mishra, N. Eds., The AAAI Press, Menlo Park, California, 321–328, 2003.

[14] Mahé, P., Ueda N., Akutsu, T., Perret, J. L., Vert, J. P., Graph Kernels for Molecular Structure-Activity Relationship Analysis with Support Vector Machines, *J. Chem. Inf. Model.*, 45, 939–951, 2005.

[15] Mauser, H., Stahl, M., Chemical Fragment Spaces for De Novo Design, *J. Chem. Inf. Comp. Sci.*, 47, 318–324, 2007.

[16] Nagamochi, H., A Detachment Algorithm for Inferring A Graph from Path Frequency, *LNCS*, 4112, 274–283, 2006.

[17] Nakano, S., Uno, T., Efficient Generation of Rooted Trees, Technical Report, NII-2003-005E, ISSN:1346-5597; National Inst. of Informatics: Tokyo, Japan, July 3, 2003.

[18] Nakano, S., Uno, T., Generating Colored Trees, *LNCS*, 3787, 249–260, 2005.

[19] Wang, J., Zhao, L., Nagamochi, H., Akutsu, T., An Efficient Algorithm for Generating Colored Outerplanar Graphs, *LNCS*, 4484, 573–583, 2007.

BSAlign: A RAPID GRAPH-BASED ALGORITHM FOR DETECTING LIGAND-BINDING SITES IN PROTEIN STRUCTURES

ZEYAR AUNG[1] JOO CHUAN TONG[1]
azeyar@i2r.a-star.edu.sg jctong@i2r.a-star.edu.sg

[1] *Institute for Infocomm Research, A*STAR (Agency for Science, Technology and Research), 1 Fusionopolis Way, #21-01 Connexis, Singapore 138632*

Detection of ligand-binding sites in protein structures is a crucial task in structural bioinformatics, and has applications in important areas like drug discovery. Given the knowledge of the site in a particular protein structure that binds to a specific ligand, we can search for similar sites in the other protein structures that the same ligand is likely to bind. In this paper, we propose a new method named "BSAlign" (**B**inding **S**ite **Align**er) for rapid detection of potential binding site(s) in the target protein(s) that is/are similar to the query protein's ligand-binding site. We represent both the binding site and the protein structure as graphs, and employ a subgraph isomorphism algorithm to detect the similarities of the binding sites in a very time-efficient manner. Preliminary experimental results show that the proposed BSAlign binding site detection method is about 14 times faster than a well-known method called SiteEngine, while offering the same level of accuracy. Both BSAlign and SiteEngine achieve 60% search accuracy in finding adenine-binding sites from a data set of 126 proteins. The proposed method can be a useful contribution towards speed-critical applications such as drug discovery in which a large number of proteins are needed to be processed. The program is available for download at: `http://www1.i2r.a-star.edu.sg/~azeyar/BSAlign/`.

Keywords: protein structure; ligand-binding site; efficient binding site detection; subgraph isomorphism; adenine-binding sites.

1. Introduction

Proteins are the physical basis of life, and perform a number of vital functions such as storage, structural lattice, movement, transport, signaling, immunity, catalysis in metabolism, etc. A ligand is a specific compound that binds to a particular receptor protein to form a complex. It can inhibit, promote, or alter the function of the receptor protein. A ligand can either be another protein or a non-protein small molecule. Drugs are examples of small molecule ligands.

A ligand-binding site is a region in a receptor protein structure to which a ligand binds. Binding site detection is a task in which, given the knowledge of the binding site in a particular protein structure a specific ligand binds to, we detect in the other protein structure(s) for the site(s) with the similar structural and physicochemical characteristics, where the same ligand is likely to bind — as illustrated in Figure 1.

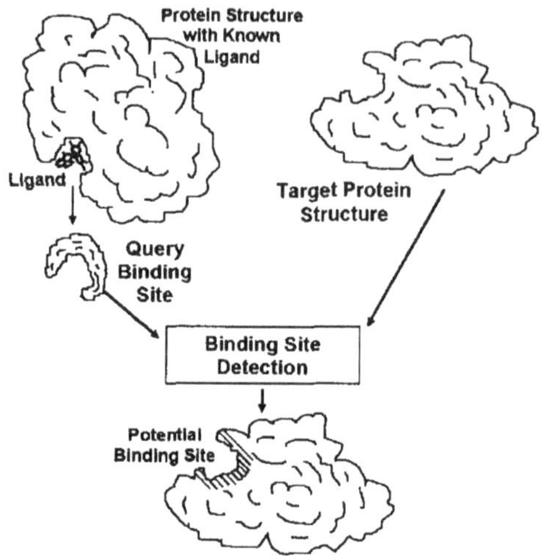

Fig. 1. Detection of a potential binding site similar to the query binding site.

This is a crucial task in structural bioinformatics, and has important applications in the area of drug discovery. In particular, binding site detection is a very useful mechanism for identifying the new drug targets and developing the targeted drug leads like inhibitors [20]. In addition to drug discovery, binding site detection is also useful for protein function prediction [14].

In this paper, we propose a new method named "BSAlign" (**B**inding **S**ite **Align**er) that detects the potential site(s) in a target protein that is/are similar to the query binding site where a specific ligand is know to bind. The method is designed to compare a query site against the similar site(s) in a single target protein, but can easily be adapted to search for potential sites in multiple target proteins.

The BSAlign method represents both the query binding site and the target protein structure as graphs. The graph representation scheme that we use captures information on both the geometrical conformations and the physicochemical properties of amino acid residues in the query and the target. Then, the method applies a subgraph isomorphism algorithm to find the maximum common subgraph(s) of the input graphs. The subgraph isomorphism problem can be effectively solved by transforming the two input graphs into an edge-product graph, and finding the maximum clique(s) or the fully-connected subgraph(s) in the edge product graph [9, 12]. From the maximum clique(s), the list(s) of maximally matching residue pairs is/are extracted. After that, those list(s) of matching residue pairs is/are refined with respect to a scoring function in order to yield the final list of optimally matching/aligned residue pairs. Depending on the size and density of the input graphs, the method automatically tunes the matching criteria of the graphs' vertices and edges on the fly so as to avoid a lengthly subgraph isomorphism process.

We tested our method by detecting the adenine-binding sites in a data set of 126 protein structures. The experimental results show that BSAlign can detect the potential binding sites for adenine-containing ligands efficiently and effectively. BSAlign is compared against another state-of-the-art binding site detection method named SiteEngine [20]. It is observed that BSAlign is 14 times as fast as SiteEngine while providing as good accuracy (60%) as SiteEngine. Since speed is a crucial factor for applications like drug discovery, which involve large quantities of ligands, ligand-binding proteins and potential target proteins [3], the efficiency of our proposed BSAlign method can be an important contribution towards such speed-critical applications.

2. Related Works

The problem of binding site detection is related to that of protein substructure alignment since both involve identifying a region similar to the query substructure in the target protein. However, the generic substructure alignment methods such as [5, 7, 19] cannot be effectively used for binding site detection, because they take only the geometrical properties of residues into account, but not their physicochemical attributes, which are essential in identifying the ligand-binding residues.

A number of algorithms dedicated to binding site detection/prediction have been proposed. The methods such as [1, 11, 14] predict potential binding sites on the surfaces of proteins without an *a priori* knowledge of a similar binding site. On the other hand, the methods such as [4, 8, 17, 18, 20] detect the target protein's potential binding site(s) which is/are similar to the query binding site.

ASSAM [4] represents residue side-chains as pseudo-atoms, and performs subgraph isomorphism to detect the side-chain patterns common to a set of binding sites. eF-site [8] and Cavbase [18] represent a binding sites as a set of detailed surface points and pseudocenters (selected atoms) in residues respectively, and apply subgraph isomorphism to find the similar binding sites. However, given the usually large quantities of objects (surface points or pseudocenters) in a query binding site and a target protein and the complexity of the subgraph isomorphism problem, which is NP-hard [15], these methods are not time-efficient.

SiteEngine [20] represents a binding site as a set of pseudocenters (as in Cavbase [18]), and applies geometric hashing to detect the binding site similarities. Being based on the efficient geometric hashing technique, it is faster than Cavbase. However, its time efficiency is still inadequate when a large amount of query binding sites and target proteins are to be processed, as usually needed in the case of drug discovery [3].

A recently proposed method, SiteAlign [17], encodes binding sites as fixed-length cavity fingerprints, and performs a time-efficient comparison on these fingerprints. No accuracy comparison of SiteAlign with those of the other methods is available. However, in general, the accuracy of fingerprint-based comparison methods tend to be lower than those of detailed comparison methods [21].

Our objective is to overcome the shortcomings, either in terms of time efficiency and accuracy, of the abovementioned methods. In order to achieve a better time efficiency, we adopt a residue-based approach, as opposed to the finer-grained approaches [8, 18, 20], which use sub-residue information like surface points or pseudocenters. On the other hand, in order to achieve the same level of accuracy as those finer-grained methods, we carefully design our residue-based graph representation scheme to encompass enough geometric and physicochemical information, and employ subgraph isomorphism for a detailed graph comparison. Our preliminary experimental results show that we have achieved our objective, and come up with a solution that is much faster than the fastest of the finer-grained methods, namely SiteEngine [20], while maintaining the same level of accuracy.

3. The BSAlign Method

3.1. *Graph Representation*

The input to the BSAlign algorithm are the query binding site and the entire target protein structure. We can define a binding site as a set of residues that are interacting with the ligand in question. A residue is considered to be interacting with the ligand if it is within 5Å radius from the ligand [13].

Both the query binding site and the target protein structure can be represented as graphs. Since the sequence order of residues is irrelevant in comparing and detecting binding sites [6], the graph representation, which is sequence-order independent, is best suited for our purpose. We use a residue-based graph representation scheme which captures information on both geometrical and physicochemical properties of the amino acid residues. Each residue is encoded as a vertex in the graph. Two vertices, representing two residues, are connected by an edge if these two residues are close enough to each other, i.e., the distance between their Cα atoms is less than or equal to 15Å (an empirically determined value). A vertex is characterized by a vertex label which comprises of the following attributes:

(1) Solvent accessibility of the residue as a percentage (0~100%) (denoted as *SA*),
(2) Physicochemical type (non-polar, polar, aromatic, positive, or negative) of the residue (*PT*), and
(3) Secondary structure type (helix, sheet, or loop) of the residue (*SS*).

An edge connecting two vertices (residues) is characterized by an edge label comprising the following attributes:

(1) Distance between the Cα atoms of the two residues (*DC*) and
(2) Angle between the Cα–Cβ vectors of the two residues (*AN*). (A Cα–Cβ vector is an imaginary line segment connecting the Cα and the Cβ atoms of a residue.)

Among these attributes, *PT*, *DC* and *AN* can be derived simply from the PDB files (`http://www.rcsb.org`), and *SA* and *SS* can be obtained by using the DSSP program (`http://swift.cmbi.kun.nl/gv/dssp/`).

3.2. *Graph Similarity*

The similarity between two graphs can be determined by finding the maximum common subgraph in them. The larger the common subgraph, the more similar the two given graphs are. The maximum common subgraph problem can be solved by transforming the two input graphs into a single edge-product graph and finding the maximum clique (fully-connected subgraph) in that edge-product graph [9, 12].

3.2.1. *Edge-product Graph Construction*

Let G be a graph of any kind defined as $G = (V, E)$ where V is the set of vertices and E is the set of edges in G respectively. We can express V as $\{v_i \mid i = 1 \ldots |V|\}$ where $|V|$ is the number of vertices in G. Similarly, was can express E as $\{e_i \mid i = 1 \ldots |E|\}$ where $|E|$ is the number of edges in G. An edge e_i can in turn be expressed as $e_i = (a_i, b_i)$ where $a_i, b_i \in V$ are the two vertices connected by e_i.

An edge-product graph GP of two input graphs $G1 = (V1, E1)$ and $G2 = (V2, E2)$ is defined as $GP = (VP, EP) = (E1 \times E2)$ in which:

- The vertex set VP of the product graph consists of all the compatible edge pairs in $E1$ and $E2$. That is, $vp_i = (e1_r, e2_s)$ if:
  - $EC(e1_r, e2_s) = \text{TRUE}$, and
  - $(VC(a1_r, a2_s) = \text{TRUE} \wedge VC(b1_r, b2_s) = \text{TRUE}) \vee$ $(VC(a1_r, b2_s) = \text{TRUE} \wedge VC(b1_r, a2_s) = \text{TRUE})$
- There exists an edge between two vertices $vp_i = (e1_r, e2_s)$ and $vp_j = (e1_t, e2_u)$ of the product graph if:
  - $(e1_r \neq e1_t) \wedge (e2_s \neq e2_u)$, and
  - Either:
    - ($e1_r$ and $e1_t$ have a common vertex $v1_{rt}$) $\wedge$ ($e2_s$ and $e2_u$ have a common vertex $v2_{su}$) $\wedge$ ($VC(v1_{rt}, v2_{su}) = \text{TRUE}$), or
    - ($e1_r$ and $e1_t$ do not have a common vertex) $\wedge$ ($e2_s$ and $e2_u$ do not have a common vertex)

The vertex compatibility function VC of the two vertices of v_i from $G1$ and v_j from $G2$ is defined as:

$$VC(v_i, v_j) = \begin{cases} \text{TRUE} & \text{if } (|v_i.SA - v_j.SA| \leq T1_{SA}) \vee \\ & \quad ((|v_i.SA - v_j.SA| \leq T2_{SA}) \wedge \\ & \quad (v_i.PT = v_j.PT) \wedge (v_i.SS = v_j.SS)) \\ \text{FALSE} & \text{otherwise} \end{cases} \quad (1)$$

where $T1_{SA}$ and $T2_{SA}$ are the two threshold values for the differences in solvent accessibility. $T1_{SA}$ is usually a very small value, and $T2_{SA}$ is a relatively larger one. The meaning of the function $VC(v_i, v_j)$ is that the two vertices (residues) v_i and v_j are regarded as compatible if either their solvent accessibility percentages are very close, or their accessibility percentages are close enough, and both of their physicochemical types and secondary structure types are respectively the same.

Similarly, the edge compatibility function EC of the two edges e_i from $G1$ and e_j from $G2$ is defined as:

$$EC(e_i, e_j) = \begin{cases} \text{TRUE} & \text{if } ((|e_i.DC - e_j.DC| \leq T_{DC}) \wedge \\ & \quad (|e_i.AN - e_j.AN| \leq T_{AN})) \\ \text{FALSE} & \text{otherwise} \end{cases} \quad (2)$$

where T_{DC} and T_{AN} are the threshold values for the differences in Cα–Cα distances and (Cα–Cβ)–(Cα–Cβ) angles of the two residues respectively. The function $EC(e_i, e_j)$ means that the two edges e_i and e_j are compatible if their distance and angle values in one edge are not very different from their counterparts in the other.

After we have constructed the edge-product graph, the next step is to detect the maximum clique(s) in it. Since maximum clique detection is an NP-hard problem [15], this will be the most time-consuming step in the BSAlign algorithm. In order to reduce the time taken for this step, we have to keep the size of the edge-product graph reasonably small. So, if required, we iterate the edge-product graph construction process up to 5 rounds with stricter threshold values for $T1_{SA}$, $T2_{SA}$, T_{DC} and T_{AN} at each time. We stop the iteration when number of edges in the edge-product graph becomes less than 1,000,000. For the first round, we use $T1_{SA} = 0.05$, $T2_{SA} = 0.30$, $T_{DC} = 2.0$ and $T_{AN} = 30$. For the second round, we use $T1_{SA} = 0.04$, $T2_{SA} = 0.25$, $T_{DC} = 1.5$ and $T_{AN} = 25$, and so on. For the last (fifth) round, we use $T1_{SA} = 0.01$, $T2_{SA} = 0.10$, $T_{DC} = 0.01$ and $T_{AN} = 10$. All of these values are empirically determined.

3.2.2. *Maximum Clique Detection*

After the final edge-product graph is obtained, we use the Cliquer program [15] to detect the maximum clique(s) in it. Cliquer is an implementation of a branch-and-bound maximum clique detection algorithm [16]. A brief description of the Cliquer algorithm as described in [15] is as follows:

The algorithm assume some order for the vertices $V = \{v_1, v_2, \ldots, v_{|V|}\}$. Let $S_i = \{v_1, v_2, \ldots, v_i\} \subseteq V$. The function $c(i)$ is defined to be the size of the maximum clique in the subgraph induced by S_i. Obviously, for every $i = 1, \ldots, |V| - 1$, we have either $c(i+1) = c(i)$ or $c(i+1) = c(i) + 1$. Moreover, $c(i+1) = c(i) + 1$ if and only if there exists a clique in S_{i+1} of size $c(i) + 1$ that includes vertex v_{i+1}. Cliquer calculates the values of $c(i)$ starting from $c(1) = 1$ up, and stores the values found. This enables a pruning strategy not found in older clique detection algorithms. Namely, when Cliquer is calculating $c(i+1)$ (that is searching for a clique of size $c(i) + 1$ within S_{i+1}), and it has formed a clique W and is considering adding vertex v_j, it can prune the search if $|W| + c(j) \leq c(i)$. Trivially, if it finds a clique of size $c(i) + 1$, it can prune the whole search and start calculating $c(i+2)$. When searching for all maximum cliques, Cliquer first determines the size of the maximum cliques, then starts the search again at the suitable position.

3.2.3. *Matching Residue Pair Generation*

The maximum clique(s) produced by Cliquer is/are mapped back into the list(s) of matching vertex pairs by using the Hungarian maximal assignment algorithm [10]. From the list of matching edge pairs, the algorithm produces the maximum possible number of matching vertex (residue) pairs — as exemplified in Figure 2. The implementation of the Hungarian algorithm is adapted from the one described in [22].

Matching Edge Pairs (query)		(target)		Matching Vertex Pairs (query)		(target)
				1	–	53
1, 2	–	53, 55		2	–	55
1, 8	–	51, 53		8	–	51
2, 3	–	55, 57	⇒	3	–	57
3, 4	–	57, 60		4	–	60
4, 5	–	58, 60		5	–	58
7, 9	–	54, 56		7	–	54
				9	–	56

Fig. 2. An example of mapping matching edge pairs into matching vertex pairs.

3.3. *Refinement and Scoring*

The two sets of matching (aligned) residue pairs are tested for their actual structural similarity using the root mean square deviation (RMSD) criterion. RMSD is calculated by superimposing the set of Cα atoms of the aligned residues in the query binding site onto their counterparts in the target protein. The smaller the RMSD, the more structurally similar the two sets of aligned residues are. However, in some cases, the RMSD values are quite large if all of the aligned residue pairs are taken into account. Therefore, we iteratively refine the initial list of aligned residue pairs by removing at each step the pair that is least fitting when superimposed. But, on the other hand, we should not remove too many pairs, because the alignment result will not be very meaningful if number of aligned residues is too small. In other words, we must balance the RMSD value the number of aligned residues in order to get the optimal alignment results. For that, we use Alexandrov and Fischer's scoring function [2], which is defined as:

$$\text{Alignment score} = \frac{3 \times \text{No. of aligned residues}}{1 + \text{RMSD}} \tag{3}$$

The refinement of the alignment is repeated until the alignment score cannot be further increased, or until the number of aligned residues is equal to one-third of the number of residues in the original query binding site. Then, the final set of aligned residues in the target protein is reported as the potential binding site. Sometimes, there are more than one maximum clique in the edge-product graph, and consequently, more than one initial lists of aligned residues exist. In such a

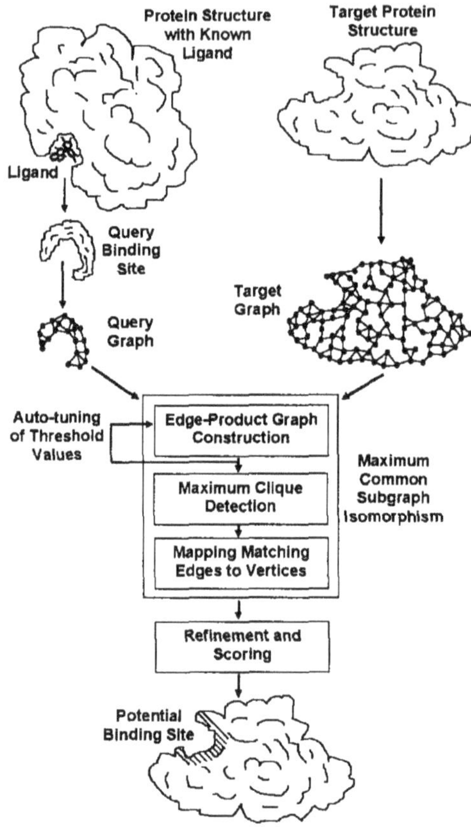

Fig. 3. Outline of the BSAlign method.

case, we refine all the available lists, and take the one that gives the highest final alignment score as the answer.

The steps taken in the BSAlign method are summarized in Figure 3.

4. Results and Discussions

Following the experiment described in [20], we test BSAlign by searching for the binding sites similar to the ATP-binding site of an adenine-binding protein "1atp" in a data set of 126 proteins listed in Table 1. The data set consists of 34 adenine-binding proteins belonging to 18 distinct SCOP Folds, and 92 proteins of other functional types from 21 distinct SCOP Folds. (SCOP – `http://scop.mrc-lmb.cam.ac.uk/scop/` is a database for structural classification of proteins. If two proteins belong to different SCOP Folds, they are very diverse in terms of their whole structures.) Adenine-binding proteins are a functional type of protein that binds to adenine-containing ligands like ATP, ANP, FAD, NAD, etc.

Table 1. The data set of 126 proteins (34 adenine-binding proteins and 92 other proteins).

Functional Type	Total	SCOP Folds	PDB IDs
Adenine-binding proteins	34	18	1a49, 1a82, 1ads, 1atp, 1ayl, 1b4v, 1b8a, 1bx4, 1byq, 1csc, 1csn, 1e2q, 1e8x, 1f9a, 1fmw, 1g5t, 1gn8, 1hck, 1hpl, 1j7k, 1jjv, 1kay, 1kp2, 1kpf, 1mjh, 1mmg, 1nhk, 1nsf, 1phk, 1qmm, 1yag, 1zin, 2src, 9ldt
Other proteins	92	21	1a27, 1a52, 1abi, 1acb, 1alq, 1arb, 1azm, 1b56, 1b6o, 1bt5, 1cbs, 1cho, 1com, 1cqq, 1cse, 1csm, 1dbf, 1dcs, 1e6w, 1ecm, 1ela, 1elc, 1equ, 1ere, 1err, 1exm, 1fby, 1fds, 1fem, 1flj, 1fnj, 1fnk, 1ftp, 1g5y, 1ghp, 1gx9, 1hah, 1har, 1hms, 1hne, 1hsg, 1hsh, 1hwr, 1ifc, 1jd0, 1jgl, 1keq, 1kop, 1kqw, 1kzk, 1l2i, 1lhu, 1lib, 1lid, 1lie, 1lvo, 1mbm, 1mdc, 1mml, 1mu2, 1oh0, 1opa, 1opb, 1pek, 1pmp, 1ppf, 1pro, 1q2w, 1qjg, 1qkt, 1rxf, 1sbn, 1sga, 1sgc, 1tgs, 1tyr, 1vrt, 1whs, 1ysc, 1znc, 2alp, 2cbr, 2ifb, 2lbd, 2lpr, 3ert, 3prk, 3sga, 3tec, 4csm, 4sgb, 4tgl
Total	126		

4.1. *Search Accuracy*

Using the BSAlign algorithm, the query ATP-binding site of 1atp is compared with every protein structure in the data set of 126 proteins in order to detect the similar binding sites in them. Then, the found binding sites are ranked by their alignment scores (Equation 3). We assess the ranking results by using the same evaluation criterion as described in [20]. We examine the 15 top ranking binding sites, and observe that 9 out of 15 (60%) belong to the adenine-binding proteins with the ligand ATP or the other adenine-containing ones (such as ANP and AP5) — as shown in Table 2. BSAlign's accuracy performance is as good as that of SiteEngine [20], which is a finer-grained method that takes the sub-residue information (namely pseudo-centers) into account. SiteEngine also ranks 9 adenine binding proteins among its top 15 answers. Among the two sets of 15 top ranking proteins by BSAlign and SiteEngine, 8 of them (1atp, 1csn, 2src, 1phk, 1chk, ajd0, 1mjh, and 1nsf) are common to both sets.

Now, let us study the details of the alignment results. We take the alignment result for the binding sites of the proteins 1atp and 1csn as an example. The ATP-binding site of 1atp consists of 13 residues: 50(G), 51(T), 52(G), 53(S), 54(F), 55(G), 57(V), 70(A), 122(Y), 123(V), 170(E), 171(N), and 184(D). Among these 13 residues, 10 are aligned with their counterparts in 1csn, with the RMSD of 0.48Å. The aligned residue pairs are: 50(G)–19(G), 52(G)–21(G), 53(S)–22(S), 55(G)–24(G), 57(V)–26(I), 70(A)–39(A), 123(V)–88(L), 170(E)–135(D), 171(N)–136(N), and 184(D)–154(D). It turns out that all of these 10 aligned residues in 1csn are within 5Å radius of the ligand ATP bound to the protein. The two ATP-binding sites of 1atp and 1csn are illustrated in Figure 4.

Table 2. The search result for the query binding site of the ligand "ATP" of the protein 1atp in the data set of 126 proteins.

Rank	PDB ID	Protein Name	SCOP Fold Name	Sequence Identity (%) [a]	Aligned Residues	RMSD (Å)	Align -ment Score	Ligand	Functional Type
1	1atp	cAMP-dependent PK, catalytic subunit	Protein kinase-like (PK-like)	100.0	13	0.00	39.00	ATP	Adenine-binding
2	1csn	Casein kinase-1, CK1, catalytic subunit	Protein kinase-like (PK-like)	17.0	10	0.48	20.24	ATP	Adenine-binding
3	2src	c-src protein tyrosine kinase	SH3-like barrel	13.4	11	0.97	16.74	ANP	Adenine-binding
4	1phk	gamma-subunit of glycogen phosphorylase kinase (Phk)	Protein kinase-like (PK-like)	24.2	8	0.58	15.18	ATP	Adenine-binding
5	1hck	Cyclin-dependent PK, CDK2	Protein kinase-like (PK-like)	19.5	9	1.06	13.12	ATP	Adenine-binding
6	3prk	Proteinase K	Subtilisin-like	2.5	6	1.10	8.55	MSU	other
7	1jd0	Carbonic anhydrase	Carbonic anhydrase	4.2	6	1.44	7.37	AZM	other
8	1mjh	"Hypothetical" protein MJ0577	Adenine nucleotide alpha hydrolase-like	15.4	6	1.47	7.27	ATP	Adenine-binding
9	1fnk	Chorismate mutase	Bacillus chorismate mutase-like	7.3	6	1.48	7.25	CSD	other
10	1zin	Adenylate kinase	P-loop containing nucleoside triphosphate hydrolases	10.1	6	1.53	7.13	AP5	Adenine-binding
11	1abi	Thrombin	Trypsin-like serine proteases	16.4	6	1.75	6.53	HMR	other
12	1hah	Eukaryotic proteases	Trypsin-like serine proteases	16.7	6	1.76	6.52	NAG	other
13	1kp2	Argininosuccinate synthetase	Adenine nucleotide alpha hydrolase-like	8.8	6	1.78	6.47	ATP	Adenine-binding
14	1dbf	Chorismate mutase	Bacillus chorismate mutase-like	3.7	6	1.79	6.45	SO4	other
15	1nsf	Hexamerization domain of N-ethylmalemide-sensitive fusion (NSF) protein	P-loop containing nucleotide triphosphate hydrolases	12.4	6	1.81	6.41	ATP	Adenine-binding

Note: [a] Calculated using EMBOSS Web Server (`http://www.ebi.ac.uk/emboss/align/`).

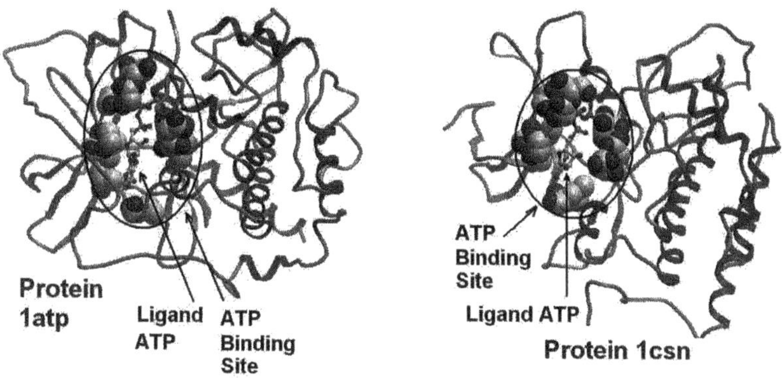

Fig. 4. ATP-binding sites of 1atp (left) and 1csn (right). Number of aligned residues = 10; RMSD = 0.48Å. The residues that involve in the alignment are shown as space-filling balls in both proteins.

4.2. *Running Time*

We compare the running times of SiteEngine and BSAlign by executing them on the same personal computer with Pentium D 3.2GHz CPU and 2GB main memory. For the aforementioned task of searching the data set of 126 proteins with the query binding site for the ligand ATP in the protein 1atp, SiteEngine takes a total of 12,010 seconds (3 hours, 18 minutes, and 10 seconds), whereas BSAlign merely takes a total of 871 seconds (14 minutes and 31 seconds). Thus, BSAlign is found to be about 14 times faster than SiteEngine while offering the same level of accuracy.

The comparable accuracy performance of the time-efficient residue-based BSAlign to that of the slower finer-grained SiteEngine can be attributed to (1) BSAlign's comprehensive graph representation scheme which captures the detailed physicochemical and geometric properties of the binding site and (2) the subgraph isomorphism process which ensures the complete matching of the two large substructures (rather than combining multiple partial matchings of the smaller substructures — as in the case of geometric hashing used by SiteEngine).

5. Conclusion

In this paper, we have presented a new ligand-binding site detection method named BSAlign, which is based on residue-based graph representation and subgraph isomorphism. Preliminary experimental results show that the method is about 14 times faster than the well-known SiteEngine method, while offering the same level of accuracy. This can be an important contribution towards the drug discovery applications where speed is critical. As a future work, BSAlign will be tested against diverse sets of protein families in order to further ascertain its accuracy and speed performances.

References

[1] Abagyan, R. and Totrov, M., High-throughput docking for lead generation, *Curr. Opin. Chem. Biol.*, 5:375–382, 2001.

[2] Alexandrov, N. N. and Fischer, D., Analysis of topological and nontopological structural similarities in the PDB: new examples with old structures, *Prot. Struct. Funct. Genet.*, 25:354–365, 1996.
[3] Alvarez, J. and Shoichet, B. (eds.), *Virtual Screening in Drug Discovery*, Taylor and Francis Ltd, 2005.
[4] Artymiuk, P. J., Poirrette, A. R., Grindley, H. M., Rice, D. W., and Willett, P., A graph-theoretic approach to the identification of three-dimensional patterns of amino acid side-chains in protein structures, *J. Mol. Biol.*, 243:327–344, 1994.
[5] Aung, Z. and Tan, K. L., MatAlign: precise protein structure comparison by matrix alignment, *J. Bioinfo. Comp. Biol.*, 4:1197–1216, 2006.
[6] Fischer, D., Wolfson, H., Lin, S. L., and Nussinov, R., Three-dimensional, sequence order-independent structural comparison of a serine protease against the crystallographic database reveals active site similarities: potential implications to evolution and to protein folding, *Protein Sci.*, 3:769–778, 1994.
[7] Holm, L. and Sander, C., Protein structure comparison by alignment of distance matrices, *J. Mol. Biol.*, 233:123–138, 1993.
[8] Kinoshita, K. and Nakamura, H., Identification of protein biochemical functions by similarity search using the molecular surface database eF-site, *Protein Sci.*, 12:1589–1595, 2003.
[9] Koch, I., Lengauer, T., and Wanke, E., An algorithm for finding maximal common subtopologies in a set of protein structures, *J. Comp. Biol.*, 3:289–306, 1996.
[10] Kuhn, H. W., The Hungarian Method for the assignment problem, *Nav. Res. Log. Quart.*, 2:83–97, 1955.
[11] Laurie, A. T. and Jackson, R. M., Q-SiteFinder: an energy-based method for the prediction of protein-ligand binding sites, *Bioinformatics*, 21:1908–1916, 2005.
[12] May, P., *Protein Structure Analysis using Contact Maps and Secondary Structure*, Ph.D. Dissertation, Free University of Berlin, 2007.
[13] Mohamad, S. B., Ong, A. L., and Ripen, A. M., Evolutionary trace analysis at the ligand binding site of laccase, *Bioinformation*, 2:369–372, 2008.
[14] Murga, L. F., Wei, Y., and Ondrechen, M. J., Computed protonation properties: unique capabilities for protein functional site prediction, *Genome Informatics*, 19:107–118, 2007.
[15] Niskanen, S. and Östergård, P. R. J., *Cliquer User's Guide, Version 1.0*, Technical Report T48, Communications Laboratory, Helsinki University of Technology, 2003.
[16] Östergård, P. R. J., A fast algorithm for the maximum clique problem, *Discrete Appl. Math.*, 120:195–205, 2002.
[17] Schalon, C., Surgand, J. S., Kellenberger, E., and Rognan, D., A simple and fuzzy method to align and compare druggable ligand-binding sites, *Prot. Struct. Funct. Bioinfo.*, 71:1755–1778, 2008.
[18] Schmitt, S., Kuhn, D., and Klebe, G., A new method to detect related function among proteins independent of sequence and fold homology, *J. Mol. Biol.*, 323:387–406, 2002.
[19] Shindyalov, I. N. and Bourne, P. E., Protein structure alignment by incremental combinatorial extension (CE) of the optimal path, *Protein Eng.*, 11:739–747, 1998.
[20] Shulman-Peleg, A., Nussinov, R., and Wolfson, H. J., Recognition of functional sites in protein structures, *J. Mol. Biol.*, 339:607–633, 2004.
[21] Sierk, M. L. and Pearson, W. R., Sensitivity and selectivity in protein structure comparison, *Protein Sci.*, 13:773–785, 2004.
[22] `http://www.public.iastate.edu/~ddoty/HungarianAlgorithm.html`

PROTEIN COMPLEX PREDICTION BASED ON MUTUALLY EXCLUSIVE INTERACTIONS IN PROTEIN INTERACTION NETWORK

SUK HOON JUNG jsh@icu.ac.kr
WOO-HYUK JANG torajim@icu.ac.kr
HEE-YUNG HUR heyrue@icu.ac.kr
BORA HYUN lovesun79ya@icu.ac.kr
DONG-SOO HAN dshan@icu.ac.rk

School of Engineering, Information and Communications University, 119, Munjiro, Yuseong-gu,Daejeon, 305-714, Korea

The increasing amount of available Protein-Protein Interaction (PPI) data enables scalable methods for the protein complex prediction. A protein complex is a group of two or more proteins formed by interactions that are stable over time, and it generally corresponds to a dense sub-graph in PPI Network (PPIN). However, dense sub-graphs correspond not only to stable protein complexes but also to sets of proteins including dynamic interactions. As a result, conventional simple PPIN based graph-theoretic clustering methods have high false positive rates in protein complex prediction. In this paper, we propose an approach to predict protein complexes based on the integration of PPI data and mutually exclusive interaction information drawn from structural interface data of protein domains. The extraction of Simultaneous Protein Interaction Cluster (SPIC) is the essence of our approach, which excludes interaction conflicts in network clusters by achieving mutually exclusion among interactions. The concept of SPIC was applied to conventional graph-theoretic clustering algorithms, MCODE and LCMA, to evaluate the density of clusters for protein complex prediction. The comparison with original graph-theoretic clustering algorithms verified the effectiveness of our approach; SPIC based methods refined false positives of original methods to be true positive complexes, without any loss of true positive predictions yielded by original methods.

Keywords: protein complex, interface, protein-protein interaction, protein interaction network

1. INTRODUCTION

Recent developments in proteomics have resulted in the increasing amount of Protein-Protein Interaction (PPI) data. Modeling PPI network with simple graphs has enabled many computational applications for the study of protein functions, one of which is the scalable method for protein complex prediction. Protein complexes generally correspond to dense sub-graphs in the PPI network because proteins in a complex are highly interactive with each other [1]. Thus, conventional network based methods have focused on the extraction of graph-theoretic clusters that are numerically determined as protein complexes. The MCODE (Molecular COmplex CEtection) algorithm utilizes connectivity values in PPIN to identify k-cores for the extraction of graph-theoretic clusters[2]. LCMA (Local Clique Merging Algorithm), which is based on local clique merging, also utilizes connectivity values in finding protein complexes[3].

However, despite the trials of graph clustering approaches for the protein complex prediction, little progress has been achieved as they are plagued by high false positive rates. Their false positive results are presumably caused by ignoring interaction dynamics. A protein complex is a group of two or more proteins formed by PPIs that are stable over time, so only dense sub-graphs excluding dynamic interactions are eligible to be protein complexes. Conventional clustering approaches can not distinguish stable protein complexes from simple PPIN that lacks in information on dynamic status of interactions, thus leading to false positive results in protein complex prediction.

In this paper, we propose an approach to predict protein complexes based on integration of PPI data and information on interfacial surfaces between protein pairs. Proposed approach is designed to reduce the false positives in prediction by excluding dynamic interactions in network clusters extracted. The basic idea is that interactions in a protein complex must be simultaneous to achieve its stability, so clusters including competitive interactions, which are incompatible at a moment, cannot be approved as protein complexes.

Competition between interactions is a type of interaction dynamics what simple PPIN cannot represent. Most proteins have a number of alternative interaction partners that may be competitive, and experimental evidences for this have been reported for several genes[4][5][6]. Many of such alternative interactions are mediated by the same or overlapping contact surface[7], so they are likely to be mutually exclusive, resulting in competition between alternative interaction partners for complex formation. Therefore, excluding competitions between mutually exclusive interactions reduces dynamics in network cluster, which may remove falsely predicted members of protein complexes.

Mutually exclusive interaction information is drawn from interfacial residue data of protein domains. More than one proteins can not physically contact the same interacting surface on a protein at a time, so utilizing interaction interface data identifies mutually exclusive interaction partners of each protein that are incompatible at a moment. PSIMAP[8] provides interface data of physical domain interactions based on tertiary structures recorded in PDB database[9]. As the domain is a sub-unit of proteins, which mediates protein interactions, we can identify the interaction interface of a protein pair by utilizing domain interface data from PSIMAP. If two or more interaction partners share the common or overlapping interface on a protein, these proteins are identified as mutually exclusive interaction partners of the protein.

The extraction of Simultaneous Protein Interaction Cluster (SPIC) is the essence of our approach. When a dense sub-graph is detected in PPIN, it is refined into SPIC by negotiating mutual exclusion among interactions. The strategy of SPIC is applicable to any simple PPI based graph-theoretic clustering methods, so we applied it to MCODE and LCMA in this research; modified methods were named SPIC_MCODE and SPIC_LCMA respectively.

Evaluation was performed on *s.cereviae* (yeast) PPIN which includes 29,683 interactions among 5,668 proteins. The results of SPIC_MCODE and SPIC_LCMA were compared with the original methods and 1,051 experimentally derived yeast protein complexes recorded in MIPS CYGD[10]. As results, SPIC_MCODE produce 135 true positives and 51 false positives, while the original method, MCODE, did 52 true positives and 88 false positives. Also, SPIC_LCMA produced 429 true positives and 1492 false positives, while LCMA did 332 true positives 1421 false positives. The comparisons showed that proposed methods adopting SPIC outperformed original graph-theoretic clustering methods. SPIC based methods refined the false positive results of original methods by achieving mutual exclusion among interactions; some of those refined clusters became true positives: 83 clusters for MCODE and 97 clusters for LCMA. Furthermore our methods did not lose any of true positive results what original methods found.

2. METHOD

2.1. Competition between Mutually Exclusive Interaction Partners

Most proteins have a number of interaction partner proteins, some of which may be cooperative or even competitive[7]. Such cooperation and competition between partners determine which one is activated among multiple functions what the host protein may serve. A Membrane protein Phospholipase D2 (PLD2) is a good example of proteins having multiple functions activated by cooperation and competition among interaction partners[11]. PLD2 catalyze the hydrolysis of Phosphatidylcholine to produce phosphatidic acid and choline, as activated by tyrosine kinase and G protein-coupled receptors among a number of interaction partners. Also it functions in regulated secretion, cytoskeletal reorganization, transcriptional regulation, and cell cycle control, which respectively are consequence of cooperation and competition between diverse interaction partners.

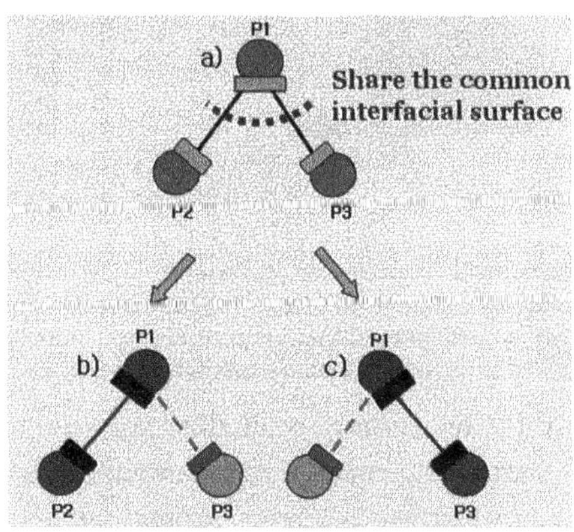

Figure 1. An example for mutually exclusive interactions: a) Two proteins, P2 and P3, bind the common surface on P1. b), c) Only one of them occurs at any given moment since the interface on protein P1 is available only for one interaction.

Among a number of interaction partners, detecting the cooperative partners for a designated function must be the essential for understanding of the protein's mechanism including protein complex formation, but only several genes are studied with great difficulty. However, the integration of current data enables exclusion of competitive interaction partners, which is an indirect method for the understating the cooperation between host protein and its partners.

The examination on physical interfaces between interaction pairs provides the information on mutual exclusiveness among partners interacting with a host protein, resulting in competition among interaction partners. If two or more interaction partners bind the common or overlapped interfacial surface on a host protein, then the surface physically available only for one interaction at given moment. Such interactions are mutually exclusive as occurrence of any one of those interactions automatically implies the non-occurrence of remaining ones. Figure 1. depicts a toy example of mutually exclusive interactions.

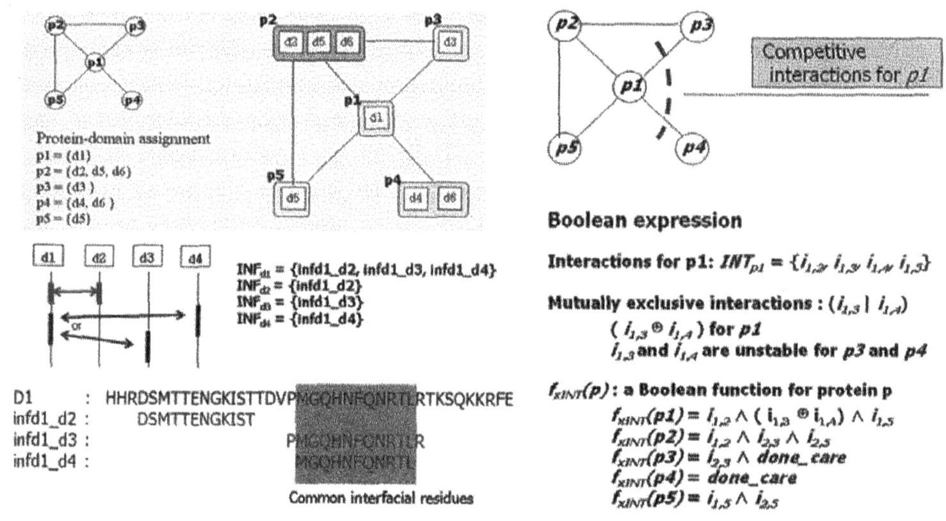

a) Detection of mutually exclusive interfaces b) Boolean expression for interaction lists on a protein

Figure 2. An example of detecting mutually exclusive interactions: a) Two interfaces, infd1_d2 and infd1_d3 are mutually exclusive since sharing common interfacial residues on d1. Therefore protein p3 and p4 are mutually exclusive for protein p1 as their interactions are mediated by interfaces infd1_d2 and infd1_d3. b) Boolean expressions for interaction list having information on mutually exclusive interactions.

The first step toward the detection of mutually exclusive interactions is identifying interface of each protein interaction, which is represented by the set of pairs of interfacial residues. In this research, an interface between protein pair is examined at the level of protein domains that are regarded as sub-units mediating protein-protein interactions. PSIMAP provides interfacial residue pairs in physical domain-domain interactions based on the analysis of crystal structures of proteins and complexes recorded in PDB database. Domain-domain interface data are extendible to many protein-protein interfaces having corresponding domain pair.

Suppose we have a simple protein interaction graph with domain information depicted in Figure 2 a), then, protein *p1* has interaction list, $INT_{p1} = \{i_{1,2}, i_{1,3}, i_{1,4}, i_{1,5}\}$

where $i_{i,j}$ means a interaction between protein *pi* and *pj*. Interfaces on protein *p1* are examined at the level of domain pair; Figure 2 a) illustrates interfaces and their notations annotated to domain *d1* for detection of mutually exclusive interactions on protein *p1*. The domain d1 has a set of interfaces, INF_{d1} = *{infd1_d2, infd1_d3, infd1_d4}*, that are provide by PSIMAP. The item *infdx_dy* denotes a set of interfacial residues on domain *x* of interface between domain *x* and *y*. Then, items in INF_{dn} are examined whether or not they have overlapped interfacial residues with each other. In the example, interfaces *infd1_d3* and *infd1_d4* overlap each other, which mediate interactions with protein *p3* and *p4*, so such interactions are mutually exclusive. Therefore, at a given moment, protein *p1* may interact with either *p3* or *p4*, and with remaining non-overlapped interaction partners. Eventually, a protein p1 have a list of non-competitive interaction partners, $xINT_{p1} = \{i_{1,2}, (i_{1,3} / i_{1,4}), i_{1,5}\}$, that contains information on mutually exclusive interactions belonging to *p1*.

The list of interaction partners is also represented by Boolean expression, for the protein *p1* $f_{xINT}(p1) = i_{1,2} \wedge (i_{1,3} \oplus i_{1,4}) \wedge i_{1,5})$ in Figure 2. b). The expression implies non-competitiveness among interactions on protein *p1*. Therefore, any sub set of original interactions which obeys the Boolean expression *xINTp1* achieves mutual exclusion, and they may occur simultaneously. Actually, protein complexes do not necessarily obey *xINT* since non-overlapped interfaces do not imply cooperation between them in nature. However, the Boolean function assumes that the interactions without mutual exclusiveness are cooperative, as following conjecture of conventional network based methods which ignore dynamics in PPIN.

2.2. *Extraction of SPIC*

A pair of mutually exclusive interactions divides a network into two possibly activated sub-networks what each of competitive interactions is contained in respectively. Therefore the number of sub-network is n^2 with the n set of mutually exclusive interactions or even more when more than two interactions are mutually exclusive each other. As we are interested in protein complexes that should be simultaneously activated, the competition between interactions are examined within a cluster.

Simultaneous Protein Interaction Cluster (SPIC) is a network cluster found in PPIN excluding interaction conflicts caused by mutually exclusive interactions, so that its interactions have a possibility to be activated at a moment. SPIC is extracted based on network clusters what any graph-theoretic clustering algorithms find. Once a graph-theoretic clustering algorithm extracts a cluster assumed to be protein complex, the cluster may have interaction conflicts causing competition between partners of each protein for complex formation. Therefore, the clusters excluding the conflicts found by the mean of mutually exclusiveness are more likely to be protein complexes than ones including conflicts.

Definition 1

A cluster in protein-protein interaction network is SPIC if and only if the cluster excludes mutual exclusiveness among interactions.

A network cluster in PPIN is examined using Boolean expression whether it is SPIC or not. All proteins with interactions in a SPIC achieve mutual exclusion, so they should obey conjunction of $f_{xINT}(pi)$ for all member proteins in the cluster. If a cluster is not a SPIC, it is refined to be several SPICs that are maximal connected sub-graphs of the cluster. (Figure 3.).

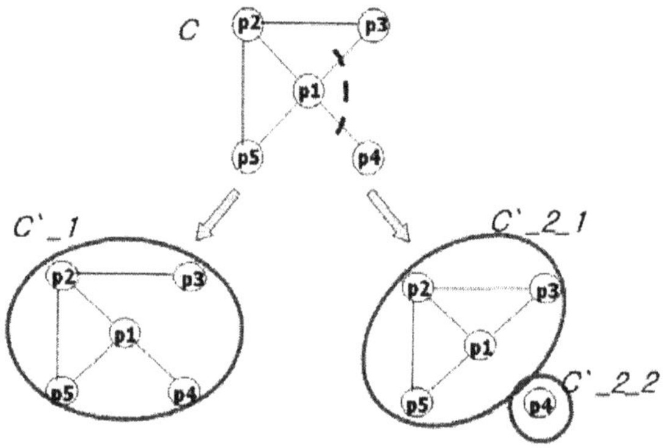

Figure 3. Extraction of SPIC from a cluster *C*. If two interactions are mutually exclusive, sub-graphs only with one of those interactions are generated(*C`_1, C`_2*). The generated sub-graphs are not necessarily clusters since elimination one of mutually exclusive interactions may disconnect proteins(*C`_2_1, C`_2_2*). In that case, each maximal connected sub-graph is SPIC respectively.

2.3. *Prediction of Protein Complex via SPIC*

The concept of SPIC focuses on the extraction of cooperative sets of proteins for prediction of protein complex as excluding competitive interactions. Therefore it additionally needs an algorithm that evaluates the density of network cluster. In this research, we adopt conventional graph-theoretic clustering algorithms, MCODE and LCMA, to evaluate the density of clusters; the methods modified using SPIC are named SPIC_MCODE and SPIC_LCMA respectively for convenience sake. The outline of our method is shown in Figure 4. SPICs are extracted from clusters what a graph-theoretic clustering algorithms detects. Then, the density of each SPIC is evaluated again since it may have lower density than original cluster has due to the elimination of mutually exclusive interactions.

3. RESULT

The evaluations are conducted twice since two conventional graph-theoretic methods, MCODE and LCMA, are modified using proposed SPIC approach; the modified methods are compared with original graph-theoretic methods. Each of results are verified by comparisons with known protein complex database, MIPS: Comprehensive Yeast Genome Database (CYGD)[10]. MIPS CYGD has 1051 complexes curated from biomedical literature. The dataset for PPIN construction was assembled from machine-readable resources: Uetz[12], Ito[13], Drees[14], Gavin[15], Tong[1], and YPD[16]. In total, PPIN consists of 29,683 experimentally determined protein-protein interactions among 5,668 yeast proteins.

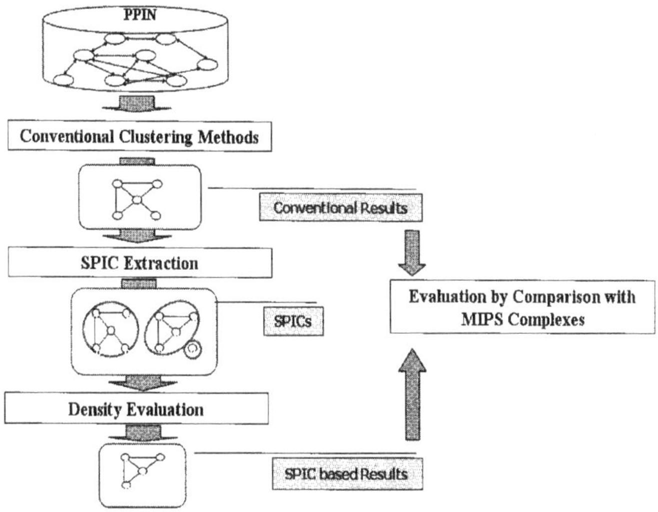

Figure 4. The structure of SPIC based methods and its evaluation plan. The conventional clustering methods are used again in the phase of density evaluation. For the evaluation of SPIC approach, the results are compared with conventional results and MIPS complexes.

We assess the precisions of results of four methods, MCODE, SPIC_MCODE, LCMA, and SPIC_LCMA, by using evaluation metric used in conventional protein complex prediction method[2][3](Equation 1), which measure the overlap score. to determine matching between a predicted complex $p \in P$ and a known complex $m \in M$, where k is the size of overlap of p and m, and $n1$, $n2$ are the sizes of p and m respectively. Given a predicted complex p and a known complex m, they are considered to be matching if $OS(p,m) \geq 0.2$, where 0.2 is an experientially determined threshold used in [2] [3].

$$OS(p, m) = \frac{k^2}{n_1 \times n_2} \quad (1)$$

Then we refer the notation in [4] to define the set of true positives (TP) as $TP = \{p / \exists m, OS(p,m) \geq 0.2,\ p \in P, m \in M\}$, and the set of false positives (FP) as $FP = P - TP$. Naturally, the set of false negatives (FN) is defined as $FN = \{m / \forall p, OS(p,m) < 0.2,\ p \in P, m \in M\}$. Then precision (specificity) is defined as $|TP|/(|TP| + |FP|)$.

3.1. *Comparative results with MCODE*

Figure 5. shows the venn diagram of the number of complexes predicted by SPIC_MCODE and MCODE, and their comparison with known complexes recorded in MIPS. The comparison of predictions with MIPS obviously depicts the effectiveness of SPIC_MCODE. The conventional MCODE method predicts 140 protein complexes ($|O|$, $O = \{b \cup d \cup g \cup f\}$), and, among them, 52 ($|d \cup g|$) complexes are shown to be correct. Meanwhile, SPIC_MCODE predicts 186 protein complexes ($|P|$, $P = \{c \cup e \cup g \cup f\}$), and, among them, 135 complexes ($|e \cup g|$) are shown to be correct. Note that $|d| = 0$ and $|e| = 83$. The SPIC_MCODE correctly predicts all of the true positives what MCODE generates, and, in addition, it obtains 83 correct complexes more.

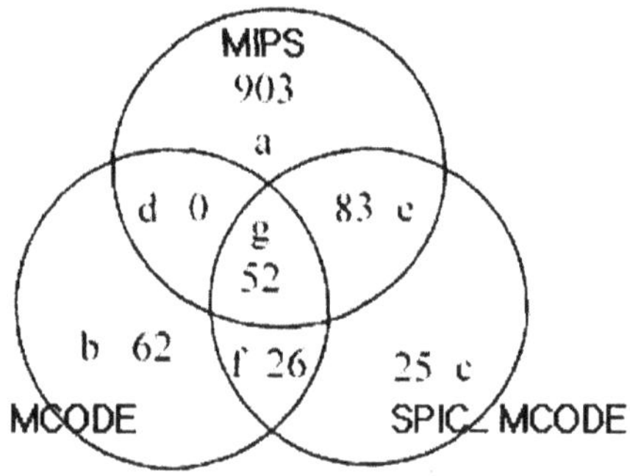

Figure 5. For MCODE, the venn diagram of the numbers of predicted and known complexes, and their comparison.

SPIC_MCODE generates 186 predictions as refining 140 original complexes what MCODE predicts. Incensement of the number of results indicates that SPIC approach splits some complexes of MCODE into several smaller and refined ones as eliminating mutually exclusive interactions within complexes. The effect of split on refinement is presented more clearly in the Table 1. as showing the number of complexes with n number of proteins and their correctness. The numbers of complexes of SPIC_MCODE tends to be greater than the ones of original MCODE for the complexes with smaller number of proteins. For complexes with 2~5 proteins, while MCODE predicts 91 complexes, SPIC_MCODE obtains 157 ones as refining complexes. The elimination of mutually exclusive interactions refines original complexes, so |TP| of SPIC_MCODE is 121 while |TP| of MCODE is 38, for complexes with 2~5 proteins.

Table 1. The number of protein complexes with n proteins, which are predicated by MCODE and SPIC_MCODE

	n	1	2 ~ 5	6 ~ 10	11 ~ 15	16 ~
I		54	378	192	205	222
II	A	0	383	449	525	396
	B	0	211	68	31	22
III	A	0	685	506	421	309
	B	0	172	163	62	32

n : The number of proteins in a complex
I : Known protein complexes in MIPS
II : LCMA Method
III : Proposed method
A : The number of protein complexes
B : The number of true positive complexes

3.2. Comparative results with LCMA

Figure 6. depicts the Venn diagram of the number of complexes predicted by SPIC_LCMA and LCMA, and their comparison with known complexes recorded in MIPS. According to Figure 6, LCMA predicts much more complexes than MCODE prediction. However, it only shows 0.19 precision caused by the large amount of false positives. The proposed method SPIC_LCMA presents the improvement of precision to 0.22. LCMA method predicts a total of 779 protein complexes ($|O|$, $O = \{ b \cup d \cup g \cup f\}$), and 332 ($|d \cup g|$) among 1753 complexes are revealed to be correct predictions. Meanwhile, SPIC_LCMA generates better results than original as predicts 1921 protein complexes ($|P|$, $P = \{c \cup e \cup g \cup f\}$), and 429 ($|e \cup g|$) complexes are correct predictions. $|d|$, is also equal to zero while $|e|$ is 97, which indicates that the SPIC_LCMA does not lose any of true positives what LCMA predicts, and, additionally, it obtains 97 correct complexes more.

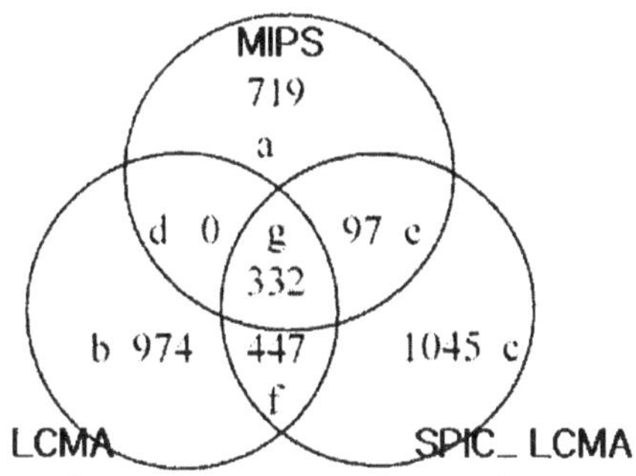

Figure 6. For LCMA, the venn diagram of the numbers of predicted and known complexes, and their comparison.

SPIC_LCMA tends to have greater number of results than LCMA method for complexes with 2~10 proteins but fewer for complexes with more than 10 proteins. That indicates that large complexes with more than 11 proteins have a tendency of being split into small complexes through the SPIC approach (Table 2.). LCMA tends to have lager number of prediction results than MCODE, especially for large complexes such as ones with more than 10 proteins. Therefore, the effect of split on refinement by SPIC approach is obvious for complexes with 6~10 proteins, contrary to MCODE having refinement effect on complexes with 2~5 proteins. For complexes with 6~10 proteins, |TP| of SPIC_LCMA is 163 while |TP| of LCMA is 68 as the elimination of mutually exclusive interactions refines original large complexes. However, the tendency of result of refinement effect on LCMA is slightly different for the complexes with 2~5 proteins as decreasing the number of additional true positives. This result is caused by the feature of local clique merging algorithm what LCMA adopts, but detail analysis reveals that vanished true positives are still valid as listed in complex group with more than 5 proteins.

Adopting SPIC approach refines clusters as eliminating unnecessary proteins that are over-predicted by conventional graph-theoretic clustering algorithms. Figure 7 illustrates the refinement effect as showing an example of MIPS complex of size 7 and

Table 2. The number of protein complexes with n proteins, which are predicted by MCODE and SPIC_MCODE

n		1	2 ~ 5	6 ~ 10	11 ~ 15	16 ~
I		54	378	192	205	222
II	A	0	383	449	525	396
	B	0	211	68	31	22
III	A	0	685	506	421	309
	B	0	172	163	62	32

n : The number of proteins in a complex
I : Known protein complexes in MIPS
II : LCMA Method
III : Proposed method
A : The number of protein complexes
B : The number of true positive complexes

two matching complexes predicted by LCMA (a) and SPIC_LCMA (b). The MIPS complex (id:15633) has a function of cell cycle, located in the nuclear of yeast. It is reported as containing seven proteins; YFL039c, YNL172w, YLR127c, YKL022c, YHR166c, YBL084c, and YGL240w. LCMA predicts (a) containing over-predicted proteins which does not appears in MIPS complex. From the complex (a), SPIC_LCMA processes the identification and elimination of mutually exclusive interactions, and then complex b) is generated. In the case of protein YDL008w, it is competitive with YGL240 for the interaction with YBL084c. Therefore one of YDL008w and YGL240 should be eliminated and density evaluation faction approves YGL240 as a winner.

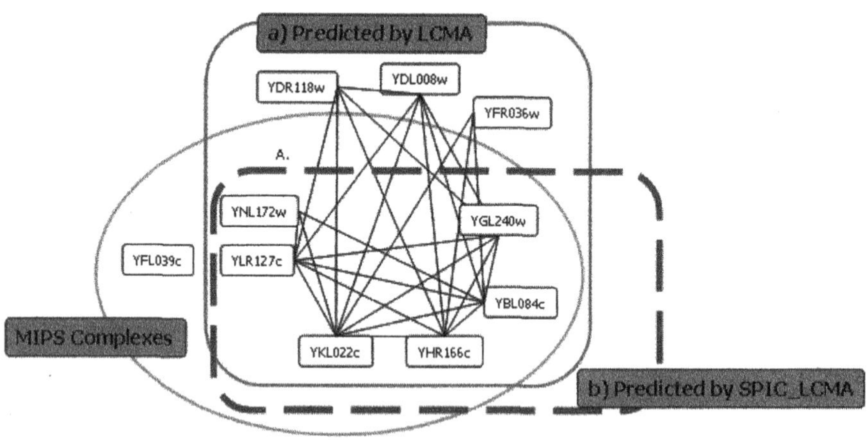

Figure 7. An comparison example for a MIPS complex (id:15633) : LCMA predicts unnecessary proteins while SPIC_LCMA excludes them.

4. CONCLUSION

In this paper, we proposed a supplementary approach to PPIN based protein complex prediction methods, which utilizes structural interface data between a protein pair. Conventional PPIN based protein complex prediction methods extract only graph-theoretic clusters without considering interaction dynamics. Interaction partners of a host protein may be mutually exclusive each other as occupying the common interfacial surface on the host protein. Even though all interactions are drawn all together in PPIN, only the clusters without those mutually exclusive interactions are eligible to be stable complexes. The concept of SPIC (Simultaneous Protein Interaction Cluster) is essential for our approach, which refines a cluster what any graph-theoretic clustering algorithms find, so that it excludes interaction conflicts caused by mutual exclusive interactions. Consequently a SPIC is likely to be closer to real protein complex than unrefined cluster is. The strategy of SPIC is applicable to any simple PPI based graph-theoretic clustering methods, so we applied it to MCODE and LCMA in this research; modified methods were named SPIC_MCODE and SPIC_LCMA respectively.

Evaluation was performed on *s.cereviae* (yeast) PPIN which includes 29,683 interactions among 5,668 proteins. The results of SPIC_MCODE and SPIC_LCMA were compared with the original methods and 1,051 experimentally derived yeast protein complexes recorded in MIPS CYGD. As results, SPIC_MCODE produced 135 true positives and 51 false positives, while the original method, MCODE, did 52 true positives and 88 false positives. Also, SPIC_LCMA produced 429 true positives and 1492 false positives, while LCMA did 332 true positives 1421 false positives. The comparisons showed that proposed methods adopting the concept of SPIC outperformed original graph-theoretic clustering methods. SPIC based methods refined the results of original methods by achieving mutual exclusion among interactions; some of those refined clusters became true positives: 83 clusters for MCODE and 97 clusters for LCMA. Especially, the fact that modified methods did not lose any of true positive results what original methods found proves that the concept of mutually exclusive interaction is quite rational and applying SPIC approach is none of the worse off.

In conclusion, the results shows that observing physical interfaces is worth consideration to improve accuracies of network based protein complex prediction methods, even thought interface and domain data are still not sufficient. Furthermore, as using structurally proved data, our approach rarely has a noise that decreases accuracy what conventional method conducts. We are sure of SPIC approach to be a firm filter for protein complex prediction, and be more useful as PPI and interface data is accumulated in future.

References

[1] A. H. Y. Tong, B. Drees, A combined experimental and computational strategy to define protein interaction networks for peptide recognition modules, *Science*, 295(5553), 321-324, 2002.

[2] G.D. Bader, C. WV. Hogue, An automated method for finding molecular complexes in large protein interaction networks, *BMC Bioinformatics*, 4:2, 2003.

[3] X. Li, S. Tan, C. Foo, S. Ng, Interaction Graph Mining for Protein Complexes Using Local Clique Merging, *Genome Informatics*, 16(2):260-269, 2005.

[4] K. Tabuchi, T. Biederer, S. Butz, T. C. Südhof, CASK Participates in Alternative Tripartite Complexes in which Mint 1 Competes for Binding with Caskin 1, a Novel CASK-Binding Protein, *The Journal of Neuroscience*, 22(11), 4264-4273, 2002.

[5] O.A. Pierrat, V. Mikitova, M.S. Bush, K.S. Browning, J.H. Doonan, Control of protein translation by phosphorylation of the mRNA 5'-cap-binding complex, *Biochemical Society Transactions*, 35, 1634-1637, 2007.

[6] R. A. Bryce, I. H. Hillier, J. H. Naismith, Carbohydrate-Protein Recognition: Molecular Dynamics Simulations and Free Energy Analysis of Oligosaccharide Binding to Concanavalin A, *Biophys J*, 81(3), 1373-1388, 2001.

[7] Hu CD, Grinberg AV, Kerppola TK, Visualization of protein interactions in living cells using bimolecular fluorescence complementation (BiFC) analysis, *Current Protocols in Cell Biology*, 21.3., 2005

[8] S. Gong, G. Yoon, I. Jang, D.M. Bolser, P. Dafas, M. Schroeder, H. Choi, Y. Cho, K. Han, S. Lee, H. Choi, M. Lappe, L. Holm, S. Kim, D. Oh, J. Bhak, PSIbase: a database of Protein Structural Interactome map (PSIMAP), *Bioinformatics*, 21, 2541-2543, 2005.

[9] H. M. Berman, J. Westbrook, Z. Feng, G. Gilliland, T.N. Bhat, H. Weissig, I. N. Shindyalov, P.E. Bourne, The Protein Data Bank, *Nucleic Acid Res.*, 28, 235-242, 2000.

[10] H. W. Mewes, D. Frishman, C. Gruber, B. Geier, et al, MIPS: A database for genomes and protein sequences, *Nucleic Acids Res.*, 28(1), 37-40, 2000.

[11] Colley WC, Sung TC, Roll R, et al. Phospholipase D2, a distinct phospholipase D isoform with novel regulatory properties that provokes cytoskeletal reorganization, *Curr. Biol*, 7 (3), 191-201, 1997.

[12] P. Uetz, L. Giot, G. Cagney, T. A. Mansfield, R. S. Judson, J. R. Knight, D. Lockshon, V. Narayan, M. Srinivasan, P. Pochart, A. Qureshi-Emili, Y. Li, B. Godwin, D. Conover, T. Kalbfleisch, G. Vijayadamodar, M. Yang, M. Johnston, S. Fields, J. M. Rothberg, A comprehensive analysis of protein-protein interactions in Saccharomyces cerevisiae, *Nature*, 403(6770), 623-627, 2000.

[13] T. Ito, T. Chiba, R. Ozawa, M. Yoshida, M. Hattori, Y. Sakaki, A comprehensive two-hybrid analysis to explore the yeast protein interactome, *Proc. Natl. Acad. Sci.*, 98(8), 4569-4574, 2001.

[14] Drees BL, Sundin B, Brazeau E, Caviston JP, Chen GC, Guo W, Kozminski KG, Lau MW, Moskow JJ, Tong A, Schenkman LR, McKenzie A 3rd, Brennwald P, Longtine M, Bi E, Chan C, Novick P, Boone C, Pringle JR, Davis TN, Fields S, Drubin DG., A protein interaction map for cell polarity development, *J. Cell Biol.*, 154(3), 549-571, 2001.

[15] A. C. Gavin, M. Bosche, R. Krause, P. Grandi, et al., Functional organization of the yeast proteome by systematic analysis of protein complexes, *Nature*, 415(6868), 141-147, 2002.

[16] M. C. Costanzo, M. E. Crawford, H. E. Hirschman, J. E. Kranz, et al., YPD, PombePD and WormPD: Model organism volumes of the BioKnowledge library, an integrated resource for protein information, *Nucleic Acids Res.*, 29(1), 75-79, 2001.

ON THE RECONSTRUCTION OF THE *MUS MUSCULUS* GENOME-SCALE METABOLIC NETWORK MODEL

LAKE-EE QUEK[1] LARS K. NIELSEN[1]
l.quek@uq.edu.au lars.nielsen@uq.edu.au

[1] *Australian Institute for Bioengineering and Nanotechnology, The University of Queensland, St Lucia Campus, Brisbane QLD 4072, Australia*

Genome-scale metabolic modeling is a systems-based approach that attempts to capture the metabolic complexity of the whole cell, for the purpose of gaining insight into metabolic function and regulation. This is achieved by organizing the metabolic components and their corresponding interactions into a single context. The reconstruction process is a challenging and laborious task, especially during the stage of manual curation. For the mouse genome-scale metabolic model, however, we were able to rapidly reconstruct a compartmentalized model from well-curated metabolic databases online. The prototype model was comprehensive. Apart from minor compound naming and compartmentalization issues, only nine additional reactions without gene associations were added during model curation before the model was able to simulate growth *in silico*. Further curation led to a metabolic model that consists of 1399 genes mapped to 1757 reactions, with a total of 2037 reactions compartmentalized into the cytoplasm and mitochondria, capable of reproducing metabolic functions inferred from literatures. The reconstruction is made more tractable by developing a formal system to update the model against online databases. Effectively, we can focus our curation efforts into establishing better model annotations and gene–protein–reaction associations within the core metabolism, while relying on genome and proteome databases to build new annotations for peripheral pathways, which may bear less relevance to our modeling interest.

Keywords: systems biology; metabolism; computational model; mouse

1. Introduction

Genome-scale metabolic network models (GSMs) are useful tools to represent and analyze the metabolism of an organism. They are information infrastructures containing chemically accurate descriptions of the cellular reactions and known gene–protein–reaction associations [11]. GSM provides a context to study cellular metabolism, not only to derive insights into the metabolic phenotypes that emerge from the system as a whole, but also to integrate heterogeneous datasets within a single modeling framework [1-3, 13]. Many organism-specific GSMs have been generated to date, ranging from microbial to multicellular organisms [5, 6, 11, 15].

Reconstruction of a metabolic network is a challenging task. For well-annotated genomes, a preliminary model can be assembled from online gene and protein databases; all that is required is an appropriate system for information storage and a consistent naming of network components. This is followed by an immense effort taken to curate the GSM such that the model reflects well-demonstrated and current knowledge of the organism's metabolism. The effort increases with the degree of content fidelity required – validating network components and their interactions using direct physical evidence in the *H. sapiens* Recon 1 model illustrate the potential challenges posed [5]. Without specialized software tools or formalized procedures, the reconstruction process is a daunting task not readily accomplished by small research groups with limited resources.

In this paper, we describe our experience with the reconstruction of the *M. musculus* GSM. We established a simple but formal approach to compile and curate a new GSM using basic software tools, namely JAVA (Sun Microsystems, Inc), Excel (Microsoft Corporation) and MATLAB (The MathWorks), which are used for information extraction, for storage and editing of the reconstructed model, and for flux simulation, respectively (Fig. 1). A new GSM is rapidly prototyped by large-scale extraction of gene, protein and reaction information from genome and proteome databases. This rudimentary GSM is then curated such that that known metabolic functions are reproduced *in silico*.

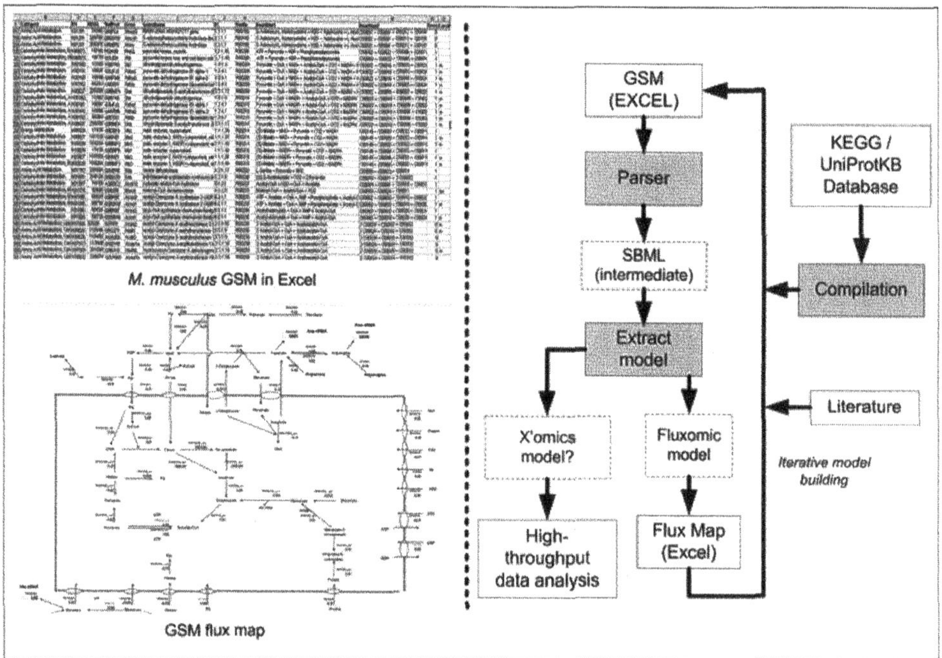

Figure 1. Workflow for the reconstruction of the *M. musculus* GSM. The GSM is compiled from online KEGG and UniProtKB databases using JAVA (grey boxes), and is stored in Excel (top-left). Gene-centric contents are parsed into reaction-centric SBML, which is a convenient intermediate for extraction of various X'omics sub-models. One instance is the fluxomic (stoichiometric) model, which is used to curate the GSM by flux balance analysis. Flux results are visualized on a flux map drawn in Excel (bottom-left). Model curation is an iterative process, whereby the GSM is consistently reconciled against biochemical literatures and new annotation data.

Our manual curation was focused on improving connectivity and annotation of the metabolic components in core metabolism, i.e., energy metabolism and anabolic reactions required for biosynthesis of major cell components (protein, DNA, RNA, lipids and carbohydrates). Main task of curation is to identify inconsistent compound name and to fill reaction gaps in the network. During metabolic simulation, the presence of an incorrect reaction is flagged by the failure to synthesize the required biomass precursors or producing unbalanced inputs and outputs. In contrast, the connectivity of peripheral pathways is progressively improved by automatically deriving new annotations from well-curated metabolic databases. All manual modifications made in core metabolism are recorded to ensure traceability, which enables the successive automated update of peripheral pathway using online databases. As a whole, this approach enabled us to make a functional *M. musculus* GSM available without significant upfront investment of resources, while still supporting continued improvements.

2. Large-scale Metabolic Reconstruction

2.1. *Gene-Reaction Assignment*

We adopted a gene-centric organization of metabolic information, in which each of the known metabolic genes is be mapped to one or many reactions. The core of the GSM was generated using the KEGG (Kyoto Encyclopedia of Genes and Genomes) genes database for *M. musculus* (Release 46) [8]. The gene–reaction mappings were derived from the four different flat files available for each pathway map: the GENE and RN (reaction) files, and their corresponding COORD (coordinate) files. These files can be downloaded from KEGG's FTP site. Each *mmu* (indicating *M. musuclus*) gene entry is mapped to a reaction entry using their positional coordinate on the pathway map, which is contained in the COORD files. This is likened to clicking the active links in KEGG's pathway maps to download the corresponding gene and reaction documents. The process is repeated for all available pathway maps. Redundant gene–reaction entries are subsequently removed. Metabolic reconstruction from KEGG's reaction database is readily performed [9]. Here, a simple JAVA script is used. The accompanying annotation attributes were included as well, namely the gene name, enzyme name, EC (Enzyme Commission) number, UniProtKB accession number, KO (KEGG Orthology) accession number and the name of the metabolic pathways where the gene entry was found.

A weakness of the above approach is that the coverage of the gene-reaction associations is limited to reactions presented on these maps. Half of the gene entries in the global gene list (in "mmu_genome" LST file) could not be mapped to a reaction entry because their corresponding coordinates do not exist. To overcome this, the EC number associated with the gene (in "mmu_enzyme" LST file) was used instead to link to one or many reaction entries using KEGG LIGAND's "reaction" file. We chose to use pathway map coordinates as the primary mapping mechanism, because genes in the maps are linked to specific reactions, whereas the use of EC number leads to mapping of genes to a broader reaction categories.

2.2. *Reaction Attributes and Compartmentalization*

The reaction attributes attached to each gene-reaction association are the reaction equation and reversibility. Reaction equations are derived from the "reaction" and "reaction_name" LST files contained in KEGG LIGAND. The original reaction formula is retained, with the compounds expressed using the full chemical name and the ID (unique entry number). As each compound ID may be associated with multiple full name aliases, it is more reliable to use the compound ID as the basis to generate the stoichiometric matrix. A full name version is kept for display purposes. The reaction reversibility (and direction) is derived from the "reaction_mapformula" LST file. Where conflicting information is encountered for the same reaction from different maps, the reaction is assumed reversible. Non-mapped reactions are automatically assumed reversible by default in absence of further information.

For reaction compartmentalization, we currently only distinguish two sub-cellular localizations: cytoplasm and mitochondria. Using the UniProtKB accession number(s) gathered for each gene entry (in "mmu_uniprot" LST file), we can interrogate the UniProtKB database for the sub-cellular localization of the corresponding protein. By default, all reactions are assigned to the cytoplasm, unless there is specific information to suggest that the reaction is localized either solely in the mitochondria or in both the cytoplasm and mitochondria.

3. Manual Curation

3.1. *Data storage*

The main objectives of curation are (a) to reproduce the known metabolic functions *in silico* by filling in network gaps and (b) to remove inconsistent naming of compounds. Metabolic modeling is performed in parallel with the curation and it is important that an appropriate data storage model is chosen that supports both curation efforts and extraction of the GSM content into structured models (i.e., stoichiometric matrix).

We used Excel as a convenient interface to curate the GSM. Contents of the GSM are easily visualized and modified. From the large-scale metabolic network reconstruction, it is relatively easy to produce a tab-delimited text file that contains a unique list of gene-reaction associations, with the accompanying gene and reaction attributes, which can be imported into Excel.

However, the GSM stored in Excel is gene-centric. For metabolic (stoichiometric) modeling, the contents must be organized into a reaction-centric form. A solution is to convert the GSM (in Excel) into SBML (System Biology Markup Language) data format as an intermediate storage medium (www.sbml.org), from which the stoichiometric model is generated. The key advantages are that (1) the gene–protein–reaction–metabolite association can be described in a hierarchical format, (2) the reaction–metabolite elements are easily transformed into a stoichiometric matrix and (3) the approach is consistent with the community's practice for storing biochemical network models. There is no specific element in SMBL allocated to store the gene–protein–reaction associations (e.g., splice-variants, isozymes, protein complex). Accordingly, additional sub-elements under the "reaction" element were created to accommodate these associations. The storage of the GSM in a hierarchical data format supports efficient interrogation and clustering of the GSM's content, especially for processing of omics datasets.

3.2. *Checking consistency of reaction equation*

Inconsistent labels used to describe the same compound are manifested as network gaps when performing stoichiometric modeling. To avoid this problem, a single candidate must be chosen to represent all other alternatives, and this chosen candidate is consistently applied throughout the GSM. The most common problem is the non-specific and specific reference to sugar stereoisomers (e.g., D-Glucose versus α-D-Glucose). Table 1 contains the modifications that were made to maintain consistent usage of compound name. Where KEGG had two identical reactions using different compound ID, only one the reaction with the desired compound name was retained.

Table 1. List of modifications made to the the compound's name and entry.

Compound name	Compound entry
D-Glucose → alpha-D-Glucose	C00031 → C00267
D-Glucose 1-phosphate → alpha-D-Glucose 1-phosphate	C00103[a]
D-Fructose → beta-D-Fructose	C00095 → C02336
D-Fructose 6-phosphate → beta-D-Fructose 6-phosphate	C00085 → C05345
D-Fructose 1,6-bisphosphate → beta-D-Fructose 1,6-bisphosphate	C00354 → C05378
N-Acetyl-alpha-D-glucosamine 1-phosphate → N-Acetyl-D-glucosamine 1-phosphate	C04501 → C04256
Electron-transferring flavoprotein → FAD	C04253 → C00016
Reduced electron-transferring flavoprotein → FADH2	C04570 → C01352
Inositol 1-phosphate → 1L-myo-Inositol 1-phosphate	C01177[a]
CMP[a]	G10621 → C00055
UDP[a]	G10619 → C00015
GDP[a]	G10620 → C00035
Dolichyl phosphate	G10622 → C00110
GDP-L-fucose	C00325 → G10615

Some reaction formulas in KEGG's were found to violate atom conservation. They are typically encountered in reactions that involve (1) the synthesis and breakdown of polymers, (2) the use of a generic atom "R", and (3) the consumption or production of H_2O, H^+, and redox equivalents (e.g., NAD(P)H, FADH2). In this GSM, polymers were described in the form of their corresponding monomers, and the use of the generic atom "R" was avoided. The active reaction set is checked when inconsistent atom balance is detected at the cellular input/output level during flux simulation. It was more difficult to close the balance for hydrogen and oxygen because metabolites like H_2O and redox units are highly connected. In recent work, automated atom mapping algorithms have been generated to validate reaction equation for these inconsistencies [7].

---

[a] No change

3.3. *Adding Membrane Transporters*

Exchange equations are used to describe the inter-compartmental exchange of metabolites: cytoplasm–extracellular and cytoplasm–mitochondria. Predominantly, the exchange equations are added to the GSM on the basis that these transporters are necessary components of normal metabolic functions. For example, the uptake of macro nutrients (e.g., amino acids, glucose), the secretion of by-products (e.g., alanine, lactate, ammonia) and the exchange of free compounds (H_2O, CO_2, O_2) are added because they represent essential cellular inputs and outputs. Similarly, the intracellular exchange of compounds between the cytoplasm and mitochondria are inferred from known mitochondrial functions, such as cellular respiration, synthesis of biomass precursors (e.g., acetyl-CoA, oxaloacetate), and oxidation of aliphatic compounds (e.g., fatty acids, branched-chain amino acids). The final GSM consists of a total of 33 and 31 inter-compartmental exchange equations added for cytoplasm–extracellular and cytoplasm–mitochondria transporters, respectively.

3.4. *Lumping Reactions*

A series of elementary reactions catalyzed by an enzyme complex should be lumped into a single overall reaction. The importance of lumping reactions is that a single flux parameter is used to describe the activity of an enzyme complex. Physiologically, this approach is used to represent the channeling of substrate to product. In KEGG for example, the pyruvate dehydrogenase complex catalyze four separate reactions: pyruvate decarboxylation (2 steps, via an intermediate thiamine pyrophosphate cofactor), dihydrolipoyl transacetylase and dihydrolipoyl dehydrogenase (Table 2). These four reactions are summed into an overall reaction by removing the intermediate metabolites. The lumping of these reactions reinforces the fact that dihydrolipoyl dehydrogenase is not shared between pyruvate dehyrogenase, oxolgutarate dehydrogenase and branched-chain oxo-acid dehydrogenase, which adopts similar reaction mechanism.

Table 2. List of elementary reactions catalyzed by the pyruvate dehydrogenase complex. These reactions are summarized into an overall reaction equation.

Reaction entry	Reactant side		Product side
R00014	Pyruvate + ThPP	=	2-Hydroxyethyl-ThPP + CO_2
R03270	2-Hydroxyethyl-ThPP + Lipoamide-E	=	S-Acetyldihydrolipoamide-E + ThPP
R02569	CoA + S-Acetyldihydrolipoamide-E	=	Acetyl-CoA + Dihydrolipoamide-E
R07618	Dihydrolipoamide-E + NAD^+	=	Lipoamide-E + NADH + H^+
R00209 (overall)	Pyruvate + CoA + NAD^+	=	Acetyl-CoA + CO_2 + NADH + H^+

Lumping is also introduced to define how NADH and $FADH_2$ contribute their redox equivalent to the electron transport chain. Without a clear description of the mechanism of the electron transport chain, and a satisfactory proton balance in both the inner membrane space (i.e., cristae) and the mitochondria matrix, it is more efficient and simpler to describe the electron transport chain with a generic oxidative phosphorylation reaction, using P/O ratio of 2.5 and 1.5 for NADH and $FADH_2$ respectively. Following this modification, dehydrogenase reactions that contain ubiquinone–ubiquinol cofactor pair are replaced by FAD–FADH2 cofactor pair (e.g., succinate dehydrogenase). This is necessary to define the entry point of the redox equivalent generated.

3.5. *Adding biomass drain equations*

Similar to the concept of adding membrane transporters to describe the efflux of by-products, the biomass drain equations are incorporated into the GSM as the accumulation terms of the biomass precursors. It is useful to describe these accumulation terms individually (e.g., "Cholesterol = Cholesterol_biomass"), in order to simplify the task of uncovering the pathway gaps in the each of the biosynthetic routes separately. For example, the pathway for cholesterol synthesis can be visualized independent of other biomass components by allowing only the production of cholesterol. A zero drain value indicates the presence of reaction gap in the cholesterol pathway. This process is iterated for all biomass precursors. Once the network gaps are filled, all biomass accumulation terms are then combined into an overall biomass synthesis equation, with the appropriate coefficients assigned to each precursor to define the composition of biomass.

For the GSM, a total of 17 biomass drain equations were added. They were used to describe the accumulation of phospholipids (7), nucleotides (8), glycogen (1) and cholesterol (1). The drains of amino acid are described via their respective amino-acyl-tRNA synthetase reactions.

3.6. *Finding Network Gaps*

Finding breaks in metabolic pathways is not an intuitive task. PathoLogic (Pathway Tools, SRI International) is an elegant program that can infer pathway gaps, from the genome annotations, using reference pathways (e.g., MetaCyc). However, Pathway Tools does not support compartmentalization and its content is not readily transformed into a stoichiometric model for flux simulations. Instead, we adopted a few elementary approaches to find the network gaps. The priority of finding network gaps is to enable the synthesis of biomass precursors *in silico*. The secondary objective is to reproduce known metabolic functions deduced from biochemical literature [10].

Visual inspection of metabolic maps from KEGG PATHWAY is a quick technique to find network gaps. Operating on a similar concept as PathoLogic, one can browse through the organismal pathway maps to deduce missing reactions using visual evidence that most of the reactions in the given pathway exist. This approach is particularly effective for tracing synthetic pathways for biomass components. These pathways are generally linear, and can also be checked against biochemical literature. Overall, this approach led to the identification of six missing reactions essential for biosynthesis (Table 3). These reactions were present in the human GSM [5].

Table 3. List of new reactions identified by visual inspection of KEGG PATHWAY.

Reaction	Reaction equation	Pathway
R01321	CDP-choline + 1,2-Diacyl-sn-glycerol = CMP + Phosphatidylcholine	Phospholipid
R01514	ATP + D-Glycerate = ADP + 3-Phospho-D-glycerate	Glycerol
R01801	CDP-diacylglycerol + sn-Glycerol 3-phosphate = CMP + Phosphatidylglycerophosphate	Phospholipid
R02029	Phosphatidylglycerophosphate + H_2O = Phosphatidylglycerol + Orthophosphate	Phospholipid
R02057	CDP-ethanolamine + 1,2-Diacyl-sn-glycerol = CMP + Phosphatidylethanolamine	Phospholipid
R07496	alpha-Methylzymosterol = Zymosterol	Cholesterol

The alternative approach is to use flux balance analysis (FBA) (i.e., linear optimization [14]) to check whether known metabolic functions can be reproduced *in silico*. To set the problem up for identifying network gaps in the biosynthetic pathways, only the uptake fluxes of nutrients that the organism is auxotrophic for are set free, while all other input nutrient fluxes are constrained to zero. An infeasible biosynthetic pathway is manifested as a zero flux value calculated for the drain of the biomass component, despite the flux being maximized. Underlying the problem may be either gap(s) in the pathway's connectivity or reversibility constraints that prevents the use of the pathway. Troubleshooting the problem, one must not only progressively trace from the end-point to the start-point to inspect potential breakage in the network connectivity, but also check whether the reversibility setting for each of the encountered reactions is realistic when compared against a given set of guidelines, such as the irreversible hydrolysis of high-energy phosphate bond [9]. The appropriate corrections are made, either by adding new reactions or by relaxing the reversibility constraint.

FBA led to the discovery of eight additional reactions essential for biosynthesis. Three of these reactions have no known gene associations, but were required to catalyze the reverse of pre-existing reactions in the GSM (Table 4). The remaining five reactions were originally compartmentalized into the mitochondria, but existing biosynthetic pathways dictate their placement in the cytoplasm (Table 5), three of which could be found in the cytoplasmic compartment of the human GSM [5].

Table 4. List of new reactions identified by FBA and their corresponding irreversible reaction that catalyze a similar reaction but in the reverse direction.

Reaction	Reaction equation	Pathway
R00841[b]	sn-Glycerol 3-phosphate + H_2O = Glycerol + Orthophosphate	Glycerol
R00847	ATP + Glycerol = ADP + sn-Glycerol 3-phosphate	
R01131[b]	ATP + Inosine = ADP + IMP	Purine
R01126	IMP + H_2O = Inosine + Orthophosphate	
R06517[b]	Acyl-CoA + Sphinganine = CoA + Dihydroceramide	Fatty acid
R06518	Dihydroceramide + H_2O = Fatty acid + Sphinganine	

[b] Reaction with no gene association

Table 5. List of pre-existing reactions added to the cytoplasmic compartment

Reaction	Reaction equation	Pathway
R00848[c]	sn-Glycerol 3-phosphate + FAD = Glycerone phosphate + FADH2	Glycerol
R01221	Glycine + THF + NAD^+ = 5,10-MethyleneTHF + NH_3 + CO_2 + NADH + H^+	Glycine
R01867[c]	(S)-Dihydroorotate + Oxygen = Orotate + H_2O_2	Pyrimidine
R02030[c]	Phosphatidylglycerol + CDP-diacylglycerol = Cardiolipin + CMP	Phospholipid
R03657	ATP + L-Leucine + tRNA(Leu) = AMP + Pyrophosphate + L-Leucyl-tRNA	Biomass drain

Apart from the reactions that were essential for the synthesis of biomass components, an additional 43 reactions with no gene associations were added to the GSM based on literature data. Overall, these modifications involve network gaps that were found in nucleotide salvage and degradation pathways, as well as in essential amino acid degradation pathways. Also, the sub-cellular localization of a further 21 reactions were corrected. From the curated GSM, there was a general sense that the degradative pathways were mostly compartmentalized into the mitochondria.

4. Metabolic Network Properties

The final version of the *M. musculus* GSM consists of 1399 genes mapped to 1757 different reactions (model available in SBML format in Supplementary). Altogether, this produced 4619 unique gene-reaction associations. A total of 52 reactions with unknown gene associations were added to the GSM. This list excludes membrane transporters (68), biomass drains (21) and auto-catalytic reactions (7) that were added on the basis that they were required. There are a total of 2037 reactions in the stoichiometric model, 387 of which are located in the mitochondria.

An interaction map of the metabolites reveals global features of the GSM (Fig. 2). Only a very small set of reactions are essential for biosynthesis of major biomass compounds. The number of essential reactions is approximately 270, although this number varies depending on the imposed input constraints. The input constraints dictate the availability of input nutrients, and therefore the biosynthetic pathways that must be activated for growth. Nodes from the essential reaction set tend to be clustered to the center of the interaction map (Fig. 2, right), suggesting that these metabolites tend to have a higher degree of connectivity. Despite a large number of reactions contained in the GSM, only 1050 reactions are considered to be active (i.e., have non-zero flux). These reactions, less the essential ones, are components of pathways that are redundant for growth. Preliminary assessment of the redundant reactions revealed that they are mostly found in parallel or cyclic pathways. Firstly, a large number of these reactions involve transhydrogenation, whereby two or more reactions can be assembled into a

[c] Cytoplasmic reactions in the human genome-scale model

pathway that produces a net transfer of redox equivalent from one cofactor to another (e.g., NAD^+, $NADP^+$, FAD, ferredoxin). In KEGG PATHWAY, the ambiguity in cofactor usage of a particular gene product often lead to a duplication of the same reaction but with different cofactors involved. Secondly, the GSM reflect a generic cell and has the potential of cells in various tissues and in varying states. For example, the network contains pathways for both biosynthesis and catabolism of a large number of biomass constituents, while in reality these pathways would be temporally and/or spatially separated. The importance of this redundancy will only be realized in an organismal-level model. Undoubtedly, the level of redundancy would be greatly reduced, if reactions were filtered based on genes actually transcribed in a given cell, e.g., using transcriptomics.

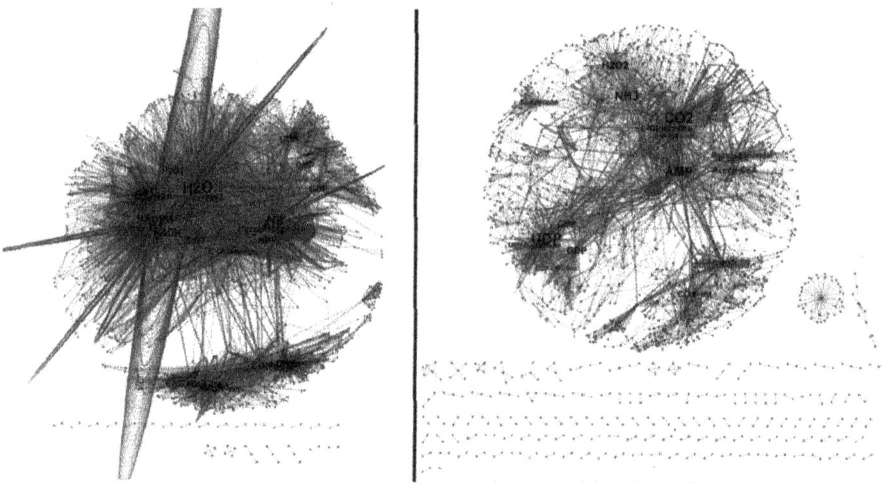

Figure 2. Visualization of the interaction network of metabolites. Metabolites are presented as nodes, while reactions are presented as edges. Metabolites with greater degree of connectivity are shown as larger node with larger label. Left figure contain all metabolites in the GSM, while the right figure has highly connected nodes (H_2O, H^+, O_2, ATP, NADPH, NADH, ADP, pyrophosphate) removed. The figures are produced using Cytoscape 2.6.0 (http://www.cytoscape.org). They are drawn using the spring-embedded layout, which tend to distribute singletons toward the peripheral space of the interaction map (see supplementary for high-resolution colour figure).

Almost half the reactions in the current model are directly or indirectly tied up to singleton (dead-end) metabolites, which account for 950 out of 2104 metabolites. Some of the singleton metabolites results from the non-specification of minor components in biomass (e.g., spermidine), which means their net synthesis must be zero. These are readily resolved by including a biomass synthesis reaction. Many singleton metabolites, however, are not connected to core metabolism in terms of carbon, only in terms of interaction with H_2O, H^+, ATP and/or redox cofactors (Fig. 2, right). This would be true of many xenobiotics metabolized in the liver, which are taken up, undergo a few reaction steps before being secreted again. Finally, some singletons are undoubtedly the result of wrongly or poorly annotated genes leading to inclusion of reactions not found in the mouse.

Using our approach, singleton metabolites will be gradually resolved as a more detailed biomass composition is considered, as better transporter annotation tools becomes available, and as secondary pathway annotation is improved. Importantly, the model remains fully functional in terms of predicting major metabolic activity despite the presence of unresolved singletons.

5. Discussion

A few critical observations were made from our experience with the large-scale reconstruction and subsequent manual curation of the *M. musculus* GSM. We demonstrated that the GSM was able to simulate basic growth and metabolic function without engaging extensive curation efforts. This validates the value of online genome and proteome databases, reinforcing the fact that the metabolic coverage by KEGG PATHWAY is, to some extent, complete, and that sub-cellular localization annotations derived from UniProtKB were accurate and readily usable. The ability to perform automated large-scale metabolic reconstruction facilitates the on-going reconciliation of our GSM with new genome annotations.

As expected, core metabolic pathways are portrayed in greater detail, and do not require extensive curation. On the other hand, curation efforts tend to be directed at singletons [5], to establish some form of network connectivity of the peripheral pathways, which are mostly discontinuous. While undoubtedly valuable, the large effort should be balanced against the returns. Where not directly linked to our needs, we are happy to let the model automatically evolve as the research community collectively improves the underlying databases. It has made possible to automate the curation tasks given a suitable reference model [12], using similar procedure outlined in our approach.

Instead we are focusing our effort at extending the existing scope of the sub-cellular compartmentalization to include nucleus, endoplasmic reticulum, peroxisome and so forth. For example, the oxidation of fatty acid should be differentiated into the peroxisomal and mitochondrial pathways. We have also commenced work on the ultimate challenge of capturing metabolic interactions between tissues and organs [5]. A feature missing from current work is metabolic regulation. The imposing regulatory network is complex, but is necessary to reflect metabolic changes for different growth conditions [4].

In conclusion, we have developed a reproducible approach for the reconstruction of *M. musculus* GSM. The approach is readily adopted because it employs generic tools for data extraction, storage and flux simulation. The development of the *M. musculus* GSM is on-going. One of the many developmental milestones is to capture and validate the gene–protein–reaction associations, and present these associations in a suitable hierarchical format [11]. This is necessary to support integration of heterogeneous datasets in future.

References

[1] Akesson, M., Forster, J., Nielsen, J., Integration of gene expression data into genome-scale metabolic models, *Metabolic Engineering*, 6(4):285-293, 2004.

[2] Cakir, T., Patil, K.R., Onsan, Z., Ulgen, K.O., Kirdar, B., Nielsen, J., Integration of metabolome data with metabolic networks reveals reporter reactions, *Molecular systems biology*, 2:50, 2006.

[3] Covert, M.W., Knight, E.M., Reed, J.L., Herrgard, M.J., Palsson, B.O., Integrating high-throughput and computational data elucidates bacterial networks, *Nature*, 429(6987):92-96, 2004.

[4] Covert, M.W., Palsson, B.O., Transcriptional regulation in constraints-based metabolic models of Escherichia coli, *Journal of Biological Chemistry*, 277(31):28058-28064, 2002.

[5] Duarte, N.C., Becker, S.A., Jamshidi, N., Thiele, I., Mo, M.L., Vo, T.D., Srivas, R., Palsson, B.O., Global reconstruction of the human metabolic network based on genomic and bibliomic data, *Proceedings of the National Academy of Sciences of the United States of America*, 104(6):1777-1782, 2007.

[6] Duarte, N.C., Herrgard, M.J., Palsson, B.O., Reconstruction and validation of Saccharomyces cerevisiae iND750, a fully compartmentalized genome-scale metabolic model, *Genome Research*, 14(7):1298-1309, 2004.

[7] Felix, L., Valiente, G., Validation of metabolic pathway databases based on chemical substructure search, *Biomol Eng*, 24(3):327-335, 2007.

[8] Kanehisa, M., Goto, S., Hattori, M., Aoki-Kinoshita, K.F., Itoh, M., Kawashima, S., Katayama, T., Araki, M., Hirakawa, M., From genomics to chemical genomics: new developments in KEGG, *Nucleic Acids Res*, 34(Database issue):D354-357, 2006.

[9] Ma, H., Zeng, A.P., Reconstruction of metabolic networks from genome data and analysis of their global structure for various organisms, *Bioinformatics*, 19(2):270-277, 2003.

[10] Michal, G., *Biochemical pathways : an atlas of biochemistry and molecular biology*, Wiley, New York, 1999.

[11] Reed, J.L., Vo, T.D., Schilling, C.H., Palsson, B.O., An expanded genome-scale model of Escherichia coli K-12 (iJR904 GSM/GPR), *Genome Biology*, 4(9):R54, 2003.

[12] Satish Kumar, V., Dasika, M.S., Maranas, C.D., Optimization based automated curation of metabolic reconstructions, *BMC bioinformatics*, 8:212, 2007.

[13] Sauer, U., High-throughput phenomics: experimental methods for mapping fluxomes, *Current Opinion in Biotechnology*, 15(1):58-63, 2004.

[14] Savinell, J.M., Palsson, B.O., Network analysis of intermediary metabolism using linear optimization. I. Development of mathematical formalism, *Journal of Theoretical Biology*, 154(4):421-454, 1992.

[15] Sheikh, K., Forster, J., Nielsen, L.K., Modeling Hybridoma Cell Metabolism Using a Generic Genome-Scale Metabolic Model of *Mus musculus*, *Biotechnology Progress*, 21(1):112-121, 2005.

PREDICTING DIFFERENCES IN GENE REGULATORY SYSTEMS BY STATE SPACE MODELS

RUI YAMAGUCHI[1] ruiy@ims.u-tokyo.ac.jp
SEIYA IMOTO[1] imoto@ims.u-tokyo.ac.jp
MAI YAMAUCHI[2] cyowako@ims.u-tokyo.ac.jp
MASAO NAGASAKI[1] masao@ims.u-tokyo.ac.jp
RYO YOSHIDA[3] yoshidar@ism.ac.jp
TEPPEI SHIMAMURA[1] shima@ims.u-tokyo.ac.jp
YOSUKE HATANAKA[1] hatanaka@hgc.jp
KAZUKO UENO[1] uepi@ims.u-tokyo.ac.jp
TOMOYUKI HIGUCHI[3] higuchi@ism.ac.jp
NORIKO GOTOH[2] ngotoh@ims.u-tokyo.ac.jp
SATORU MIYANO[1] miyano@ims.u-tokyo.ac.jp

[1]*Human Genome Center, Institute of Medical Science, University of Tokyo, 4-6-1 Shirokanedai, Minato-ku, Tokyo 108-8639, Japan*
[2]*Division of Systems Biomedical Technology, Institute of Medical Science, University of Tokyo, 4-6-1 Shirokanedai, Minato-ku, Tokyo 108-8639, Japan*
[3]*Institute of Statistical Mathematics, 4-6-7 Minami-Azabu, Minato-ku, Tokyo 106-8569, Japan*

We propose a statistical strategy to predict differentially regulated genes of case and control samples from time-course gene expression data by leveraging unpredictability of the expression patterns from the underlying regulatory system inferred by a state space model. The proposed method can screen out genes that show different patterns but generated by the same regulations in both samples, since these patterns can be predicted by the same model. Our strategy consists of three steps. Firstly, a gene regulatory system is inferred from the control data by a state space model. Then the obtained model for the underlying regulatory system of the control sample is used to predict the case data. Finally, by assessing the significance of the difference between case and predicted-case time-course data of each gene, we are able to detect the unpredictable genes that are the candidate as the key differences between the regulatory systems of case and control cells. We illustrate the whole process of the strategy by an actual example, where human small airway epithelial cell gene regulatory systems were generated from novel time courses of gene expressions following treatment with(case)/without(control) the drug gefitinib, an inhibitor for the epidermal growth factor receptor tyrosine kinase. Finally, in gefitinib response data we succeeded in finding unpredictable genes that are candidates of the specific targets of gefitinib. We also discussed differences in regulatory systems for the unpredictable genes. The proposed method would be a promising tool for identifying biomarkers and drug target genes.

Keywords: differentially regulated genes; state space model; time-course gene expression data; differences of regulatory systems.

1. Introduction

In the last decade, gene expression data analysis allowed us to explore differentially expressed genes in case-control studies, e.g., cells before and after drug dosing, normal and disease tissues, *etc.* The identified genes have been utilized as biomarkers. The case-control studies applied to gene expression data so far, however, only focused on the differences of expression profiles in each gene, it is difficult to elucidate the reason for the differences, if we do not have *a priori* knowledge about underlying regulatory system. More specifically, classical fold-change analysis and statistical group comparison method like SAM [13] can not distinguish the following two situations: Situation 1) a gene is differentially expressed between case and control samples but the regulatory systems, which regulate the gene, are the same; Situation 2) a gene is differentially expressed between case and control samples and the regulatory systems are different.

We show a schematic diagram for such situations with examples of regulatory networks among five genes ($g_1, \cdots, g_5$) in Figure 1. The top and bottom two panels represent the situations 1 and 2, respectively. The both of right panels show differentially expressed profiles of g_5 from case and control systems. The top and bottom left panels illustrate examples of case and control systems for the situation 1 (with identical regulations) and the situation 2 (with different regulations), respectively. It is presumable that such differentially expressed profiles can be generated from the systems in both situations due to following reasons. If we observe a differential expressions for g_5 in the situation 1, where the case and control systems are identical, it is possibly due to different expression patterns of regulator genes ($g_1, \cdots, g_4$) in the case and control ones. On the other hand, in the situation 2, it is due to differences of regulatory relationships from the regulators, e.g., deletion or addition of regulation, changes of strength or direction of regulation (up or down). An example of such differentially regulating systems for the situation 2 are shown in the bottom left panel. The case system equips inherently different regulations from the control one. If we consider that the control system is for normal cells, the case one may correspond to abnormal cells such as tumor cells. Another kind of differential regulations can be also considered, that is, a case system affected by drug dosing as shown in the center panel of Figure 2. If the drug represses g_5 to a constant level regardless of expression levels of the regulators, it means a structural change in regulatory relationships. An equivalent regulations can be drawn in view of gene networks as shown in the right panel of Figure 2. For identifying drug targets and understanding differences of drug efficacy for patients, the genes in the situation 2 are more important than those in the situation 1.

Therefore, the identified differentially expressed genes need to be interpreted in terms of the underlying regulatory systems. Such regulatory relationships, however, are not known in most of the cases and it is necessary to estimate them from data. As for estimation of gene regulatory systems from gene expression data, a number of probabilistic methods were proposed during the last decade, e.g., graphical

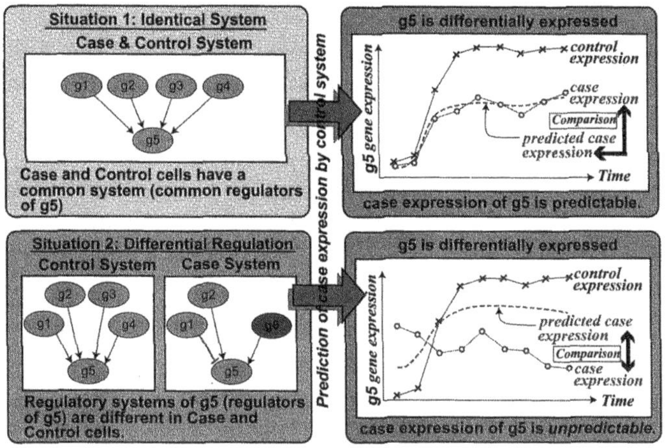

Fig. 1. Two kinds of situations for differentially expressed genes between case and control samples. Situation 1 (top panels): The gene regulatory systems are identical for the case and control. Situation 2 (bottom panels): The gene regulatory systems are different between the case and control.

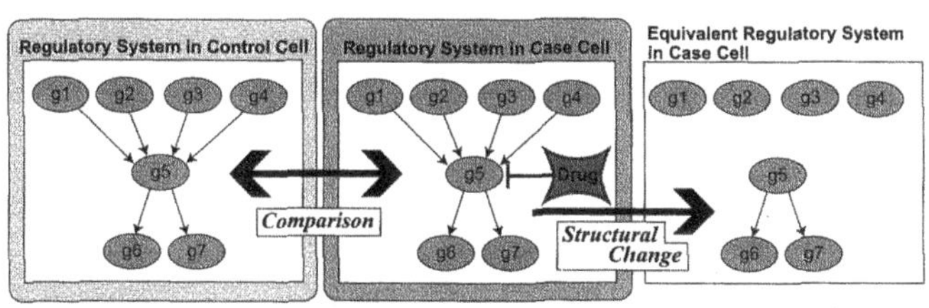

Fig. 2. A conceptual diagram for differential regulations by a drug dosing.

Gaussian models [12], dependence networks [9], boolean networks [1, 11], Bayesian networks [3, 6, 7], state space models [5, 16, 17], *etc.* Although many success stories for inferring gene regulatory systems exist, only few studies have been conducted to elucidate and utilize the differences of the regulatory systems. Recently, Qiu *et al.* [9] proposed a method to identify informative biomarkers by comparing topological changes in estimated networks for case and control of mass-spectrometry or microarray data. However, to our knowledge, there has been no study comparing dynamic systems inferred from time-course gene expression data.

Here we propose a new computational strategy for predicting differentially expressed genes under different regulations. The proposed method is to identify differentially expressed genes from time-course gene expression data of case and control samples by leveraging unpredictability of the patterns from the underlying regulatory system inferred by using a time series model, i.e., state space model. It consists of three steps. Firstly, a gene regulatory system is inferred by a state space model with the control data. Secondly, the obtained model for the underlying regulatory system of the control sample is used to predict time-course gene expressions of the case data. Finally, we explore genes whose expressions are not well predicted in the case data, but well predicted in the control data by using the model of control system. Such genes are considered unpredictable due to differential regulations in the case and control systems.

The organization of this paper is as follows. In Section 2, we describe our proposed strategy to find genes under differential regulations by using state space models. There we explain more about such differentially regulated genes identifiable as unpredictable genes and introduce state space models for modeling dynamic gene regulatory systems. In Section 3, an application result for real data, that is, time-course gene expression data of normal human small airway epithelial cells treated with epidermal growth factor (control) and ones treated with epidermal growth factor and gefitinib (case) is shown. Finally, discussion and conclusion are given in Section 4.

2. Strategy for Finding Differentially Regulated Genes

In this section, we explain our proposed strategy to identify differentially expressed genes under different regulations (i.e., differentially regulated genes) from case and control time-course gene expression data. In the method, the differentially regulated genes are identified in terms of unpredictability of the case data from a dynamic regulatory system inferred with the control data.

2.1. *Differentially regulated genes*

Here we clarify differentially regulated genes which our proposed method aims to identify. The method predicts differentially regulated genes between case and control systems by utilizing predictive abilities of state space models for time-course gene expression data. It assumes that if time-course expressions of a gene generated from one system is not producible from the other system, the regulations for the gene are different between the two systems. Such unpredictable genes are the targeted differentially regulated genes. In other words, if a time-course of a gene from one system is predictable by the other system, both systems are considered to equip the same or an equivalent regulations for the gene. Therefore, our task is to differentiate such predictable and unpredictable genes.

A conceptual diagram for identifying predictable and unpredictable genes is illustrated in Figure 1. If we see differentially expressed profiles for the case and control observations as shown in the right panels of Figure 1, it is probably hard to discriminate those of the situation 1 from those of the situation 2 by only comparing the observed expression profiles. However, if we have a dynamic model of the control system, we may differentiate those situations by comparing the observed case data (circles) and the predicted case data from the control model (dashed line). In the situation 1, the profile of g_5 in the case data will be predicted well, if the profiles of the regulators are given, since the system is the same as the control one. Meanwhile, in the situation 2, probably g_5 in the case data is not predicted well by the control model, since the control system may not generate the profile with the observation values of the regulators (see discrepancies between the circles and the dashed line in the bottom right panel).

If we can identify such differentially regulated genes, those may provide insightful information regarding systematic differences between cells, e.g., normal and cancer cells, cells with different drug susceptibility, *etc.* Especially, estimating regulatory relationships around the identified genes with a drug dosing (e.g., the right panel of Figure 2), we can expect to obtain keys for elucidating action mechanism of the drug (e.g., the center panel of Figure 2), and searching novel points of action for drug resistance cells.

2.2. *State Space Models*

We use state space models (SSMs) to estimate gene regulatory systems. SSMs have been used in a wide variety of areas for modeling time-course data and dynamic systems [8, 14]. Those have also been applied, in order to learn gene regulatory networks [2, 5, 10, 16, 17]. We briefly describe the method in the following (cf. [5] for details).

Let $y_n \in \mathcal{R}^p$, $n \in \mathcal{N}_{\text{obs}} \subseteq \mathcal{N}$ be a series of vectors containing observed expression levels of p genes at the nth time step. The set of entire time points, $\mathcal{N}$, consists of the observed time set $\mathcal{N}_{\text{obs}}$ and the unobserved one $\mathcal{N}^{\text{c}}_{\text{obs}}$. In order to model such time-course data, we use linear Gaussian state space models which are often simply called state space models. In SSMs, a sequence of the observation vectors $Y_{\mathcal{N}_{\text{obs}}} = \{y_n\}$, $n \in \mathcal{N}_{\text{obs}}$ is modeled by assuming that at each time step, y_n was generated from k-dimensional hidden state variable denoted by x_n. A basic model of SSM is shown as follows:

$$\begin{aligned} x_n &= F x_{n-1} + v_n, \; n \in \mathcal{N}, \\ y_n &= H x_n + w_n, \; n \in \mathcal{N}_{\text{obs}}, \end{aligned}$$

where the first and second equations are called the system and observation models, respectively, $F \in \mathcal{R}^{k \times k}$ is the state transition matrix, $H \in \mathcal{R}^{p \times k}$ is the observation matrix, $v_n \sim N_k(0_k, Q)$ and $w_n \sim N_p(0_p, R)$ are the system noise and the observation noise respectively. The initial state vector x_0 is assumed to be a Gaussian random vector with mean vector μ_0 and covariance matrix Σ_0, i.e., $x_0 \sim N_k(\mu_0, \Sigma_0)$. To estimate a gene regulatory system underlying the observation data, we need to estimate unknown parameters $\theta = \{H, F, Q, R, \mu_0\} \in \Theta$ characterizing the dynamic system and state vector x_n in the model. The dimension of state vector (k) is also unknown and thus needs to be determined for the optimal one.

The parameters θ in an SSM with a fixed k can be estimated by applying an EM algorithm. However, the parameters are not uniquely determined by the ordinal estimation procedure, since there are infinite number of parameter values which can yield the same likelihood. Namely the model lacks an identifiability. In order to obtain the identifiability, they proposed the following constraints on the parameters: (i) $Q = I$, (ii) $H^T R^{-1} H = \Lambda \equiv \text{diag}(\lambda_1, \cdots, \lambda_k)$ where $\lambda_1 > \lambda_1 > \cdots > \lambda_k$, and (iii) An arbitrary sign condition is imposed on the elements of the first row of H. Therefore the parameters in the model becomes $\theta = \{H, F, R, \mu_0\}$. We utilize the EM algorithm with the constraints for the parameter estimation. We denote an SSM with the parameters θ by SSM(θ).

By converting the estimated parameters and the model, a parsimonious representation of the first order vector autoregressive model is obtained as

$$R^{-1/2}(y_n - w_n) = \Psi R^{-1/2}(y_{n-1} - w_{n-1}) + R^{-1/2} H v_n, \tag{1}$$

where the autoregressive coefficient matrix is given by $\Psi \equiv D^T \Lambda F D$ with $D = \Lambda^{-1} H^T R^{-1/2}$. Since Ψ represents magnitude of interactions between genes, we can estimate a gene regulatory network with it.

2.3. *Prediction of differentially regulated genes by SSM*

If the parameters θ and observation data $Y_{\mathcal{N}_{\text{obs}}}$ are given, SSM(θ) can predict the observation y_n with the one-step-ahead prediction estimator

$$y_{n|n-1} = H x_{n|n-1},$$

where $x_{n|n-1} = E(x_n | Y_{(n-1)})$ with $Y_{(n-1)} \subseteq Y_{\mathcal{N}_{\text{obs}}}$ which is the set of observations obtained before the nth time step. Namely the prediction estimator predicts future observation with the previous observations in time course. The estimators are calculated sequentially by utilizing Kalman filter algorithm.

To identify differentially regulated genes, we search genes which have unpredictable profiles in the case data by using a model for underlying dynamic system of the control data. The procedure of the method consists of three steps as follows:

(i) SSM(θ) is applied to the time-course gene expression data of the control $Y^{\text{CTRL}}_{\mathcal{N}_{\text{obs}}} = \{y^{\text{CTRL}}_n\}$, $n \in \mathcal{N}_{\text{obs}}$ and the parameters are estimated. As a result, a model for dynamic system of the control data, SSM($\hat{\theta}^{\text{CTRL}}$), is obtained.

(ii) SSM($\hat{\theta}^{\text{CTRL}}$) is applied to predict time-course gene expression data of the case $Y^{\text{CASE}}_{\mathcal{N}_{\text{obs}}} = \{y^{\text{CASE}}_n\}$, $n \in \mathcal{N}_{\text{obs}}$.

(iii) In order to identify differentially regulated genes, we search genes whose expressions are not well predicted for the case data but well predicted for the control data by using the control model SSM($\hat{\theta}^{\text{CTRL}}$).

In the step (ii), we utilize Kalman filter and smoother algorithms twice: the first one is to estimate the initial value of the state vector x^{CASE}_0 and the second one is to obtain prediction values. At the first time, to estimate x^{CASE}_0, SSM($\hat{\theta}^{\text{CTRL}}$) is applied to $Y^{\text{CASE}}_{\mathcal{N}_{\text{obs}}}$ with the initial state $\hat{x}^{\text{CTRL}}_0 = \hat{\mu}^{\text{CTRL}}_0$. Using these algorithms, we obtain the smoother estimate of $x_{0|N} = E(x_0 | Y^{\text{CASE}}_{\mathcal{N}_{\text{obs}}})$ and use it as $\hat{x}^{\text{CASE}}_0$. At the second time, with the estimated initial state vector, we again predict the case data using one-step-ahead prediction estimator.

In the step (iii), in order to identify differentially regulated genes, we employ a statistical test. Although we can use prediction errors between the observation values and the estimators to identify differentially regulated genes, the errors do not consider the variances of the estimators. Therefore we propose a statistical test utilizing such uncertainty of the estimations. It may be more suitable for that purpose. We use a statistical testing procedure called Meta Gene Profiler (MetaGP) which evaluate the significance of a set of tests by integrating the p-values from the individual tests and yields an integrated p-value [4, 15]. In our case, an individual p-value of the gene expression data for the ith gene and at the observation time point n ($p_{i,n}$) is calculated based on the Gaussian (marginal) predictive distribution of the data, i.e., $N(y_{i,n|n-1}, \sigma_{i,n|n-1})$, where $y_{i,n|n-1}$ is the ith element of the prediction estimator $y_{n|n-1}$ and $\sigma_{i,n|n-1}$ is the ith diagonal element of the covariance matrix of prediction estimator

$$\Sigma_{n|n-1} = HV_{n|n-1}H^T + R,$$

with

$$V_{n|n-1} = E\{(x_n - x_{n|n-1})(x_n - x_{n|n-1})^T|Y_{(n-1)}\}.$$

To calculate the individual p-value, we use the two-sided test. Integrating $\{p_{i,n}\}$, $n \in \mathcal{N}_{\rm obs}$ by MetaGP, we obtain an integrated p-value p_i for the ith gene, measuring significance of the prediction errors.

3. Experimental Result

3.1. *Time-course gene expression data*

We applied the proposed method to time-course gene expression data of normal human small airway epithelial cells (SAECs) from an individual. As a control sample, we used the SAECs treated by epidermal growth factor (EGF). On the other hand, as a case sample, we used the SAECs treated by not only EGF but also gefitinib (GFT) which was extracted from tablets of Iressa (AstraZeneca). The control and case samples are labeled as "EGF" and "EGF-GFT", respectively. GFT is a selective inhibitor of epidermal growth factor receptor's (EGFR) signaling pathway. Therefore we can expect that the underlying regulatory networks for both samples are different due to the different drug dosing and also we can expect that not all of the regulations are different since samples are taken from the same cell line.

For the both samples, we took gene expression data at 19 time points during 48 hours (i.e., $\mathcal{N}_{\rm obs} = \{0, 0.5, 1, 2, 3, 4, 9, 12, 15, 18, 21, 24, 27, 30, 33, 36, 39, 43, 48\}$ [hour]) after starvation to synchronize the cell cycle by Agilent Whole Human Genome Oligo Microarray (G4112F). GFT was dosed at two hours before the 0 hour to the case sample. EGF was dosed at the 0 hour to the both samples.

3.2. *Preprocessing*

The preprocessing procedures to extract time-course gene expression signal values, for SSM analysis, from the above obtained gene expression data are described below. Each microarray has more than 40000 probes. For each probe, a raw signal value and a quality frag (i.e., present, marginal, or absent) are obtained. We applied a median shift normalization for the signals from each microarray, i.e., the median of the processed signals on a microarray is one. In the following analysis, the normalized signals after log-transformation with the basis of two were used.

We then selected a unique probe for each gene, since a gene is often measured by multiple probes for scanning different parts of the sequence. The procedure is as follows. We first counted the number of the present frags assigned to 38 data points in EGF and EGF-GFT data for each probe. Then we selected probes with the largest number of the present frags for each gene. If only one probe was chosen after the comparison of the number of the frags, we added the probe for the gene to a list. If multiple probes with the same number of the frags existed, then we chose one probe with the highest mean expression value during the interval. As a result, 19633 probes were listed for the same number of unique genes.

We reduced the number of genes to be analyzed in the following analysis, since it is not feasible to estimate a network for all of the genes in the list and also hard for interpretation. Therefore, we selected 500 genes in the list by using coefficient of variation (CV). Genes with higher CV are included in the selected gene list. Note that we only used CVs calculated from EGF data, since such a selection allowed to include genes showing different levels of the variations between the two data sets.

Finally, we obtained the two time-course gene expression data sets to be analyzed by SSMs, i.e., $Y^{\mathrm{EGF}}_{\mathcal{N}_{\mathrm{obs}}} = \{y^{\mathrm{EGF}}_n\}$ and $Y^{\mathrm{EGF\text{-}GFT}}_{\mathcal{N}_{\mathrm{obs}}} = \{y^{\mathrm{EGF\text{-}GFT}}_n\}$, $n \in \mathcal{N}_{\mathrm{obs}}$. y^{EGF}_n and $y^{\mathrm{EGF\text{-}GFT}}_n$ are 500 dimensional vectors containing the gene expression values at time n for EGF and EGF-GFT treatments, respectively. We note that the gene expression values in the data sets were from those obtained by the abovementioned normalization and transformation, but the mean expression value during the time course of each gene was shifted to be 0 for each data set, i.e., $(\sum_{n\in\mathcal{N}_{\mathrm{obs}}} y^{\mathrm{EGF}}_{i,n})/|\mathcal{N}_{\mathrm{obs}}| = (\sum_{n\in\mathcal{N}_{\mathrm{obs}}} y^{\mathrm{EGF\text{-}GFT}}_{i,n})/|\mathcal{N}_{\mathrm{obs}}| = 0$, $(i = 1, \cdots, 500)$.

3.3. *Parameter estimation*

We applied SSM(θ) to $Y^{\mathrm{EGF}}_{\mathcal{N}_{\mathrm{obs}}}$ and $Y^{\mathrm{EGF\text{-}GFT}}_{\mathcal{N}_{\mathrm{obs}}}$, respectively, in order to estimate parameters $\theta = \{H, F, R, \mu_0\}$ representing underlying gene regulatory systems. Since the minimum interval of the observation is 0.5 hour, the index of the time points are renumbered as $\mathcal{N}_{\mathrm{obs}} = \{1, 2, 3, 5, 9, 13, 19, 25, 31, 37, 43, 49, 55, 61, 67, 73, 79, 87, 97\}$ and $\mathcal{N} = \{1, \cdots, 97\}$. The dimension of state vector k was ranged $1, \cdots, 10$. In order to obtain the maximum likelihood estimator of the parameter vector for each k, the EM algorithm were applied 100 times with different initial values of the parameters. We discarded the estimated parameter vectors if the estimated time-course profiles of $x_{n|n-1}$ clearly showed unstable high frequency components by visual inspection.

Searching parameters yielding higher likelihood and reasonable time-course profiles for the state vector, we obtained the maximum likelihood estimates of parameters $\hat{\theta}_k^{\text{EGF}}$ and $\hat{\theta}_k^{\text{EGF-GFT}}$ for each k.

To determine an optimal dimension of the state vector, i.e., $\hat{k}$, we used prediction errors of data sets which were not used for the parameter estimation. Here a model yielding smaller prediction error is better in terms of generalization capability. Since there are time-course data from technical replicates at the first 11 observation time points in $\mathcal{N}_{\text{obs}}$ for both data sets, we applied SSMs with the maximum likelihood parameters for each k. As a result, $\hat{k}$ was determined to 9 for both EGF and EGF-GFT data sets since the prediction errors became the smallest for each data set. We set $\hat{\theta}^{\text{EGF}} = \hat{\theta}_9^{\text{EGF}}$ and $\hat{\theta}^{\text{EGF-GFT}} = \hat{\theta}_9^{\text{EGF-GFT}}$ in the following analysis. Thus we obtained the dynamic models for the control data and the case data, respectively, i.e., SSM($\hat{\theta}^{\text{EGF}}$) and SSM($\hat{\theta}^{\text{EGF-GFT}}$).

3.4. *Differentially regulated genes*

In order to predict differentially regulated genes between control and case systems, following the procedures explained in Section 2.3, we applied the dynamic model for the control system SSM($\hat{\theta}^{\text{EGF}}$) to the control data $Y_{\mathcal{N}_{\text{obs}}}^{\text{EGF}}$ and the case data $Y_{\mathcal{N}_{\text{obs}}}^{\text{EGF-GFT}}$.

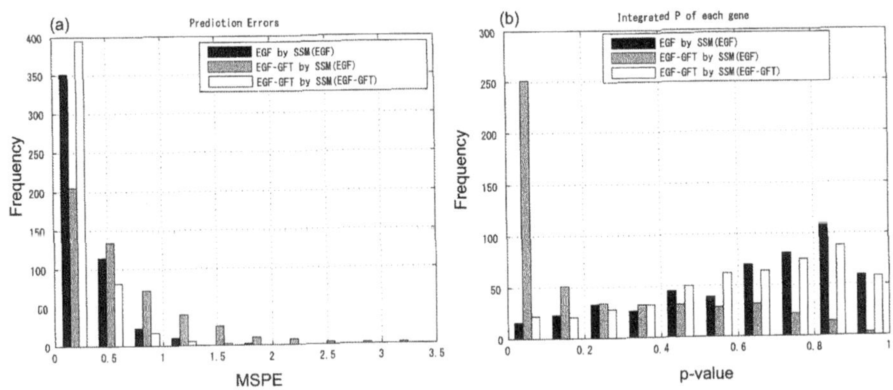

Fig. 3. Distributions of (a) the mean of squared prediction errors (MSPEs) and (b) the integrated p-values. The black (gray) bars are for EGF (EGF-GFT) data predicted by SSM($\hat{\theta}^{\text{EGF}}$). The white ones are for EGF-GFT data predicted by SSM($\hat{\theta}^{\text{EGF-GFT}}$).

Figure 3(a) shows histograms of the mean of squared prediction errors (MSPE) for each of the genes in each data obtained by using one-step-ahead prediction estimator. The black (gray) bars represent MSPEs from the EGF (EGF-GFT) data predicted by SSM($\hat{\theta}^{\text{EGF}}$). The white bars for MSPEs from the EGF-GFT data predicted by SSM($\hat{\theta}^{\text{EGF-GFT}}$) are also shown. Comparing the distributions of these bars, we can see that there exist genes in the case data which could not be predicted well by the control system as shown by the gray ones.

Figure 3(b) shows the integrated p-values of the prediction error for each gene. The colors of the bars are the same as in Figure 3(a). We can see that a number of p-values for predicted errors of EGF-GFT data by SSM($\hat{\theta}^{\mathrm{EGF}}$) (gray) accumulate in the smallest p-value group which is for significantly unpredictable genes.

In order to identify significantly differentially regulated genes, we select genes with the integrated p-values for the EGF-GFT data predicted by SSM($\hat{\theta}^{\mathrm{EGF}}$) (gray) are less than 0.01 and those for the EGF data predicted by the same model (black) are larger than 0.5.

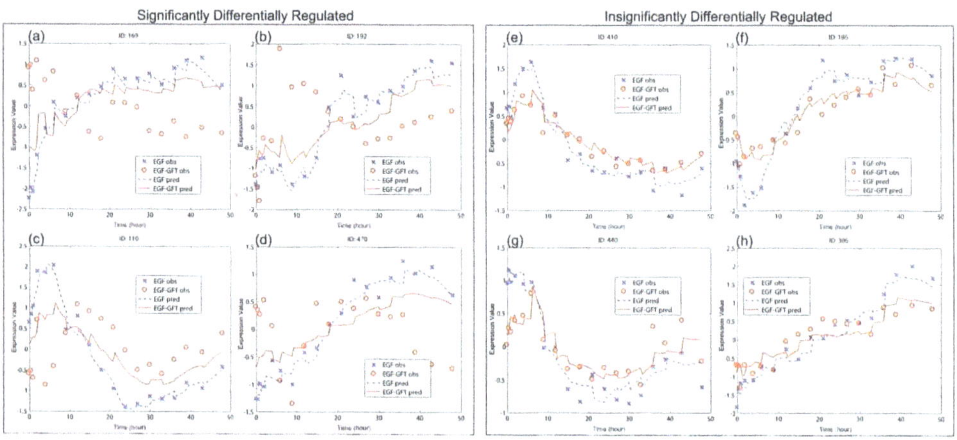

Fig. 4. The left panel: Time-course profiles of significantly differentially regulated genes, (a) g_{169}, (b) g_{192}, (c) g_{110}, and (d) g_{470}. The right panel: those of insignificantly differentially regulated genes, (e) g_{410}, (f) g_{186}, (g) g_{440}, and (h) g_{306}. The observed data points for EGF (EGF-GFT) are represented by $\times$ (o). The predicted observations for EGF (EGF-GFT) by SSM(θ^{EGF}) are represented by dashed (solid) lines.

The left panel of Figure 4 shows the top four significantly differentially regulated genes in the selected gene list sorted by the integrated p-value. Here we can see that although the prediction for EGF data (dashed line) could trace the data ($\times$) well, those for EGF-GFT (solid line) clearly deviate from the observations (o). It means the control system can not generate the pattern in the case data. This is probably due to the differences in the regulatory systems of the case and the control. In the next section, we discuss that point with estimated gene networks.

As a comparison, the right panel of Figure 4 shows genes well predicted in both data, where the integrated p-values are larger than 0.5 for both EGF and EGF-GFT. Interestingly, the control system could trace the case data which does not show very similar pattern to the control one.

4. Discussion and Conclusion

In this paper, we proposed a novel strategy to predict differentially regulated genes. There we utilized predictability and unpredictability for gene expression profiles generated from a gene regulatory system by using a dynamic gene regulatory model for another system.

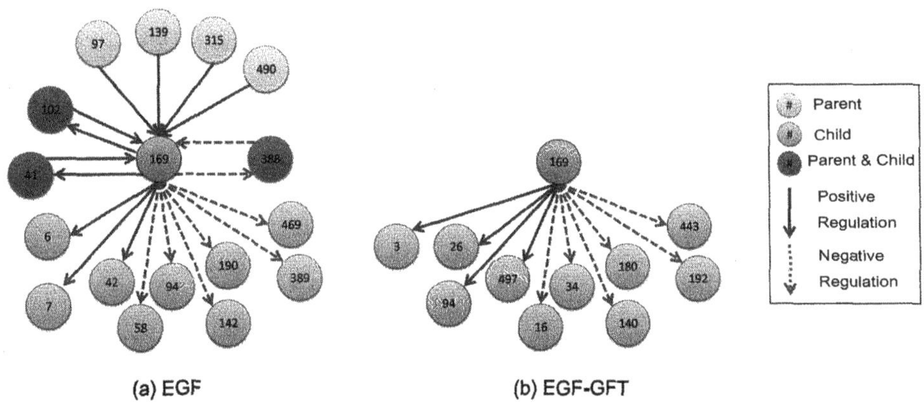

Fig. 5. Resulting gene networks around a significantly differentially regulated gene g_{169} (see Figure 4(a)) estimated from (a) EGF data, and (b) EGF-GFT data using SSMs. The edges are with the weights $|\psi_{ij}| > 0.015$.

According to the proposed strategy, we identified such unpredictable genes as candidates for differentially regulated genes from a real data set of SAEC treated by EGF and GFT. Figure 5 shows gene regulatory networks around the identified gene g_{169} (Figure 4(a)). Those were estimated from gene-gene interaction matrix $\Psi = \{\psi_{ij}\}$ (Equation 1) for $\hat{\theta}^{\text{EGF}}$ and $\hat{\theta}^{\text{EGF-GFT}}$, respectively. We drew an edge from the jth gene to the ith gene if $|\psi_{ij}| > 0.015$. The threshold value was chosen somewhat arbitrarily to obtain sparse networks. With the threshold value, there exist 6321 (10673) edges in EGF (EGF-GFT) network for the 500 genes, which is about 2.5% (4.3%) of possible ones. An alternative to select edges with more objective manner is a permutation test [5], though the method requires high computational power. Comparing the resultant networks, we can see that after dosing GFT, parents genes regulating g_{169} disappear and the children nodes are altered except for g_{94}. The regulatory relationship to g_{94} are switched from the negative regulation to the positive one. Since the disappearance of the parents genes looks like a predicted change of regulations by a drug dosing depicted in Figure 2, it supportively shows an ability of our strategy to find differentially regulated genes, although further inspection will be needed for validation. It is also interesting that g_{192} which is one of the newly attached children nodes for the EGF-GFT network is also predicted as a differentially regulated gene (Figure 4(b)).

We note that topological changes in gene networks estimated by some method with the case and control data can be utilized to search for differentially regulated genes. In our method, we can also search such topological changes by comparing Ψ matrices for the case and control data. However, we emphasize here that a characteristic of our strategy is the capability to elucidate differences of dynamic behaviors of gene regulatory systems for such differentially regulated genes, which are hard to be clarified from a static view of the networks.

Finding genes under different regulations is important task, since such genes are involved in substantial differences between the systems and, thus, potentially become good biomarkers and give clues for searching drug targets. Our proposed strategy probably become a promising tool to discover such differentially regulated genes. We are planning to conduct further experiments to validate the predicted differential regulations.

References

[1] Akutsu, T., Kuhara, S., Maruyama, O., and Miyano, S., A system for identifying genetic networks from gene expression patterns produced by gene disruptions and overexpressions, *Genome Informatics*, 9:151–160, 1998.

[2] Beal, M. J., Falciani, F., Ghahramani, Z., Rangel, C., and Wild, D. L., A Bayesian approach to reconstructing genetic regulatory networks with hidden factors, *Bioinformatics*, 21(3):349–356, 2005.

[3] Friedman, N., Linial, M., Nachman, I., and Pe'er, D., Using Bayesian network to analyze expression data, *J. Comput. Biol.*, 7:601–620, 2000.

[4] Gupta, P. K., Yoshida, R., Imoto, S., Yamaguchi, R., and S. Miyano, Statistical absolute evaluation of gene ontology terms with gene expression data, *Lecture Notes in Bioinformatics*, 4463:146–157, 2007.

[5] Hirose, O., Yoshida, R., Imoto, S., Yamaguchi, R., Higuchi, T., Charnock-Jones, D. S., Print, C., and Miyano, S., Statistical inference of transcriptional module-based gene networks from time course gene expression profiles by using state space models, *Bioinformatics*, 24(7):932–942, 2008.

[6] Imoto, S., Goto, T., and Miyano, S., Estimation of genetic networks and functional structures between genes by using Bayesian networks and nonparametric regression, *Pacific Symp. Biocomput.*, 7:175–186, 2002.

[7] Imoto, S., Kim, S., Goto, T., Aburatani, S., Tashiro, K., Kuhara, S., and Miyano, S., Bayesian network and nonparametric heteroscedastic regression for nonlinear modeling of genetic network, *J. Bioinform. Comput. Biol.*, 1:231–252, 2003.

[8] Kitagawa, G. and Gersch, W., *Smoothness Priors Analysis of Time Series*, Springer-Verlag, 1996.

[9] Qiu, P., Wang, Z. J., Liu, K. J. R., Hu, Z., and Wu, C. H., Dependence network modeling for biomarker identification, *Bioinformatics*, 23(2):198–206, 2007.

[10] Rangel, C., Angus, J., Ghahramani, Z., Lioumi, M., Sotheran, E., Gaiba, A., Wild, D. L., and Falciani, F., Modelling T-cell activation using gene expression profiling and state space models, *Bioinformatics*, 20(9):1361–1372, 2004.

[11] Shmulevich, I., Dougherty, E. R., Kim, S., and Zhang, W., Probabilistic Boolean networks: a rule-based uncertainty model for gene regulatory networks, *Bioinformatics*, 18(2):261–274, 2002.

[12] Toh, H. and Horimoto, K., Inference of a genetic network by a combined approach of cluster analysis and graphical Gaussian modeling, *Bioinformatics*, 18(2):287–297, 2002.

[13] Tusher, V. G., Tibshirani, R., and Chu, G., Significance analysis of microarrays applied to the ionizing radiation response, *Proc. Natl. Acad. Sci. USA*, 98(9):5116–5121, 2001.

[14] West, M. and Harrison, J., *Bayesian Forecasting and Dynamic Models*, 2nd edition, Springer-Verlag, 1997.

[15] Yamaguchi, R., Yamamoto, M., Imoto, S., Nagasaki, M., Yoshida, R., Tsuiji, K., Ishige, A., Asou, H., Watanabe, K., and Miyano, S., Identification of activated transcription factors from microarray gene expression data of kampo medicine-treated mice, *Genome Informatics*, 18:119–129, 2007.

[16] Yamaguchi, R., Yoshida, R., Imoto, S., Higuchi, T., and Miyano, S., Finding module-based gene networks with state-space models - mining high-dimensional and short time-course gene expression data, *IEEE Signal Processing Magazine*, 24(1):37–46, 2007.

[17] Yoshida, R., Imoto, S., and Higuchi, T., Estimating time-dependent gene networks from time series microarray data by dynamic linear models with Markov switching, In *Proc. IEEE Comput. Sys. Bioinform. Conf.*, 289–298, 2005.

Exploratory simulation of cell ageing using hierarchical models

Marija Cvijovic[1]⋆ cvijovic@molgen.mpg.de
Hayssam Soueidan[2]⋆ soueidan@labri.fr
David James Sherman[3] david.sherman@labri.fr
Edda Klipp[1] klipp@molgen.mpg.de
Macha Nikolski[2] macha@labri.fr

⋆ *These authors contributed equally to this work*
[1] *Max-Planck Institute for Molecular Genetics, Ihnestraße 63, D-14195 Berlin, Germany*
[2] *LaBRI, Université Bordeaux 1, 351 cours de la Libération, 33405 Talence, France*
[3] *INRIA Team Magnome, 351 cours de la Libération, 33405 Talence, France*

Thorough knowledge of the model organism *S. cerevisiae* has fueled efforts in developing theories of cell ageing since the 1950s. Models of these theories aim to provide insight into the general biological processes of ageing, as well as to have predictive power for guiding experimental studies such as cell rejuvenation. Current efforts in *in silico* modeling are frustrated by the lack of efficient simulation tools that admit precise mathematical models at both cell and population levels simultaneously. We developed a novel hierarchical simulation tool that allows dynamic creation of entities while rigorously preserving the mathematical semantics of the model. We used it to expand a single-cell model of protein damage segregation to a cell population model that explicitly tracks mother-daughter relations. Large-scale exploration of the resulting tree of simulations established that daughters of older mothers show a rejuvenation effect, consistent with experimental results. The combination of a single-cell model and a simulation platform permitting parallel composition and dynamic node creation has proved to be an efficient tool for in silico exploration of cell behavior.

Keywords: Systems biology, Ageing and senescence, Hybrid hierarchical modeling, Yeast

1. Introduction

A recurring challenge for *in silico* modeling of cell behavior is that hand-tuned, accurate models tend to be so focused in scope that it is difficult to repurpose them. *Hierarchical modeling* [2] is one way of combining specific models into networks. Effective use of hierarchical models requires both formal definitions of the semantics of such compositions, and efficient simulation tools for exploring the large space of complex behaviors. In this study, we propose the use of a hierarchical model to reduce the complexity of analysing cell ageing phenomena such as cell rejuvenation. To this end, we extend a single-cell model of inheritance of protein damage to a structured population where mother-daughter relations are tracked. This requires definition and implementation of an exploratory simulation software system. Using this system we validate the model, discover a cell rejuvenation effect consistent with the experimental literature, and derive testable hypotheses on cell ageing.

Unlike most microorganisms or cell types, the yeast *Saccharomyces cerevisiae* undergoes asymmetrical cytokinesis, resulting in a large mother cell and a smaller daughter cell. The mother cells are characterized by a limited replicative potential accompanied by a progressive decline in functional capacities, including an increased generation time [15]. Accumulation of oxidized proteins, a hallmark of ageing, has been shown to occur also during mother cell-specific ageing, starting during the first G1 phase of newborn cells [1]. Both asymmetric and symmetric division exist in different yeast species. In particular, *S. cerevisiae* is known to divide asymmetrically, although symmetrical division is observed in about 30% of cells at the end of their replicative lifespan [9]. Another yeast model organism, *Schizosaccharomyces pombe*, divides symmetrically by fission (see [11] for review). The following is a mathematical model we have developed to explain how the accumulation of damaged proteins influences fitness and ageing in yeast. In this paper we consider the two theoretically possible scenarios, namely asymmetrically and symmetrically dividing cells in different damaging environments. To explore any and all branches of the pedigree tree of a cell population, we will use a hierarchical model that allows us to track mother-daughter relations. We can therefore explore lineage-specific properties, such as the rejuvenation property.

2. From single cell to population model

Single cell model A minimal single-cell model of inheritance of damaged proteins can be formalized by the following three equations:

$$\frac{dP_{\text{int}}}{dt} = \frac{k_1}{k_s + P_{\text{int}} + P_{\text{dam}}} - k_2 P_{\text{int}} - k_3 P_{\text{int}} \tag{1}$$

$$\frac{dP_{\text{dam}}}{dt} = k_3 P_{\text{int}} - k_4 P_{\text{dam}} \tag{2}$$

$$\frac{dP}{dt} = \frac{k_1}{k_s + P_{\text{int}} + P_{\text{dam}}} - k_2 P_{\text{int}} - k_4 P_{\text{dam}} \tag{3}$$

The size of the cell is the sum (P) of intact (P_{int}) and damaged (P_{dam}) proteins, $P = P_{\text{int}} + P_{\text{dam}}$. Protein temporal dynamics are determined by five rate constants k_1, k_2, k_3, k_4, and k_s. Protein production rate, k_1, has been adjusted by hand allowing for a steady state to be reached and has been assigned a final value of 10^7. We choose values of k_2 and k_4, the degradation rates of P_{int} and P_{dam}, respectively, so that $k_2 < k_4$. Degradation rates are computed using the half-life formula $t_{1/2} = \ln 2/k$, where k is the degradation rate; setting the half-life of intact proteins to be 1 time unit, $k_2 = \ln 2$. Since degradation of damaged proteins is faster, k_4 needs to be greater than k_2 and it has been set to $\ln 5$. To simulate different rates of conversion, k_3 has been given a range of values, from 0.1 to 2.3. Finally, k_s is a half-saturation constant in the model, not used in this study. We assume that cells grow until they have attained a critical cell size, P_{div}, which triggers a cell division. A cell may divide symmetrically (halving its mass) or asymmetrically, as defined by size coefficients s_{mother} and $s_{daughter}$. These different types of divisions are modeled by varying the two coefficients $s_{mother}$and $s_{daughter}$. In the case of symmetrical division, the size of both progeny and progenitor is equal, so $s_{mother} = s_{daughter} = 0.5$. In asymmetrical division, cells in the next generation will have different sizes such as when $s_{mother} = 0.75$ and $s_{daughter} = 0.25$. In this study, rate constants k_1, k_2, and k_4 received fixed values, k_3 was given a range of values with step size 0.1, and s_{mother} and $s_{daughter}$ were given two pairs of values representing symmetric and asymmetric growth strategies, namely $\langle s_{mother}, s_{daughter}\rangle$ being $\langle 0.5, 0.5\rangle$ or $\langle 0.75, 0.25\rangle$.

The proteins distribution between the progenitor and the progeny after division is described by the following set of transition assignments:

$$P_{\text{int}} := P_{\text{int}} \cdot s_P \tag{4}$$

$$P_{\text{dam}} := P_{\text{dam}} \cdot s_P \tag{5}$$

$$P := P_{\text{int}} \cdot s_P + P_{\text{dam}} \cdot s_P, \tag{6}$$

where s_P is s_{mother} for the progenitor and $s_{daughter}$ for the progeny.

Population Model Based on this single-cell model, we first define a hierarchical model of a structured population where complete mother-daughter relations are recorded, using the BioRica formalism. BioRica [16] is a high-level modeling framework integrating discrete and continuous multi-scale dynamics within the same semantic domain. It is in this precise sense of mixing different dynamics that BioRica models are *hybrid* following the classical definitions [2]. Moreover, BioRica models are built hierarchically. In [2] two types of hierarchy are defined: *architectural* and *behavioral.* While BioRica admits both, in this paper we are only concerned with the former. This type of hierarchy allows for both concurrency and parallel composition.

In this work each cell is encoded by a BioRica node that has a 2-level hierarchy: a discrete controller and a continuous system. The former determines the distribution of proteins at division time using the discrete transition assignments (4–6), while the latter determines the evolution of protein quantities during one cell cycle and is realized by the equations (1–3). More precisely, the discrete controller is encoded by a *constraint automaton* [6] defining the discrete transitions between states. A *state* of a cell c^i is a tuple $\langle P^i_{\text{int}}, P^i_{\text{dam}}, D^i\rangle$, where P^i_{int} and P^i_{dam} are protein quantities as before, and D^i is a single dimension array of integers representing the identifiers of every daughter of c^i. A *transition* between states is a tuple $\langle G, e, A\rangle$, where G is a guard, e is an event, and A is a parallel assignment. For this model, the discrete transitions are atomic operations and consequently take zero time. In our case we have: for mitosis (event e), if the threshold of the cell size is attained $P_{\text{int}} = 1500$ (guard G), then create a new BioRica node c^j for the daughter of the current cell c^i, append c^j to the vector D^i, and perform the assignments (4–6) (assignments A of state variables). A second discrete event representing clonal senescence is triggered whenever protein production reaches zero, that is $\partial P_{\text{int}} < 0$.

The cell population is encoded by a BioRica node using the mechanism of parallel composition. This node contains the population array *Pop*, the root of the lineage tree R and the parameter vector $\vec{\mathcal{P}}$. Since our model focuses on the division strategy,

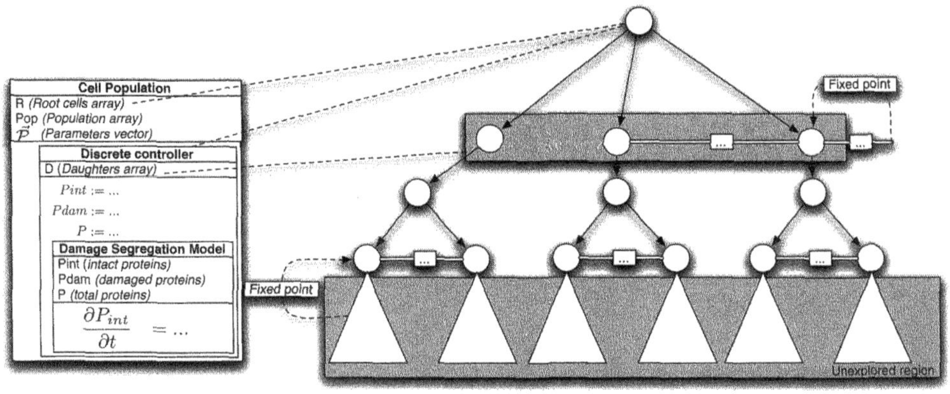

Fig. 1. Three level hierarchical model, showing the discrete cell population and cell division controllers, and the continuous single-cell model. This model generates pedigree trees during simulation, instantiating new single-cell models for each cell division. Infinite width and depth are represented finitely by relaxing the tree constraints to permits loops from the leaves. These *fixed points* represents immortal cells or immortals lineages.

we consider the growth medium as a non limiting factor, and consequently we do not account for cell to cell interactions. This absence of interaction is directly modeled by parallel composition of independently evolving cell nodes. For illustration see figure 1. The algorithmic challenges related to dealing with multiple time scales and event detection, and our solutions, are described in section 3.

3. Algorithm

We now describe our method for efficient simulation of the cell population model (section 2), starting with an overview of the general simulation schema (algorithm 3.1) followed by a concrete specialization for damage segregation. The simulation schema for a given BioRica node is given by a hybrid algorithm that deals with continuous time and allows for discrete events that **roll back** (see figure 2) the time according to these discrete interruptions. Time advances optimally either by the maximal stepsize defined by an adaptive integration algorithm [12], or by discrete jumps defined by the minimal delay necessary for firing a discrete event. As shown in algorithm 3.1, the simulation advances in a loop that is interrupted when either the simulation time expires, or the *alive* flag indicates that this node has died in the current or previous state. The node evolves continuously by calling *advance_numerical_integration*, after which we check whether any guard G of some event $\langle G, e, A\rangle$ was satisfied. In which case a number of updates is performed: the time is set to the firing time of e, e is stored in the trace database, the current state S is set according to the algebraic equation A, and the numerical integrator is reset to take into the account the discontinuity.

As illustrated in figure 2, the step size proposed by the numerical integrator

Algorithm 3.1 General simulation schema

Require: current state S, current simulation time t, maximal simulation time $t_{\max}$

$S' = S$
while alive$(S, S') = 1$ and $t < t_{\max}$ **do**
    $S' = S$
    $t, S =$ advance_numerical_integration()
    if $e =$ discrete_events() **then**
        $t =$ get_discrete_event_time()
        store_event(e)
        $S =$ update(S, e)
        reset_numerical_integrator()
    end if
    store_state(S)
end while

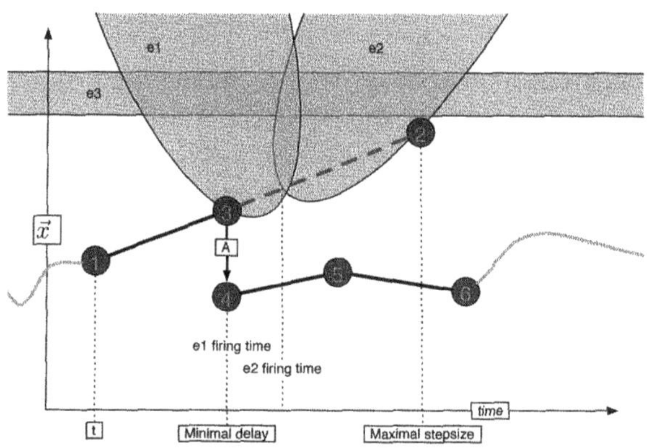

Fig. 2. The numerical integrator advances between t (point 1) and the maximal stepsize (2). The guards of events e_1, e_2 are satisfied. The regions where these guards are satisfied are shaded. The firing time of e_1 (3) is used to reset the simulator after the discrete transition A (4).

guarantees that the continuous function is linear between the current time t and the maximal step size. In this way the location of discrete events whose guards have been satisfied in this interval is reduced to computing the **first intersection** (see figure 2). It is the event e with the smallest firing time that is retained for the next discrete transition. After this transition the numerical integrator must restart from the point defined by A.

Correction of the stepping algorithm The numerical stability of the stepper described in algorithm 3.1 has been checked by evaluating the stiffness of the single cell ODE system by comparing the accumulated integration error for several explicit and implicit integration methods. For the final implementation the embedded Runge-Kutta-Fehlberg method was retained because of its good tradeoff between efficiency and precision.

Failure of an event detection can be caused when a guard is falsely assumed to be monotonic between two successive simulation steps. In BioRica, the guard of an event can not refer to the current simulation time. For this reason, given a BioRica node with n state variables, detecting the occurrence of an event is reduced to an intersection test between an $n+1$ dimensional segment and a n dimensional region or polytope. In the specific case of guards in our cell population model, this test is accelerated since it is sufficient to evaluate the guard at the last integration step due to the guards region convexity.

Specialization The generic simulation algorithm 3.1 was specialized for the damage segregation study. In particular, *alive* and *update* had to be redefined in a specific way. The *alive* predicate verifies three conditions. First, the cell is checked for immortality, which is realized by fixed point detection. Second, we verify whether the cell is in the state of clonal senescence, by evaluating the two guards described in section 2. *Update* has the role of managing new cell creation. For the current cell c it updates its state variables, according to the algebraic equations (4-6 for progenitors), and its statistics (fitness, generation time, etc). It creates a new cell node (daughter of c) according to the equations (4-6 for progeny) and inserts it into the population array *Pop*.

Population simulation On top of this specific stepping algorithm, another algorithm drives the whole population simulation by selectively starting simulations for pending cells in *Pop*. Given a depth n, a root cell c and an extent value e, this algorithm first selects pending nodes required to get a complete binary pedigree tree of depth n rooted at cell c. Afterwards, e leftmost and e rightmost leaves are used as root cells in recursive calls of this algorithm with a decremented value of e. Fix points are detected by testing before simulation if a candidate cell's initial values P_{int} and P_{dam} are equal to a previously simulated cell, in which case we get a pedigree graph by adding a loop edge.

Determining parameter values that exhibit optimal population fitness is based on averaging individual cell statistics (defined in section 4) to compute the mean fitness. In fact, this averaging maps a real value denoting the population fitness to each parameter vector. A coarse computation of this mapping is then built by varying the parameter vector using fixed step size. This coarse estimation is used to determine initial guess of optima position, that are then established by using Brent Principal Axis method on the mapping.

4. Results

Initial calibration To calibrate and validate the system, complete simulations were run to depth 4 in the pedigree tree for a large range of parameter values. Rate constants k_1, k_2, and k_4 received fixed values, k_3 was given a range of values with step size 0.1, and s_{mother} and $s_{daughter}$ were given two pairs of values representing symmetric and asymmetric growth strategies. A total of 625 simulations were run, summing to 9375 different initial conditions and parameters values. Sample results

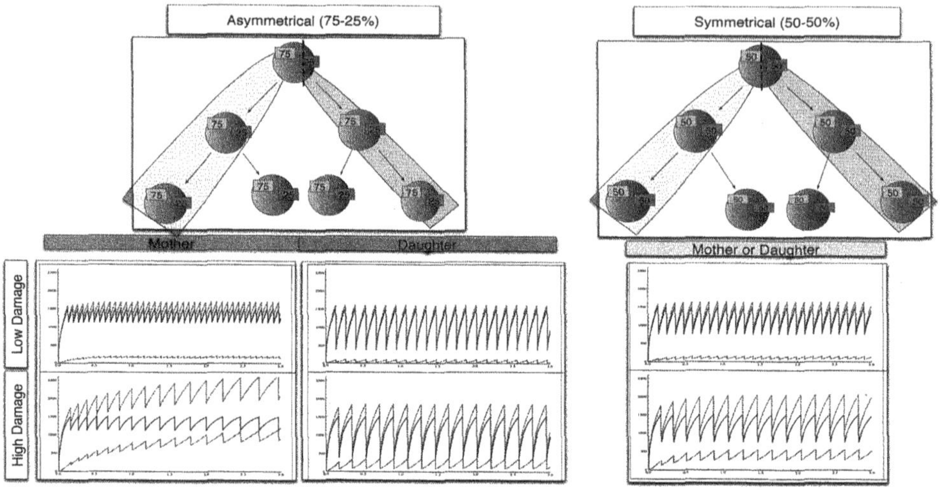

Fig. 3. Sample pedigree tree results for asymmetrical (left) and symmetrical (right) division strategies. Pedigree tree (Top) showing mother-daughter relations; and simulation results (Bottom) showing single cell protein amounts over time: normal proteins (blue), damaged proteins (pink), total proteins (dashed). For example, in the asymmetrical case with high damage (lowest left plot), from time zero, the amount of normal proteins (blue) in the mother eventually crosses the division threshold at time approx. 0.15. At this time, the proteins repartition is approx. 250 damaged proteins for 1750 total proteins. Once division is triggered, the progeny separates, and a new simulation is started for the mother (resp. the daughter) with initial normal proteins set at $1500 \times 0.75 = 1125$ (resp. $1500 \times 0.25 = 375$) and damaged proteins set at $250 \times 0.75 = 187.5$ (resp. $250 \times 0.25 = 62.5$). Since in this case the mother accumulates damage during divisions (compare damaged proteins amount between time 0.15 and 3.0), it will eventually reach a senescence point after 27 divisions. Comparison of its life span with the life span of each of its daughters and the life span of each daughter of its daughters (as tracked by the pedigree tree) shows a rejuvenation effect.

for pedigree tree are illustrated on figure 3. Successful comparisons with a small number of experimental cell growth results were also performed (data not shown).

Parameter Exploration. Using parameter exploration (section 3) we identified sets of parameters that exhibited a given emerging high-level behaviour, both at the single-cell and whole pedigree tree levels. For example, for the former we are interested in detecting cells that have a certain number of daughters (say, 24), and for the latter we are looking for parameters giving high rejuvenation value across the whole population. These two values are computed by a trace simulation analysis script.

Thus, for each of scenarios studied here, a representative simulation was chosen by inspecting properties of the initial mother. From the whole parameter space, we selected simulations where the mother cell produces a number of daughters that is both finite and large enough (20-24 divisions depending on the case, since the average life span of wild type budding yeast is 24 divisions). For each of these simulations, the pedigree tree was calculated up to depth 30, and for each cell in the tree we calculated five values: *initial damage* and *terminal damage* levels (corresponding respectively to the amounts of damage P_{dam} at the beginning of cell cycle, and at the end of the cycle when division is about to occur), *generation time* (time between two divisions), *absolute date of birth* (in arbitrary time units, measured from the moment when mother starts its first division) and the *fitness* (defined as number of divisions during first time unit).

Model analysis. The hierarchical model we have defined explicitly tracks mother-daughter relations in pedigree trees of simulations. This allows us to study lineage-specific properties, which are properties associated with connected subgraphs of the pedigree tree. Pedigree trees and typical simulation results are shown in Figure 3.

In the pedigree tree, a given mother cell generates a series of daughter cells; these siblings are ordered in time, and the younger a sibling, the older the mother at the time of division. We observe in simulation results that younger siblings have higher damage, consistent with inheritance from an older mother that has accumulated more damaged proteins, and these younger siblings are thus born "prematurely old." This increase in damage accumulation is reflected in the decrease of fitness values, shown in the first level of Figure 4.

Extending this analysis one level further in the pedigree tree shows, expectedly, that daughters born early to the same mother have low damage, and their daughters have normal fitness. Daughters born late to the same mother have high damage and lower fitness, but remarkably, in simulations with asymmetric division, their own daughters are born with lower damage and higher fitness. This increase in fitness in the second generation is a *rejuvenation effect*, in part explaining how populations maintain viability over time despite inheritance of protein damage.

The testable hypothesis is thus that there exists a mechanism for segregation of damaged proteins during cell division, that attenuates the accumulation of such proteins in descendants, and that the asymmetry coefficients (s_{mother} and $s_{daughter}$) in the model determines the scale of the rejuvenation effect.

These predictions are consistent with *in vivo* experimental results reported in the literature: Kennedy, et al. [9] report that daughter cells of an old mother cell are born prematurely old, with lower replicative potential, but that the daughters of these daughters have normal life spans.

However, this rejuvenation effect is not present in symmetric division case, since inheritance of damaged proteins should be proportional in both mother and daughter cells, and indeed is what is observed.

Finally, in simulations we observe that fitness and viability are sensitive to precise values of k_3, the rate by which proteins are damaged (see figure 4). This provides a series of testable hypotheses that could be investigated experimentally in different damaging environments, such as oxidative damage or radiation damage.

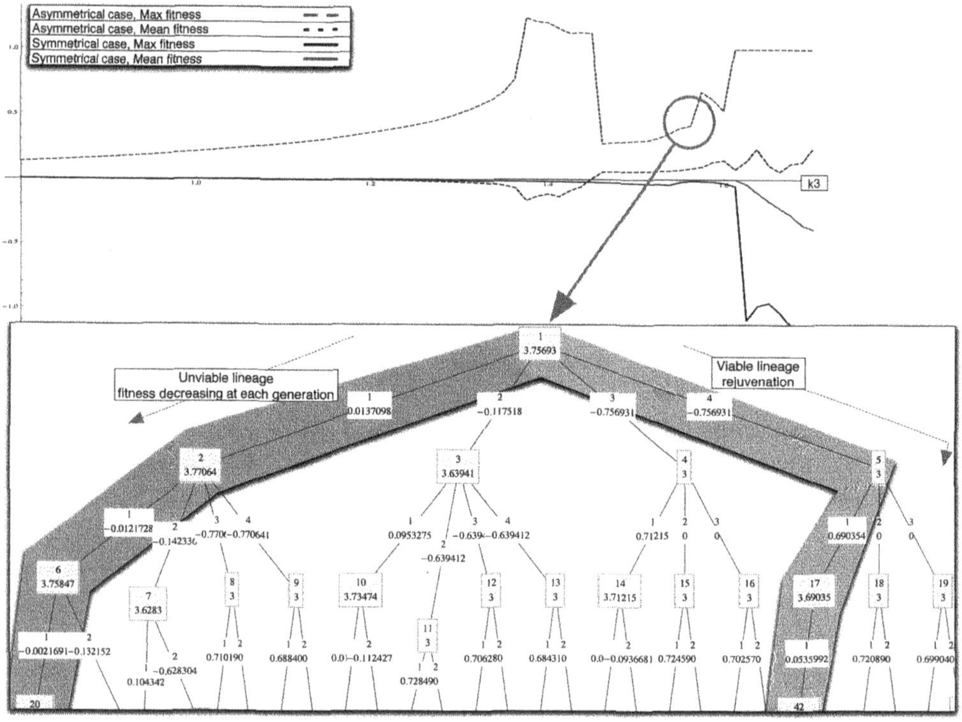

Fig. 4. Sample parameter exploration result showing the high sensitivity and non linearity of the rejuvenation effect w.r.t. precise values of the damage rate k_3. Only some ranges of the parameter value (approx. between 1.53 and 1.59) exhibit an increase of both the maximal and the mean fitness difference between every mother and daughter in a population. **Top**: Estimation of the maximal and mean rejuvenation amounts for damage rate ranging from 1.2 to 1.7 for the asymmetrical and symmetrical case. The higher the values, the fitter some cells of the lineage are compared to their direct mother. **Bottom:** Close up of a lineage tree used to compute a single point in the previous rejuvenation plot. Each node (yellow box) is labeled with a numeric id and the floating-point fitness value of the cell, and each edge (white label) is labeled with the index of the daughter relatively to its mother and with the difference of fitness between daughter and mother. For such a tree, we compute the mean and max values of the edges, which is represented as one point on the rejuvenation plot. The rejuvenation effect in young daughters of old mothers (right blue colored branch) is consistent with the experimental results of [9].

5. Conclusions

Although purely continuous systems such as ODEs have long been used for quantitative modeling and simulation of biological systems (for example [10]) and are commonly thought to be powerful enough, they do not suffice for highly structured models where emerging properties result from dynamic changes to the model.

For this study, the BioRica hybrid formalism and the related framework proved to be powerful enough to model, simulate and analyze the rejuvenation property of a hierarchical damage segregation process, by extending an existing continuous cell model to a population model. Since our hybrid formalism allows a BioRica node to describe and import an ODE system, we maintain the low computational cost and biological soundness of ODEs.

Hybrid simulation in BioRica scheme enjoys the traditional advantages of numerical integration, since the computational overhead for the hybrid stepper is proportional to the number of discrete events in the model; thus hybrid simulation of a purely continuous model is as fast as integrating it. For hybrid models mixing both continuous and discrete dynamics, our clear semantics permits concise description and reproducibility of simulation results in other simulation frameworks. For example, while the division strategy could be described in a continuous model by adjusting sigmoid functions, it is more naturally described by algebraic equations and their description in the model ought to be kept algebraic. The resulting gain in clarity has been observed elsewhere, for example in the complete cell cycle model of [4]. While most existing simulation tools admit a programming interface that allows for the modeler to simulate discrete events, the lack of a precise semantics renders the simulation predictions questionable and merely reproducible, since allowing such discrete events in a model has semantics issues[a]. Indeed, two discrete events can be enabled at the same time, but nothing defines whether in such cases the simulator should fire neither event, both events, or some random choice; and different strategies imply radically different simulation results.

In BioRica, we use the mathematical definition of non-determinism, standard in discrete formalisms, thus giving any BioRica model an unambiguous precise mathematical semantics. Furthermore, while not explicitly used in this study, BioRica relies on and extends the compositional operators initially defined in the AltaRica language family [3]. This enables parallel, partially synchronous and data sharing composition of hybrid, stochastic, multi-models as well as composition with external abstract processes. The underlying formalism class encompasses the range from constraint automata to hybrid stochastic differential systems. Furthermore, since such compositions are mathematically defined in BioRica, we can exactly identify subclasses of models admitting modern model analysis such as model checking, compositional reasoning, functional module decomposition and automatic simplification; all of which were spotted as grand challenges for modeling and simulation in system biology [13]. For compatibility with other systems biology software, BioRica imports SBML files through libSBML [8]. In addition to SBML support, BioRica exports the model as software independent C++ code, that can be compiled on any POSIX compliant system. This approach allows initial model prototyping in user friendly workbenches such as xCellerator [14], followed by use of optimized command line simulators for large scale analyses.

More specifically for population studies, since discrete variables and dynamic node creation are allowed in BioRica, our cell model can explicitly track a dynamic mother-daughter relationship. A realistic population model needs such a dynamic topology. Even when restricting ourselves to the biologically realistic case of dying cells, the number of daughters that any cell can have is *a priori* unbounded; thus,

[a]See for example http://www.sys-bio.org/sbwWiki/compare/themysterysolved

simply replicating the ODE equations to get a continuous population model as in [7] is not scalable. Furthermore, when simulations were carried up to depth 30, approximately 2^{30} cells were evolving in parallel, adding up to a 2^{32}-variable differential system that is untractable using a classical ODE approach. Instead, our population model clearly separates each cell behavior from the population by using hierarchical composition, and uses this modularity to provide a hierarchical simulation scheme, thus ensuring that each individual cell continuous part will be integrated with the most efficient step size.

Furthermore, the properties of parallel composition render study of population model with up to 2^{30} individuals still partially tractable by our scheme since we can linearize this population tree to simulate each cell independently. This approach is efficient since the cost of simulating a population is linearly proportional to the cost of simulating an individual, while flat and unstructured models have a quadratic complexity [5]. Finally, since we use discrete variables to track the mother-daughter relationships, we can directly estimate the rejuvenation effects, which would otherwise be buried in a flat and unstructured model.

Large scale exploration to detect the rejuvenation effect required a tree coverage that is out of reach of naive exploration algorithms such as breadth-first or depth-first. In fact, neither the population tree width nor its height are bounded, and thus these algorithms do not terminate. An *ad hoc* exploration algorithm partially solves this problem by alternating evaluation of first born daughters and evaluation of late born daughters, but does not provide the required coverage to detect significant rejuvenation. However, substantial acceleration is provided by the fix point detection scheme encoded in our tree visitor pattern, whose soundness is ensured by the deterministic nature of a cell behavior. In the continuous model, initial values of $P_{\text{int}}$and $P_{\text{dam}}$for given parameter values entirely determine a unique single cell behavior, and we proved that this property is preserved in the hybrid model. In fact, this is a special case of a more general result stating needed and sufficient conditions for this determinism to be preserved when adding discrete transitions to a continuous model. These conditions, beyond the scope of this paper, are not stringent and thus this acceleration can be used in most dynamic populations models built upon individuals.

Funding

This work was supported in part by the European Commission FP6 programme "Yeast Systems Biology Network" (YSBN), LSHG-CT-2005-018942; and by the French ANR project ANR-05-BLAN-0331-03 (GENARISE). MC is funded by the EU Marie Curie Early Stage Training (EST) Network "Systems Biology".

Acknowledgement

The authors thank Pascal Durrens of the CNRS for helpful discussions and Thomas Nyström (Göteborg University, Sweden) for initial discussion and valuable ideas.

References

[1] H Aguilaniu, L Gustafsson, M Rigoulet, and T Nystrom. Asymmetric inheritance of oxidatively damaged proteins during cytokinesis. *Science*, 299(5613):1751–1753, 2003.

[2] R Alur, F Ivancic, J Kim, I Lee, and O Sokolsky. Generating embedded software from hierarchical hybrid models. In *LCTES*, pages 171–182, 2003.

[3] A Arnold, A Griffault, G Point, and A Rauzy. The altarica formalism for describing concurrent systems. *Fundamenta Informaticae*, 40:109–124, 2000.

[4] KC Chen, L Calzone, A Csikasz-Nagy, FR Cross, B Novak, and JJ Tyson. Integrative analysis of cell cycle control in budding yeast. *Mol Biol Cell*, 15(8):3841–62, Aug 2004.

[5] J Esposito and V Kumar. An asynchronous integration and event detection algorithm for simulating multi-agent hybrid systems. *TOMACS*, Jan 2004.

[6] L Fribourg and M Veloso Peixoto. Concurrent constraint automata. In *ILPS*, page 656, 1993.

[7] M Henson, D Muller, and M Reuss. Cell population modelling of yeast glycolytic oscillations. *Biochem J*, Jan 2002.

[8] M Hucka, A Finney, HM Sauro, JC Bolouri, and H Kitano et al. The systems biology markup language (sbml). *Bioinformatics*, 19(4):524–31, 2003.

[9] BK Kennedy, NR Austriaco Jr, and L Guarente. Daughter cells of saccharomyces cerevisiae from old mothers display a reduced life span. *J Cell Biol*, 127(6 part 2):1985–93, 1994.

[10] P Mendes. Biochemistry by numbers: simulation of biochemical pathways with gepasi 3. *Trends Biochem. Sci.*, 22(9):361–3, 1997.

[11] P Nurse. Fission yeast morphogenesis–posing the problems. *Mol Biol Cell.*, 5(6):613–616, June 1994.

[12] WH Press, SA Teukolsky, WT Vetterling, and BP Flannery. *Numerical Recipes 3rd Edition: The Art of Scientific Computing.* Cambridge University Press, third edition, 2007.

[13] HM Sauro, D Harel, M Kwiatkowska, CA Shaffer, AM Uhrmacher, M Hucka, P Mendes, L Stromback, and JJ Tyson. Challenges for modeling and simulation methods in systems biology. *Winter Simulation Conference: Proceedings of the 38 th conference on Winter simulation*, 3(06):1720–1730, 2006.

[14] B Shapiro, A Levchenko, E Meyerowitz, and B Wold. Cellerator: extending a computer algebra system to include biochemical arrows for signal .... *Bioinformatics*, Jan 2003.

[15] DA Sinclair, K Mills, and L Guarente. Molecular mechanisms of yeast aging. *Trends Bioch. Sci.*, 23(4):131–4, April 1998.

[16] H Soueidan, M Nikolski, and D Sherman. Biorica: A multi model description and simulation system. In F Allgöwer and M Reuss, editors, *Proceedings of Foundations of Systems Biology in Engineering (FOSBE)*, Stuttgart, Germany, September 2007. ISBN 978-3-8167-7436-5.

INFERRING DIFFERENTIAL LEUKOCYTE ACTIVITY FROM ANTIBODY MICROARRAYS USING A LATENT VARIABLE MODEL

JOSHUA W.K. HO[1,6] RAJEEV KOUNDINYA[2] TIBÉRIO S. CAETANO[5,6]
joshua@it.usyd.edu.au raj@anatomy.usyd.edu.au tiberio.caetano@nicta.com.au
CRISTOBAL G. DOS REMEDIOS[2] MICHAEL A. CHARLESTON[1,3,4]
crisdos@anatomy.usyd.edu.au mcharleston@it.usyd.edu.au

[1] *School of Information Technologies, The University of Sydney, NSW 2006, Australia*
[2] *Bosch Institute, The University of Sydney, NSW 2006, Australia*
[3] *Sydney Bioinformatics, The University of Sydney, NSW 2006, Australia*
[4] *Centre for Mathematical Biology, The University of Sydney, NSW 2006, Australia*
[5] *RSISE, Australian National University, ACT 2601, Australia*
[6] *NICTA, Australia*

Recent development of cluster of differentiation (CD) antibody arrays has enabled expression levels of many leukocyte surface CD antigens to be monitored simultaneously. Such membrane-proteome surveys have provided a powerful means to detect changes in leukocyte activity in various human diseases, such as cancer and cardiovascular diseases. The challenge is to devise a computational method to infer differential leukocyte activity among multiple biological states based on antigen expression profiles. Standard DNA microarray analysis methods cannot accurately infer differential leukocyte activity because they often fail to take the cell-to-antigen relationships into account. Here we present a novel latent variable model (LVM) approach to tackle this problem. The idea is to model each cell type as a latent variable, and represent the class-to-cell and cell-to-antigen relationships as a LVM. Once the parameters of the LVM are learned from the data, differentially active leukocytes can be easily identified from the model. We describe the model formulation and assumptions which lead to an efficient expectation-maximization algorithm. Our LVM method was applied to re-analyze two cardiovascular disease datasets. We show that our results match existing biological knowledge better than other methods such as gene set enrichment analysis. Furthermore, we discuss how our approach can be extended to become a general framework for gene set analysis for DNA microarrays.

Keywords: antibody microarray, latent variable model, Bayesian network, EM algorithm

1. Introduction

Leukocytes (white blood cells) play a critical role in the human immune system. Several subtypes of leukocytes exist, including granulocytes, lymphocytes (T, B and NK cells), monocytes and others. These leukocyte subtypes can be characterized by different subsets of cell surface proteins, called cluster of differentiation (CD) antigens. The activity (in terms of absolute cell count, or density of expressed CD antigens) of each leukocyte subtype is associated with inflammation, particularly in cardiovascular diseases [10, 15]. Therefore efficiently quantifying leukocyte activity is important. Our laboratory has been developing a cell-captured antibody microarray platform that enables concurrent quantification of many CD antigens [1, 2]. This array platform has been successfully used to identify changes in the immunophenotype of various human diseases, such as leukemia [1, 2], heart failure [8, 9], and coronary artery disease [4].

Standard DNA microarray analysis methods, such as differential expression (DE) analysis, clustering and classification, are used to analyze these antigen expression profiles. However, essentially none of them can directly infer differential activity of leukocyte subpopulations as they only focus on mining "interesting" antigen expression patterns. Currently we rely on manual inspection of the list of DE antigens and their associated leukocyte subtypes to infer cellular activity. This approach is subject to human bias, and does not scale to analyzing larger expression profiles. Therefore the challenge is to devise a computational method that can accurately infer differential leukocyte activity from a set of antigen expression profiles.

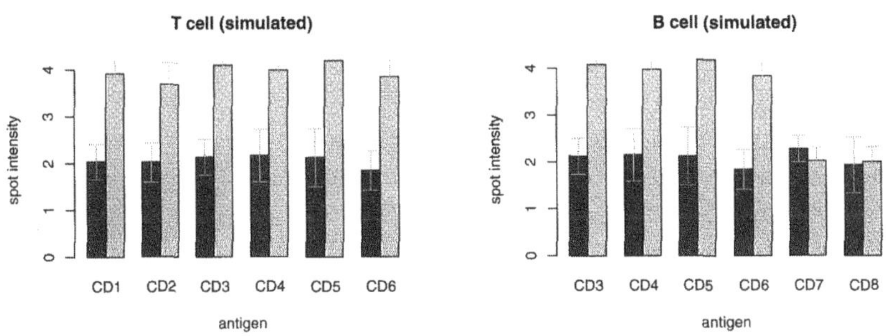

Fig. 1. The mean expression values of a simulated "toy" dataset. The dark and light bars represent antigen expression from normal individuals and diseased patients respectively.

We initially tackled this problem by gene set enrichment analysis (GSEA) [19]. In this case, gene sets correspond to leukocyte subpopulations. However, we soon discovered that the small number of genes and gene sets, and the large amount of overlap among gene sets leads to incorrect inference of leukocyte activity. To illustrate these problems, we constructed a simple "toy" dataset of two cell types - T and B cells, where each expresses six CD antigens, and four of these are expressed by both cell types (Figure 1). T cells were simulated to have elevated activity in the diseased patients, while B cells activity remain unchanged. GSEA indicated that neither T nor B cells was significantly enriched in DE antigens, based on the false discovery rate (FDR) of 0.70 and 0.69 respectively. Since the FDR calculated by GSEA is related to the distribution of enrichment score of all the gene sets, a large overlap between the two gene sets renders both gene sets insignificant. These statistical problems are likely to be shared by other gene set analysis methods which

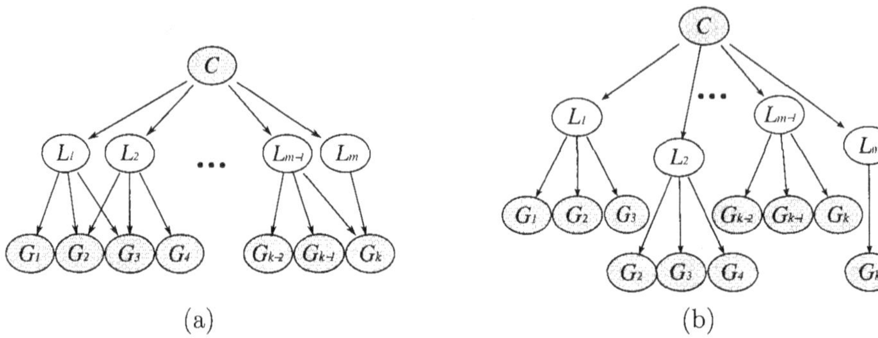

Fig. 2. An exemplary latent variable model (LVM). The shaded nodes are observed variables while the clear nodes represent latent variables. (a) The full model. (b) The effective decomposition of the full model.

use a hypothesis-testing approach [12]. Clearly we need an alternative approach that takes the cell-to-antigen relationships into account.

In this paper, we present a probabilistic graphical modeling approach to solve the problem. The main idea is to encode the observed data (antigen expression and class labels) and unobserved entities (leukocyte activity) as a special type of Bayesian network, called a latent variable model (LVM). The structure of the LVM is determined by the cell-to-antigen relationships, and the model parameters can be learned from the data (see Figure 2(a) for an exemplary LVM). Once the model parameters are learned, differential cellular activity can easily be obtained by performing probabilistic inference on the model.

2. Methods

2.1. *Model specification*

Our LVM consists of a set of variables $\mathbf{X} = \{C, \mathbf{L}, \mathbf{G}\}$, which includes an observed class variable, C, a set of m latent variables, $\mathbf{L} = \{L_1, L_2, ..., L_m\}$, and a set of k observed antigen expression variables, $\mathbf{G} = \{G_1, G_2, ..., G_k\}$. In this paper, the latent variables represent the cellular activity status. These variables are connected as a Bayesian network such that each L_i is the immediate parent of a subset of $\mathbf{G}$ corresponding to the antigens that are expressed by that cell type, and each L_i is an immediate child of C (Figure 2(a)). For convenience, we denote the set of $w(X)$ parents and $d(X)$ descendents of any variable X as $\pi(X) = \{\pi_1^X, \pi_2^X, ..., \pi_{w(X)}^X\}$ and $\delta(X) = \{\delta_1^X, \delta_2^X, ..., \delta_{d(X)}^X\}$. We further denote the set of possible realizations, or the *state space*, of each variable X to be $S(X)$. Each node X is associated with a conditional probability distribution (CPD), which is the probability distribution of X given the state of its parents, i.e., $P(X|\pi(X))$. Using the standard Bayesian network approach, the joint probability distribution (JPD) of the LVM can be decomposed as the product of local CPDs:

$$p(\mathbf{X}) = p(C)\left[\prod_{i=1}^{m} p(L_i|C)\right]\left[\prod_{i=1}^{k} p(G_i|\pi(G_i))\right] \quad (1)$$

We observe that the terms $p(G_i|\pi(G_i))$ do not generally decompose because each G_i can have many parents. Full parameterization of this CPD can result in a large computational burden during parameter estimation. Therefore we introduce an assumption here to further simplify the JPD:

$$p(G_i|\pi(G_i)) = p(G_i|\pi_1^{G_i}, \pi_2^{G_i}, ..., \pi_{w(G_i)}^{G_i}) = p(G_i|\pi_1^{G_i})p(G_i|\pi_2^{G_i})...p(G_i|\pi_{w(G_i)}^{G_i}) \quad (2)$$

The above assumption decomposes the poly-tree structure of the LVM into a tree by effectively duplicating those G_i with $|\pi(G_i)| > 1$ (Figure 2(b)). This leads to the following effective decomposition of the JPD:

$$p(\mathbf{X}) = p(C)\prod_{i=1}^{m}\left[p(L_i|C)\prod_{j=1}^{d(L_i)} p(\delta_j^{L_i}|L_i)\right] \quad (3)$$

The LVM is associated with a set of model parameters, which are used to specify the CPDs. We model $P(C)$ as a multinomial distribution, where $S(C)$ is the set of distinct class labels. Since $P(C)$ is the relative frequency of each distinct class label, it can be directly estimated from the dataset. We model $P(L_i|C)$ as a binary variable where $S(L_i) = \{\text{inactive}, \text{active}\}$. Each $P(L_i|C)$ is associated with $2|S(C)|$ parameters, each specifying the probability of L_i being active or inactive in each of the $|S(C)|$ classes. $P(G_i|\pi(G_i))$ is modeled as a Gaussian distribution with means $\{\mu_{G_i,1}, .., \mu_{G_i,w(G_i)}\}$, and variances $\{\sigma^2_{G_i,1}, .., \sigma^2_{G_i,w(G_i)}\}$.

One major consequence of the decomposition of the JPD in Equation 3 is that data of some nodes are duplicated. In general, such duplication of data may lead to bias in parameter learning. To alleviate this problem, we down-play the contribution of each duplicated antigens G_i by scaling up the set of variances $\{\sigma^2_{G_i,1}, .., \sigma^2_{G_i,w(G_i)}\}$. The basic idea is that antigens that are expressed by more than one cell type should have higher expression variability compared to antigens that are expressed by only one cell type. Therefore, we fix the variance of antigen expression per cell to be proportional to the number of parents, i.e., $\sigma^2_{i,j} = w(G_i)^r \times \sigma^2$, where we use $r = 3$ in this study since it works well in practice. The more parents G_i has, the higher is its expression variance. Using this formulation, the set of parameters that have to be estimated from the model is $\Theta = \{\theta_{L_1|C}, ..., \theta_{L_m|C}, \theta_{G_1|\pi(G_1)}, ..., \theta_{G_k|\pi(G_k)}\}$ where $\theta_{L_i|C} = p(L_i|C)$, and $\theta_{G_i|\pi(G_i)} = \{\mu_{G_i,1}, .., \mu_{G_i,w(G_i)}\}$.

2.2. *Parameter learning using EM algorithm*

Here we describe an efficient algorithm to obtain an approximate maximum likelihood estimate (MLE) of the LVM parameters from data. Since our model contains latent variables, we learn parameters by the expectation maximization (EM) approach [5]. The main idea of the EM algorithm is to iteratively calculate the expected distribution of the latent variables (E-step) and then use the results from E-step to re-estimate the MLE (M-step). Since the log likelihood of the model increases after each iteration of E- and M-step, the algorithm terminates when the expected log-likelihood of the model converges.

Given the observed data of array u, the E-step finds the expected distribution of the latent variables, $q_u(\mathbf{L}|C, \mathbf{G}, \Theta^{(t-1)})$, based on the current parameters. Using the effective decomposition in Equation 3, we can decompose q_u as well:

$$\begin{aligned} q_u(\mathbf{L}|C, \mathbf{G}, \Theta^{(t-1)}) &= p(L_1, L_2, ..., L_m|G_1, G_2, ..., G_k, C) \\ &= \frac{p(L_1, L_2, ..., L_m, G_1, G_2, ..., G_k, C)}{\sum_{S(\mathbf{L})} p(L_1, L_2, ..., L_m, G_1, G_2, ..., G_k, C)} \\ &= \frac{p(L_1|C)p(\delta(L_1)|L_1)}{\sum_{S(L_1)} p(L_1|C)p(\delta(L_1)|L_1)} \cdots \frac{p(L_m|C)p(\delta(L_m)|L_m)}{\sum_{S(L_m)} p(L_m|C)p(\delta(L_m)|L_m)} \\ &= q_u(L_1|C, \mathbf{G}, \Theta^{(t-1)}) ... q_u(L_m|C, \mathbf{G}, \Theta^{(t-1)}) \end{aligned}$$

Such decomposition implies that expected distribution of $\mathbf{L}$ is the product of the expected marginal distribution of each L_i, which can be computed by:

$$q_u(L_i|C, \mathbf{G}, \Theta^{(t-1)}) = \frac{p(L_i|C) \prod_{j=1}^{d_i} p(\delta_j^{L_i}|L_i)}{\sum_{S(L_i)} p(L_i|C) \prod_{j=1}^{d(L_i)} p(\delta_j^{L_i}|L_i)}$$

The M-step re-estimates Θ using the set of q_u calculated from the previous E-step. Given the antigen expression data $\{e_{1,1}, ..., e_{n,k}\}$, and the class labels $\{c_1, ..., c_n\}$, where n is the number of arrays and k is the number of antigens, we can calculate the MLEs as follows:

$$p(L_i|C) = \frac{\sum_{u=1}^{n} q_u(L_i|C, \mathbf{G})}{n} \qquad \mu_{L_i,j} = \frac{\sum_{u=1}^{n} e_{u,j} \times q_u(L_i|C, \mathbf{G})}{\sum_{u=1}^{n} q_u(L_i|C, \mathbf{G})}$$

Since there is no known efficient way to obtain the best initial parameters, we turn to a heuristic approach. The idea is to iteratively try out different random initial parameters. We select the parameter set that produces a model with the highest likelihood score after two iterations of EM. In general, the more random initial parameter sets being tested, the higher chance of finding the optimal one.

2.3. *Model analysis*

Once the model parameters are estimated, differential cellular activity can be obtained by inspecting the set of $P(L_i|C)$. To quantify the extent of differential cellular activity, we use total correlation $C_{\text{tot}}(L_i, C)$ [20] to measure the extent of dependency between L_i and C. Total correlation can be calculated by:

$$C_{\text{tot}}(L_i, C) = \sum_{l \in S(L_i)} \sum_{c \in S(C)} p(l, c) \log \left[\frac{p(l, c)}{p(l)p(c)} \right]$$

where $p(L_i)$ and $p(C)$ are the marginal distributions of L_i and C respectively. If L_i and C are statistically independent, C_{tot} becomes 0. In general, a higher C_{tot} implies that L_i is more strongly dependent on C, and therefore more differentially active.

3. Results

3.1. *Analysis of the toy example*

Here we analyze the toy example described in the introduction (Figure 1) using our LVM approach. We performed 20 iterations of search heuristics (described in Section 2.2) to obtain the initial values, then performed 20 iterations of EM procedures. By inspecting the set of $P(L_i = \text{active}|C)$, we observe an increase in T cell activity in patients with disease compared to healthy individuals (Figure 3(a)). The C_{tot} of T and B cells are 1.0 and 0.052 respectively (Figure 3(b)), which correctly implies that T cell is the only cell type that is differentially active. Moreover, we observed that the total correlation results converge after the first two EM iterations, implying that our results are stable.

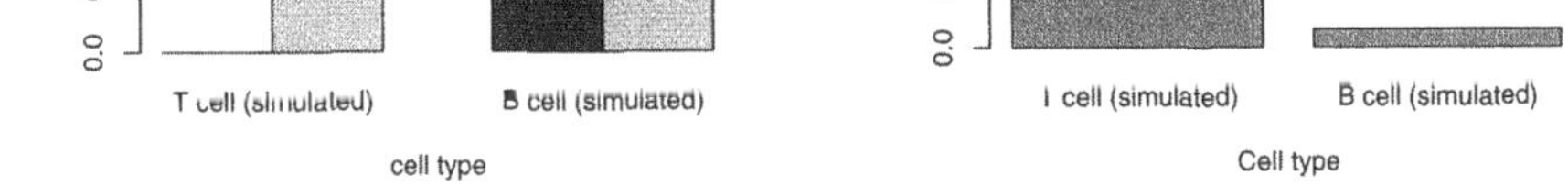

(a) (b)

Fig. 3. Results of LVM analysis of the simulated toy example shown in Figure 1. (a) A plot showing the probability of each cell type being active in each condition. The dark and light bars represent the healthy and diseased individuals respectively. (b) The C_{tot} of T and B cells.

3.2. *Re-analysis of two cardiovascular disease datasets*

Two cardiovascular disease datasets [4, 9] were re-analyzed using our LVM approach. All data were generated in our laboratory using an 82 spot antibody array platform.

In the original studies, only peripheral blood mononuclear cells (PBMCs), which include T cells (T), natural killer cells (NK), B cells (B) and monocytes (M), were investigated. The set of CD antigens being expressed by each leukocyte subpopulation is shown in Table 1. The set of CD antigens that are not expressed by any PBMCs are also listed here under the category *Others*, and should be regarded as a negative control for the analysis since it should not be differentially active. After data filtering and normalization (discussed in the original studies), the datasets were analyzed by our LVM approach. For each dataset, we performed 100 iterations of heuristic search to obtain the initial parameters, then performed 20 iterations of EM procedures to obtain the model parameters.

Table 1. A list of all CD antigens expressed by each type of PBMC.

Leukocyte	CD antigens[a]
T cell (T)	TCR a/b TCR g/d CD1a CD2 CD3 CD4 CD5 CD7 CD8 CD9 CD11a CD11b CD11c CD16 CD25 CD28 CD29 CD31 CD37 CD38 CD43 CD44 CD45 CD45RA CD49d CD49e CD52 CD54 CD56 CD57 CD60 CD62L CD80 CD86 CD95 CD102 CD103 CD120a CD122 CD126 CD128 CD130 CD134 CD154
B cell (B)	CD1a CD2 CD5 CD9 CD11a CD11b CD11c CD19 CD20 CD21 CD22 CD23 CD24 CD25 CD29 CD31 CD32 CD37 CD38 CD40 CD44 CD45 CD45RA CD45RO CD49d CD52 CD54 CD62L CD77 CD79a CD79b CD80 CD86 CD95 CD102 CD120a CD122 CD126 CD130 CD138 HLA-DR l FMC7 k
Monocyte (M)	CD1a CD4 CD9 CD11a CD11b CD11c CD13 CD14 CD15 CD16 CD29 CD31 CD32 CD33 CD36 CD37 CD38 CD40 CD43 CD44 CD45 CD45RA CD45RO CD49d CD49e CD52 CD54 CD60 CD61 CD62L CD64 CD65 CD80 CD86 CD88 CD95 CD102 CD120a CD122 CD126 CD128 CD130 HLA-DR
Natural Killer (NK)	CD2 CD7 CD8 CD11a CD11b CD11c CD16 CD25 CD29 CD31 CD38 CD43 CD44 CD45 CD45RA CD45RO CD49d CD49e CD52 CD56 CD57 CD62L CD95 CD102 CD120a CD122 CD128 CD130
Others	CD10 CD34 CD41 CD42a CD62E CD62P CD66c CD71 CD117 CD135 CD235a

Note: [a]These relationships were extracted from the official poster of the Eight International Workshop on Human Leukocyte Differentiation Antigens.

Brown *et al.* [4] studied two major coronary artery diseases (CAD): stable angina pectoris (SAP), and unstable angina pectoris (UAP). The dataset consists of antigen expression profiles from 15 SAP patients, 19 UAP patients and 19 healthy donors. Brown *et al.* manually mapped 19 DE antigens with the leukocytes that express them, and concluded that the observed patterns support a drop in T cell activity and an elevation in monocyte activity. Our results support their conclusion. Additionally we observe a drop in NK cell activity in CAD patients (Figure 4(a)-(b)). Unlike the original analysis by Brown *et al.* [4], we excluded granulocytes from our analysis since they are not PBMC. As noted by Brown *et al.*, the presence of granulocytes specific CD antigens may be an experimental artefact.

Lui *et al.* [9] studied two major aetiologies of heart failure (HF): ischemic heart disease (IHD), and idiopathic dilated cardiomyopathy (IDCM). Their dataset consists of antigen expression profiles from 22 IHD patients, 15 IDCM patients and 19 healthy donors. Our results (Figure 5(a)-(b)) show that HF patients have decreased NK cell activity and elevated monocyte activity. Further, we found that T cells are down-regulated in IHD patients but not in IDCM patients.

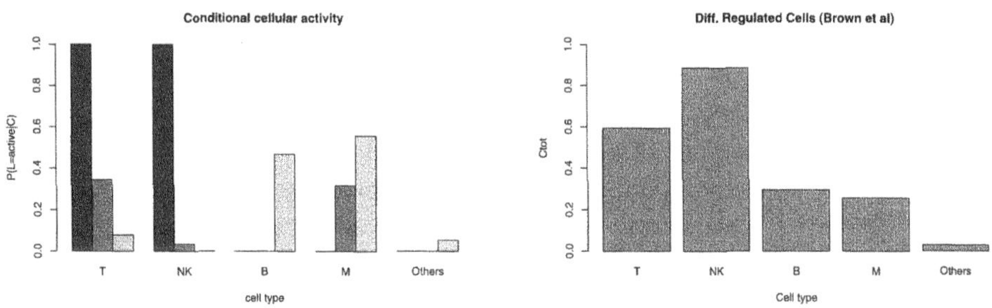

(a) (b)

Fig. 4. The LVM analysis result of Brown *et al.*'s data. (a) The conditional cellular activity plot. (b) The C_{tot} of various leukocyte populations.

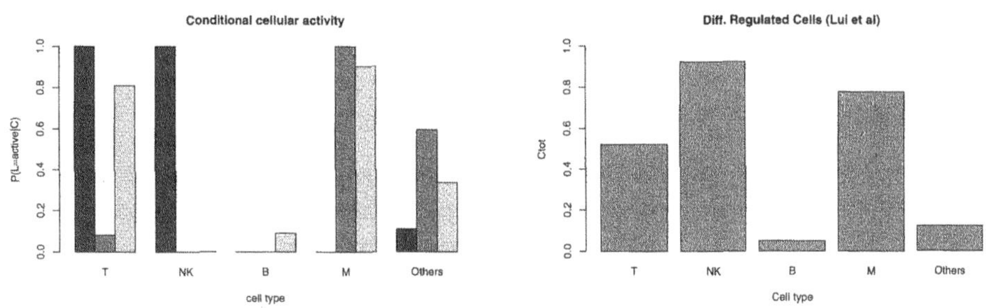

(a) (b)

Fig. 5. The LVM analysis result of Lui *et al.*'s data. (a) The conditional cellular activity plot. (b) The C_{tot} of various leukocyte populations.

In general, our approach indicates that there are decreased T and NK cell activity and increased monocyte activity in cardiovascular patients compared to healthy donors. An increase in monocyte count is known to be linked to various cardiovascular conditions [13, 14, 21]. In our arrays, all CD antigens in NK cells represented in our arrays are also expressed by other leukocytes in this study (primarily because NK cells are a sub-lineage of T cells). None of the original studies found differential activity of NK cells, since their changes are attributed to other classes of leukocytes. However, our model detected a strong signal for decrease in NK cells activity in both CAD and HF compared to healthy donors. This drop in NK cell activity is supported by the literature [7]. T cell activity is down-regulated in CADs and IHD, but not in IDCM. This is again consistent with previous findings which link decreased T cell count with myocardial infarction [3]. Our results correctly indicate no differential activity for the *Others* category in both studies.

In addition to our LVM analysis, we performed GSEA [19] on the two datasets. We used version 2 of the Java GSEA program [23]. Default parameters were used for all analyses. Only half of those true differentially active leukocyte subtypes (according to known biology and visual inspection of the data) are considered significantly enriched with DE antigens by GSEA (Table 2). The significant enrichment of B cells in Lui *et al.*'s dataset contradicts the results from manual data inspection and known biological knowledge. The results indicate that our LVM approach is superior to GSEA in terms of identifying biologically meaningful differential leukocyte activities. We note that general conclusion holds even when a nominal P-value is used to determine statistical significance.

Table 2. Results from GSEA. Gene sets with FDR$\leq$ 0.25 are deemed significant (in **bold**).

Analysis	Up-regulated in control (FDR)	Up-regulated in disease (FDR)
control vs. SAP	**T (0.11)** , B (0.66), NK (0.49)	M (0.33), Others (0.87)
control vs. UAP	T (0.27), NK (0.54)	M (0.62), Others (0.95), B (0.9)
control vs. IHD	**T (0.051)** , **B (0.17), NK (0.17)**	M (0.64), Others (0.63)
control vs. IDCM	**T (0.25), B (0.15), NK (0.15)**	M (0.34), Others (0.67)

4. Discussion

There has been a great interest in applying probabilistic graphical modeling (PGM) techniques to analyzing microarray data. Applications of PGM include pathway discovery [17], regulatory gene modules discovery [16], inferring alternative splice variants [18], and inferring gene network structures [6]. One advantage of PGM is that it allows structural information (relationships between variables) and systems dynamics (expression values) to be integrated under a simple yet theoretically sound framework.

There are two main contributions in this paper. The first is the application of PGM to the inference of differential leukocyte activity using antigen expression profiles. The re-analysis of the two real datasets clearly demonstrates the applicability of our approach to discover biological knowledge. With an increasing number of arrayed antibodies and more reliable experimental protocols, this cell-captured antibody array technology should become increasingly useful in both basic biological investigations and clinical diagnostic applications.

To demonstrate the merit of our approach, let us consider the mean expression value of all the CD antigens expressed by T cell in the Brown *et al.* dataset as an example (Figure 6). The changes in expression patterns across all these antigens differ a lot since many antigens are expressed by other leukocytes. We notice that the expression patterns of those cell specific antigens are much more informative in elucidating the cellular activity. However, removing those antigens expressed by multiple leukocytes is not desirable since some leukocytes do not express, or express only one or two, cell specific CD antigens (like NK cells in this study). Therefore our LVM model provides a general framework for such inference.

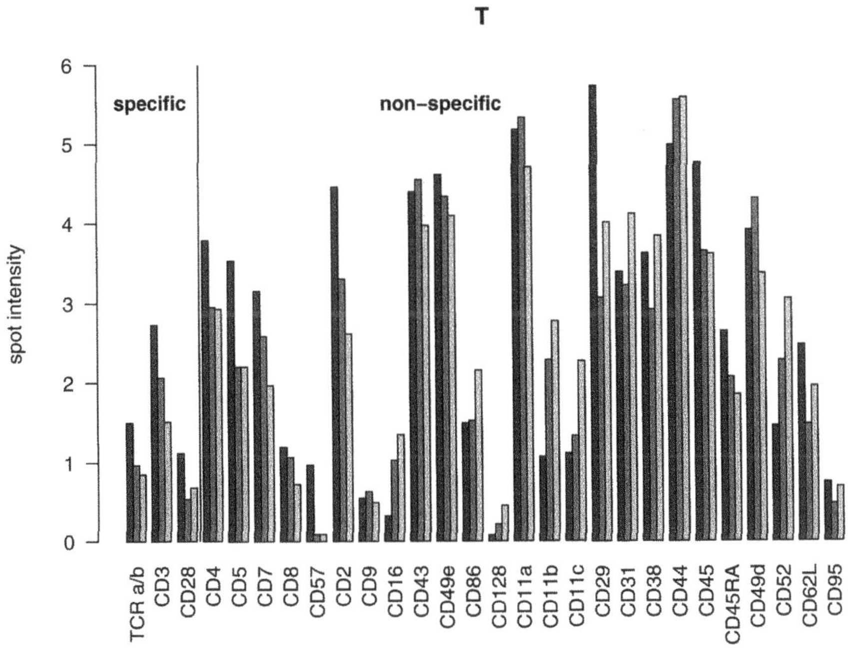

Fig. 6. The mean antigen expression levels of the CD antigens associated with the T cells from the Brown *et al.* dataset. The CD antigens in the barplots are sorted according to the number of different cell types that express it. The antigens on the left of the vertical line represent T cell specific CD antigens. The dark, gray and light bars represent healthy donors, SAP and UAP patients respectively.

Our second contribution is to introduce a novel LVM approach for microarray gene set analysis. Our model is similar to the hierarchical naïve Bayes model proposed by Zhang *et al.* [22], except that our LVM consists of strictly one level of latent variables, and our LVM network structure is known *a priori*. Since the network structure of the LVM is given by biological knowledge, our method eliminates the need to perform computationally expensive structural learning.

In this work we also present a computationally efficient method to learn the conditional probabilities associated with the latent variables. The computational efficiency is achieved by the product assumption in Equation 2, which leads to the decomposition of the JPD (Equation 3). To avoid losing the antigen overlapping information, we made the second assumption that antigens which are expressed by multiple cell types have higher expression variability. This assumption effectively gives more weight to cell type specific CD antigens. As a result, the antigen overlapping information is retained without increasing the computational complexity in parameter learning. The effectiveness of our approach is demonstrated by the analyses of a simulated and two real datasets.

We propose that our LVM approach can be used as a general framework for finding differentially expressed gene sets in DNA microarrays. Since the initial publication of GSEA [11, 19], many gene set analysis methods emerged [12]. All of them use a hypothesis testing approach to define interesting gene sets. However, as indicated by our toy example, the correctness of the results depends on meeting a set of assumptions which may be biologically or technically unrealistic. Our LVM approach is not based on hypothesis testing, so the aim of our method is not to find significantly differentially expressed gene sets, but to map the gene expression profiles into the hidden gene set expression space.

In general, there are many possible formulations of the CPDs in our model. We are currently investigating the CPD formulation that is most suitable for general gene set analysis. Moreover, we will investigate the use of other learning techniques to achieve more robust estimates of the model parameters. Nonetheless, this paper presents a conceptually new approach to perform gene set analysis.

Acknowledgement

JWKH is supported by an Australia Postgraduate Award and a NICTA Research Project Award. We thank Angus Brown and Rodney Lui for providing the antibody microarray data.

References

[1] Belov, L., de la Vega, O., dos Remedios, C.G., Mulligan, S.P., and Christopherson, R.I., Immunophenotyping of leukemias using a cluster of differentiation antibody microarray. *Cancer Res.*, 61:4483-4489, 2001.

[2] Belov, L., Huang, P., Barber, N., Mulligan, S.P., and Christopherson, R.I., Identification of repertoires of surface antigens on leukemias using an antibody microarray, *Proteomics*, 3:2147-2154, 2003.

[3] Blum, A., Sclarovsky, S., Rehavia, E., and Shohat, B., Levels of T-lymphocyte subpopulations, interleukin-1 beta, and soluble interleukin-2 receptor in acute myocardial infarction, *Am. Heart J.*, 127:1226-1230, 1994.

[4] Brown, A., Lattimore, J.-D., McGrady, M., Sullivan, D., Dyer, W., Braet, F., and dos Remedios, C.G., Stable and unstable angina: Identifying novel markers on circulating leukocytes. *Proteomics Clin. Appl.*, 2:90-98, 2008.

[5] Dempster, A.P., Laird, N.M., and Rubin. D.B., Maximum likelihood from incomplete data via the EM algorithm, *J. R. Statist. Soc. B.*, 39:1-38, 1977.

[6] Friedman, N., Inferring cellular networks using probabilistic graphical models. *Science*, 303:799-805, 2004.

[7] Jonasson, L., Backteman, K., and Ernerudh, J., Loss of natural killer cell activity in patients with coronary artery disease. *Atherosclerosis*, 183:316-321, 2005.

[8] Lal, S., Lui, R., Nguyen, L., Macdonald, P.S., Denyer, G., and dos Remedios, C.G., Increases in leukocyte cluster of differentiation antigen expression during cardiopul-

monary bypass in patients undergoing heart transplantation, *Proteomics*, 4:1918-1926, 2004

[9] Lui, R., Macdonald, P.S., Hayward, C., and dos Remedios, C.G., Proteomics analysis of leukocyte membrane proteins from human heart failure patients using an antibody microarray platform, *J. Mol. Cell. Cardiol.*, 42:S146, 2007.

[10] Madjid, M., Awan, I., Willerson, J.T., and Casscells, S.W., Leukocyte count and coronary heart disease. *J. Am. Coll. Cardiol.*, 44:1945-1956, 2004.

[11] Mootha, V.K., Lindgren, C.M., Eriksson, K.F., Subramanian, A., Sihag, S., Lehar, J., Puigserver, P., Carlsson, E., Ridderstrle, M., Laurila, E., Houstis, N., Daly, M.J., Patterson, N., Mesirov, J.P., Golub, T.R., Tamayo, P., Spiegelman, B., Lander, E.S., Hirschhorn, J.N., Altshuler, D., and Groop. L.C., PGC-1alpha-responsive genes involved in oxidative phosphorylation are coordinately downregulated in human diabetes, *Nat. Genet.*, 34:267-273, 2003.

[12] Nam, D., and Kim, S.-Y., Gene-set approach for expression pattern analysis, *Brief. Bioinform.*, 9:189-197, 2008.

[13] Nasir, K., Gaullar, E., Navas-Acien, A., Criqui, M.H., and Lima, J.A.C., Relationship of monocyte count and peripheral arterial disease: Results from the national health and nutrition examination survey 1999-2002, *Arteroscler. Thromb. Vasc. Biol.*, 25:1966-1971, 2005.

[14] Olivares, R., Ducimetiére, P., and Claude, J.R., Monocyte count: A risk factor for coronary heart disease, *Am. J. Epidemiol.*, 137:49-53, 1993.

[15] Ommen, S.R., Hodge, D.O., Rodeheffer, R.J., McGregor, C.G.A., Thomson, S.P., and Gibbons, R.J., Predictive power of the relative lymphocyte concentration in patients with advanced heart failure, *Circulation*, 97:19-22, 1998.

[16] Segal, E., Shapira, M., Regev, A., Peer, D., Botstein, D., Koller, D., and Friedman, F., Module networks: identifying regulatory modules and their condition-specific regulators from gene expression data, *Nat. Genet.*, 34:166-176, 2003.

[17] Segal, E., Wang, H., and Koller, D., Discovering molecular pathways from protein interaction and gene expression data. *Bioinformatics*, 19:i264-i272, 2003.

[18] Shai, O., Morris, Q.D., Blencowe, B.J., and Frey, B.J., Inferring global levels of alternative splicing isoforms using a generative model of microarray data, *Bioinformatics*, 22:606-613, 2006.

[19] Subramanian, A., Tamayo, P., Mootha, V.K., Mukherjee, S., Ebert, B.L., Gillette, M.A., Paulovich, A., Pomeroy, S.L., Golub, T.R., Lander, R.S., and Mesirov., J.P., Gene set enrichment analysis: A knowledge-based approach for interpreting genome-wide expression profiles, *Proc. Natl. Acad. Sci. U.S.A.*, 102:15545-15550, 2005.

[20] Watanabe, S., Information theoretical analysis of multivariate correlation, *IBM J. Res. Dev.*, 4:66-82, 1960.

[21] Zalai, C.V., Kolodziejcyk, M.D., Pilarski, L., Christov, A., Nation, P.N., Lundstrom-Hobman, M., Tymchak, W., Dzavik, V., Humen, D.P., William, K., Jablonsky, G., Pflugfelder, P.W., Brown, J.E., and Lucas, A., Increased circulating monocyte activation in patients with unstable coronary syndromes, *J. Am. Coll. Cardiol.*, 38:1340-1347, 2001.

[22] Zhang, N.L., Nielsen, T.D., and Jensen, F.V., Latent variable discovery in classification models, *Artif. Intell. Med.*, 30:283-299, 2004.

[23] `http://www.broad.mit.edu/gsea/`

Assessing and Predicting Protein Interactions Using Both Local and Global Network Topological Metrics

Guimei Liu[1] Jinyan Li[2] Limsoon Wong[1]
liugm@comp.nus.edu.sg jyli@ntu.edu.sg wongls@comp.nus.edu.sg

[1] *School of Computing, National University of Singapore, Singapore*
[2] *School of Computer Engineering, Nanyang Technological University, Singapore*

High-throughput protein interaction data, with ever-increasing volume, are becoming the foundation of many biological discoveries. However, high-throughput protein interaction data are often associated with high false positive and false negative rates. It is desirable to develop scalable methods to identify these errors. In this paper, we develop a computational method to identify spurious interactions and missing interactions from high-throughput protein interaction data. Our method uses both local and global topological information of protein pairs, and it assigns a local interacting score and a global interacting score to every protein pair. The local interacting score is calculated based on the common neighbors of the protein pairs. The global interacting score is computed using globally interacting protein group pairs. The two scores are then combined to obtain a final score called *LGTweight* to indicate the interacting possibility of two proteins. We tested our method on the DIP yeast interaction dataset. The experimental results show that the interactions ranked top by our method have higher functional homogeneity and localization coherence than existing methods, and our method also achieves higher sensitivity and precision under 5-fold cross validation than existing methods.

Keywords: protein-protein interaction; network topology

1. Introduction

Protein-protein interactions play a critical role in most cellular processes and form the basis of biological mechanisms. Protein interactions have been traditionally studied on an individual basis, which is accurate but is often slow and laborious. In the past several years, high-throughput experimental techniques—such as yeast two-hybrid assay, mass spectrometry, protein chip and phage display—have been introduced to detect a large number of interactions simultaneously, which enables the study of protein-protein interactions at the proteome scale. However, high-throughput protein interaction data are often associated with high false positive and false negative rates due to the limitations of the associated experimental techniques and the dynamic nature of protein interaction maps. It is therefore desirable to develop computational methods to identify these errors.

Many computational approaches have been proposed to assess the reliability of high-throughput protein interaction data or predict new protein interactions. Various information has been used in these approaches, including protein primary structures and associated physicochemical properties [1], 3D structures of protein complexes [10], gene fusion [18], protein domains [13, 14], literature [23], co-localization information [8] and co-evolution information [11, 22]. Every method for protein interaction assessment and prediction is limited by the availability and reliability of the information it uses, and methods using different information sources are complementary to one another. Some work integrates multiple information sources to achieve better performance [12, 20].

Recent screening techniques have made large amounts of protein-protein interaction data available, which makes it possible to assess or predict protein interactions using solely the topology of the protein interaction networks [4, 5, 24, 25, 29]. Saito et al. [24, 25] introduced two measures called IG1 and IG2 which use the local topological structure of protein pairs to assess their reliability, and they do not consider topological properties beyond the candidate protein pair and their neighbors. Chen et al. [4] proposed a more global measure called IRAP, which is defined as the collective reliability of the strongest alternative path between two proteins. The authors later improved the IRAP measure by iteratively removing low-confidence interactions from the network and adding high-confidence new interactions to the network [5]. Yu et al. [29] proposed a method to predict new protein interactions by completing defective cliques. Chua et al. [6] proposed a measure called FSweight which exploits indirect neighbors to predict protein functions. The same group of authors later showed that FSweight could also be used to assess and predict protein interactions and it outperformed IG1, IG2 and IRAP on large interaction datasets [3].

FSweight is still a local measure. In this paper, we propose a computational method which uses both local topological information of protein pairs and global topological structures discovered from the whole interaction network to assess and predict protein interactions. The local interacting score of a protein pair is calculated based on the neighbors of the two proteins, and the reliability of the interactions between these two proteins and their neighbors is also taken into consideration. The global interacting score is obtained based on the observation that if one group of proteins interact with another group of proteins, then it is likely that the interaction between these two protein groups is mediated by an underlying complementary binding domain/motif pair. The above observation has been used to discover interacting motif pairs [16, 19, 27]. We call such protein group pairs interacting protein group pairs. If a protein pair participates in an interacting protein group pair, that is, the two proteins belong to different groups of the interacting protein group pair, then the interaction between the two proteins is likely to be true.

To calculate global interacting scores, we first generate groups of proteins that have common interacting partners from the interaction network using frequent itemset mining techniques, and then for every pair of discovered protein groups, we calculate their interacting scores. The global interacting score of a protein pair is computed based on the interacting score of the interacting group pairs it participates in and the degree of its participation. We studied the performance of our method on the DIP yeast interaction dataset. Our experiment results showed that our method outperforms FSweight and CD-distance, especially for predicting new interactions.

The rest of the paper is organized as follows. Section 2 describes our method, and the experiment results on the DIP yeast interaction dataset are presented in Section 3. Section 4 discusses and concludes the paper.

2. Method

In this section, we first describe how to calculate local interacting scores and global interacting scores of protein pairs, and then discuss how to combine them together to get the final score. The following notations are used in this section. A protein interaction network can be modeled as an undirected graph $G = (V, E)$ where vertex set V is the set of proteins and edge set E is the set of interactions between proteins. We use u, v, x to denote individual vertices (proteins), V_1, V_2 to denote vertex sets (protein groups), and (u, v) to denote the edge between u and v. The neighbor set of a vertex u in G, denoted as N_u, is defined as $N_u = \{v | (u, v) \in E\}$.

2.1. *Local interacting score*

The local interacting score is defined based on the observation that if two proteins have many common neighbors, then these two proteins are likely to interact with each other. We use a variant of the CD-distance measure to calculate local interacting score of protein pairs. The CD-distance measure was originally proposed by Brun et al. [2] for function prediction, and later was shown to be very effective in assessing the reliability of high-throughput interaction data [3].

It has been estimated that more than half of current high-throughput data are spurious [15, 26, 28], and these spurious interactions usually have a low score. To alleviate the impact of spurious interactions, we iteratively apply the scoring method on the weighted interaction network. The local interacting score of a protein pair in the k-th ($k > 0$) iteration, denoted as $w_L^k(u, v)$, is defined as follows:

$$w_L^k(u,v) = \frac{\sum_{x \in N_u \cap N_v} w_L^{k-1}(x,u) + \sum_{x \in N_u \cap N_v} w_L^{k-1}(x,v)}{\sum_{x \in N_u} w_L^{k-1}(x,u) + \sum_{x \in N_v} w_L^{k-1}(x,v) + \lambda_u^k + \lambda_v^k} \quad (1)$$

where $w_L^{k-1}(x, u)$ is the score of (x, u) in the (k-1)-th iteration, $w_L^0(x, u)$=1 if $(x, u) \in E$ and $w_L^0(x, u)$=0 if $(x, u) \notin E$. The two terms, λ_u^k and λ_v^k, are used to penalize proteins with very few neighbors (as in [6]), and they are defined as follows:

$$\lambda_u^k = \max\{0, \frac{\sum_{x \in V} \sum_{v \in N_x} w_L^{k-1}(v,x)}{|V|} - \sum_{v \in N_u} w_L^{k-1}(v,u)\} \quad (2)$$

When k=1, the local interacting score is similar to the CD-distance score except that it uses λ_u^1 and λ_v^1 to penalize proteins with very few neighbors. In our experiments, we have found that the local interacting score reaches the best performance when k=2, and the subsequent iterations do not improve the performance further.

2.2. *Global interacting score*

The global interacting score is based on the observation that if one group of proteins interact with another group of proteins, then it is likely that the interaction between these two protein groups is mediated by an underlying complementary binding domain/motif pair. We call such protein group pairs *interacting protein group pairs.* Given a protein pair (u, v) and an interacting protein group pair (V_1, V_2), we say (V_1, V_2) *contains* (u, v) if $u \in V_1$ and $v \in V_2$. We also say that (u, v) *participates in* interacting protein group pair (V_1, V_2). If a protein pair participates in an interacting protein group pair whose two groups are densely connected, then the interaction between these two proteins is likely to be true.

Proteins on one side of an interacting group pair are expected to have some common domains or motifs, so we expect that they have some common interacting partners. Also it is not desirable to have very few proteins on either side of an interacting group pair, because otherwise, the underlying interacting domain/motif pair may not be significant. Here we use two thresholds *min_sup* and *min_size* to restrict the minimum number of common neighbors and the minimum size of a protein group. We call *min_sup* the minimum support threshold and *min_size* the minimum size threshold. For an interacting protein group pair, each of its two protein groups must has at least *min_sup* common neighbors and contains at least *min_size* proteins.

The calculation of global interacting scores of protein pairs consists of three steps. In the first step, protein groups that have at least *min_sup* common interacting partners and contain at least *min_size* proteins are generated. In the second step, the interacting score of every pair of discovered protein groups is calculated. In the last step, the global interacting score of a protein pair is computed.

2.2.1. *Generating protein groups*

The protein groups that have at least *min_sup* common interacting partners and contain at least *min_size* proteins are generated using frequent itemset mining techniques. The adjacency matrix of an undirected graph can be regarded as a transaction database where each adjacency list is a transaction and each vertex (protein) is an item. The support of an itemset (protein group) is defined as the number of transactions (adjacent lists) containing it, which is equal to the number of common partners of the corresponding protein group. Finding protein groups that have at least *min_sup* common interacting partners and contain at least *min_size* proteins is equivalent to finding frequent itemsets occurring in at least *min_sup* transactions and containing at least *min_size* items. Frequent itemset mining algorithms use the anti-monotone property of itemsets to prune the search space, that is, if an itemset appears in less than *min_sup* transactions, then all of its supersets also appear in less than *min_sup* transactions, thus the itemset can be pruned. Given that the adjacency matrix of a protein interaction network is usually sparse, frequent itemset mining algorithms can generate the desired protein groups within several minutes.

In this paper, we use the AFOPT algorithm [17] to generate the protein groups.

2.2.2. *Calculating interacting confidence score of protein group pairs*

Let V_1 and V_2 be two protein groups generated in the first step. The interacting confidence score of V_1 and V_2, denoted as $conf(V_1, V_2)$, is defined as the ratio of the number of interactions between V_1 and V_2 to the total number of distinct protein pairs contained in (V_1, V_2):

$$conf(V_1, V_2) = \frac{|\{(u,v)|(u,v) \in E, u \in V_1, v \in V_2\}|}{|V_1| \cdot |V_2| - |V_1 \cap V_2| \cdot (|V_1 \cap V_2| + 1)/2} \tag{3}$$

When calculating the total number of distinct protein pairs contained in (V_1, V_2), we need to consider the situation that V_1 and V_2 may contain some common proteins. In the simple case that the two protein groups contain no common proteins, the total number of distinct protein pairs contained in (V_1, V_2) is simply $|V_1| \cdot |V_2|$. Otherwise, among the $|V_1| \cdot |V_2|$ protein pairs, there are $|V_1 \cap V_2|$ self-interactions and $|V_1 \cap V_2| \cdot (|V_1 \cap V_2| - 1)/2$ duplicated protein pairs, and they should be discarded. Therefore, the total number of distinct protein pairs contained in (V_1, V_2) is $|V_1| \cdot |V_2| - |V_1 \cap V_2| - |V_1 \cap V_2| \cdot (|V_1 \cap V_2| - 1)/2 = |V_1| \cdot |V_2| - |V_1 \cap V_2| \cdot (|V_1 \cap V_2| + 1)/2$.

2.2.3. *Calculating global interacting score of protein pairs*

The global interacting score of a protein pair is computed based on the interacting confidence score of the interacting group pairs it participates in and the degree of its participation, and it is defined as follows:

$$w_G(u,v) = \max\{conf(V_1, V_2) \cdot \frac{2|N_u \cap V_2|}{|V_2| + |N_u|} \cdot \frac{2|N_v \cap V_1|}{|V_1| + |N_v|} |u \in V_1, v \in V_2\} \tag{4}$$

where $\frac{2|N_u \cap V_2|}{|V_2|+|N_u|}$ and $\frac{2|N_v \cap V_1|}{|V_1|+|N_v|}$ are the participation degree of protein u and v respectively.

2.3. *The final interacting score of protein pairs*

The final interacting score of a protein pair is simply defined as the sum of its local interacting score and its global interacting score. For local interacting scores, we set $k = 2$.

$$LGTweight(u,v) = w_L^2(u,v) + w_G(u,v). \tag{5}$$

The higher the interacting score is, the more likely the two proteins interact with each other. After the interacting scores of the protein pairs are calculated, we rank the protein pairs in descending order of their score.

3. Results

In this section, we study the performance of our method and compare it with FSweight [6] and the original CD-distance [2]. We used the DIP (`http://dip.doe-mbi.ucla.edu/`) yeast interaction dataset dated 10/07/2007 in our experiments, which contains 17491 interactions. After removing duplicate interactions and self-interactions, the dataset contains 4932 distinct proteins and 17201 interactions. The DIP yeast core dataset contains 6459 interactions that were validated according to the criteria described in [9], and it is used as golden standard in our experiments.

3.1. *Functional homogeneity and localization coherence*

By the "guilt-by-association" principle [21], true interacting proteins usually share some common functional roles or are in the same cellular components. Hence we use the degree of functional homogeneity and localization coherence of protein pairs as one of the measures to evaluate our method. The interacting score of a protein pair indicates the interacting possibility of the protein pair. The higher the score is, the more likely the two proteins interact with each other. If we use a cut-off value *min_score* to select the protein pairs with score no less than *min_score* as interacting protein pairs, we expect that the proportion of the protein pairs sharing common functions or localizations in the selected protein pairs increases with the increase of *min_score*.

We use the annotations in Gene Ontology (GO) (`http://www.geneontology.org/`) to calculate functional homogeneity and localization coherence. The Gene Ontology comprises three orthogonal taxonomies or aspects that hold terms describing molecular functions, biological processes and cellular components of a gene product. We use the terms in the first two taxonomies for functional homogeneity calculation, and the terms in the last taxonomy for localization coherence calculation. The GO terms are organized hierarchically. Two different GO terms may share a common parent or a common child in the hierarchy. GO terms at high levels may occur in many proteins, and they are too common to be useful. GO terms appearing in very few proteins are also not very useful. In our experiments, we select only those informative GO terms. A GO term is informative if itself occurs in at least 30 proteins, but none of its children appears in at least 30 proteins. Using the proteins in the DIP yeast dataset, 50 molecular function terms, 110 biological process terms and 42 cellular component terms are selected.

Among the 4932 proteins in the DIP yeast dataset, 3251 proteins have functional annotations. There are 11229 interactions whose two proteins both have functional annotations, and among them 3660 interactions have common function annotations between its two proteins. We consider only those protein pairs whose two proteins both have functional/localization annotations when calculating the degree of functional homogeneity and localization coherence. Thus the degree of functional homogeneity of the DIP yeast interaction dataset is 32.6% (3660/11229). The overall functional homogeneity of all the possible protein pairs is 3.4%. There are 1615 proteins with cellular component annotations and 4246 interactions whose two proteins both have localization annotations. Among them, 2321 interactions have common localization annotations between its two proteins, so the degree of localization coherence of the DIP yeast dataset is 54.7%. The localization coherence over all possible protein pairs is 4.9%.

3.1.1. *The effect of the number of iterations on local interacting scores*

Our first experiment is to study the effect of the number of iterations on the performance of local interacting scores. Figure 1(a) shows the degree of functional homogeneity of the interactions in the DIP yeast dataset ranked using local interacting scores under different k values. It shows that the local interacting score reaches the best performance when k=2. The subsequent iterations do not improve the performance much. We use local interacting scores to rank the protein pairs that are not in the DIP dataset and select the top ranked protein pairs as predicted new interactions. Figure 1(b) shows the degree of functional homogeneity of these new interactions ranked under different k values. Again, the performance of the local interacting score reaches the best when k=2. We also observed the same trend using localization coherence. In the following experiments, we set k=2.

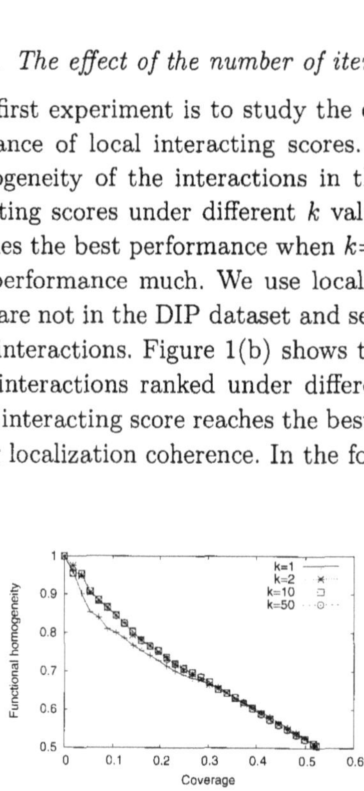

(a)

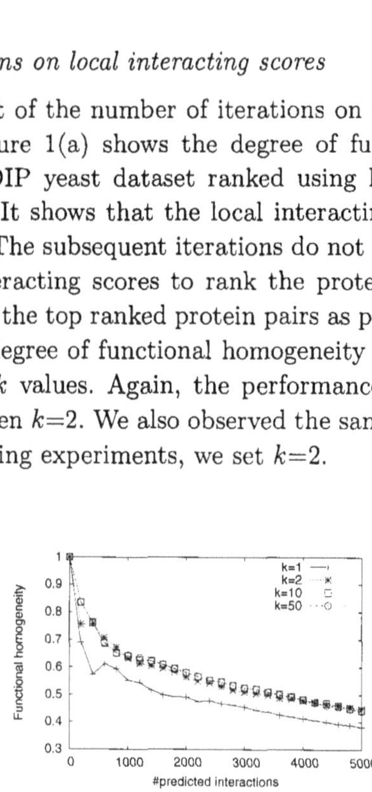

(b)

Fig. 1. The effect of the value of k (a) interactions in the DIP yeast dataset (b) New interactions predicted.

3.1.2. *Assessing and predicting interactions*

Our second experiment is to compare the performance of our method with that of FSweight and CD-distance in terms of functional homogeneity and localization coherence. When calculating global interacting scores, we set min_sup=1 and min_size=5. More specifically, the generated protein groups have at least one common neighbors and contains at least five proteins. Frequent itemset mining algorithms use the minimum support threshold min_sup to prune the search space. Here the value of min_size is larger than that of min_sup, so we swapped the values of the two thresholds and used min_size as the minimum support threshold to first find the partner groups of the desired groups, and then generated the desired protein groups in a post-processing step. The time used for generating the protein groups is less than one minute on a PC with 2.33GHz CPU.

In our experiments, we retained only those protein group pairs with a confidence score no less than 0.1. We assessed the significance of these retained protein group pairs as follows. For a protein group pair (V_1, V_2), we randomly generate 1000 protein group pairs (V_1', V_2') such that $|V_1'| = |V_1|$, $|V_2'| = |V_2|$ and $|V_1' \cap V_2'| = |V_1 \cap V_2|$. We then calculate the interacting confidence score of these random protein group pairs, and use the percentage of these random group pairs whose confidence score is no less than $conf(V_1, V_2)$ to approximate the p-value of (V_1, V_2). We have found that the p-value of all of the retained protein group pairs is no larger than 0.005.

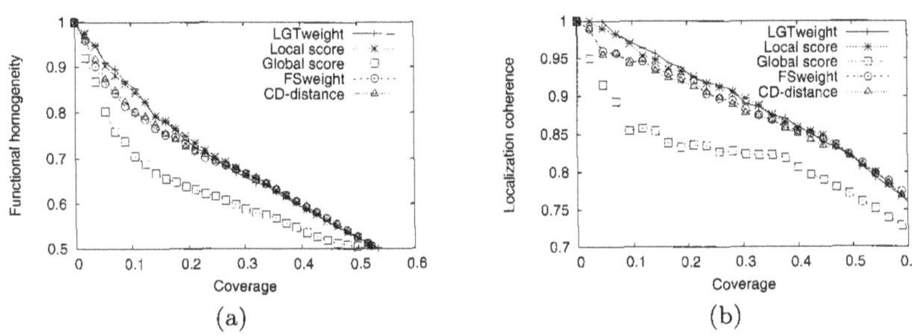

Fig. 2. Performance of the five methods in assessing reliability of interactions (a) functional homogeneity (b) localization coherence

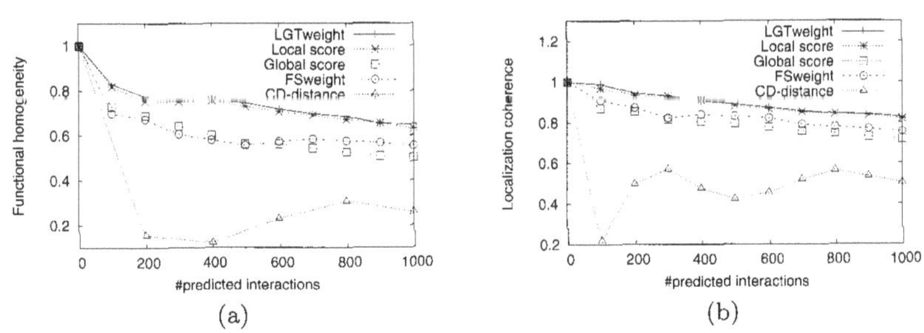

Fig. 3. Performance of the five methods in predicting new interactions (a) functional homogeneity (b) localization coherence

Figure 2(a) shows the functional homogeneity and localization coherence of the interactions in the DIP datasets ranked using five methods: LGTweight, local interacting score, global interacting score, FSweight and CD-distance. Protein pairs ranked by global interacting score show lower functional homogeneity and localization coherence than those ranked by other methods. The reason being that local interacting score, FSweight and CD-distance rank protein pairs based on their level-1 neighbors, and proteins are more likely to share common functions or localizations with their level-1 neighbors than with other proteins. The global interacting score ranks protein pairs based on interacting protein group pairs. A protein pair contained in an interacting group pair may have no common neighbors at all. The local interacting score performs better than FSweight and CD-distance, and its performance can be improved when combined with global interacting score.

Figure 3 shows the functional homogeneity and localization coherence of the new interactions predicted by the five methods. CD-distance performs the worst among the five methods. The global interacting score still performs worse than FSweight, but the gap between it and FSweight becomes smaller. Local interacting score and LGTweight perform significantly better than FSweight. LGTweight performs better than local interacting score due to the use of global interacting score.

3.2. *Five-fold cross validation*

Our last experiment is to study the performance of our method using five-fold cross validation. Here we use the DIP yeast core dataset as the golden standard. We divide the proteins into five disjoint groups. For each group, we remove the interactions between proteins in that group, and use the remaining interactions as the training dataset. The testing dataset contains all the possible pairs of proteins in the group. The removed interactions that are in the DIP yeast core dataset are regarded as the correct answers. The number of proteins in each of the five groups is 986, the average number of interactions in the five training datasets is 16723, the number of testing interactions in each of the five testing datasets is 486591 and the average number of correct-answer interactions is 307.

Sensitivity and specificity are two commonly used measures to assess prediction algorithms, and they are defined as follows.

$$sensitivity = \frac{TP}{TP + FN} \tag{6}$$

$$specificity = \frac{TN}{TN + FP} \tag{7}$$

where TP (True Positive) is the number of true interacting protein pairs that are also predicted to be interacting, FN (False Negative) is the number of true interacting protein pairs that are predicted to be non-interacting, TN (True Negative) is the number of non-interacting protein pairs that are predicted to be non-interacting, and FP (False Positive) is the number of non-interacting protein pairs that are predicted to be interacting.

In our testing data, the number of non-interacting protein pairs is orders of magnitude larger than the number of interacting protein pairs. Only 0.063% testing protein pairs are considered as truly interacting. In this case, the specificity of an

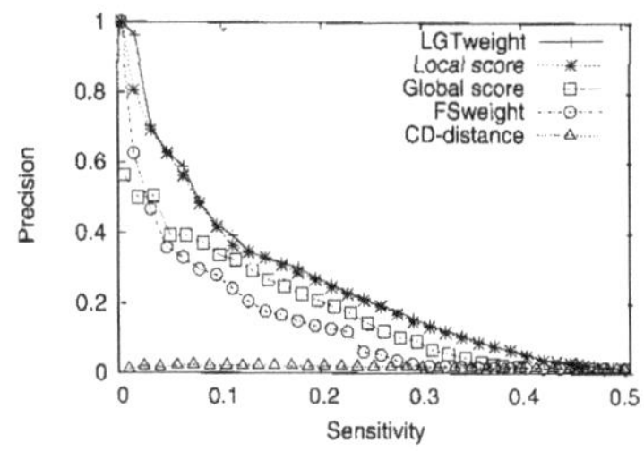

Fig. 4. Sensitivity vs. precision.

algorithm can be always very high. In our experiments, the specificy of all the algorithms is no less than 97.8% when they reach their maximal sensitivity. To have a clearer comparison of the algorithms, here we use another measure called precision to assess the algorithms, and it is defined as the percentage of true interactions among all the predictions made by the algorithms.

$$precision = \frac{TP}{TP + FP} \tag{8}$$

Figure 4 shows the precision of the five methods with respect to their sensitivity. CD-distance shows very poor performance. FSweight performs worse than the other three methods. Under the same sensitivity, the precision of FSweight is lower than that of the other methods. It indicates that FSwegith makes more false positive predictions than other methods. However, the maximal sensitivity achieved by FSweight is 50.5%, which is higher than the other methods. The maximal sensitivity achieved by LGTweight is 49.9%, which is higher than that of local interacting score (46.3%) and global interacting score (40.0%). Under the same sensitity, the precision of LGTweight is also higher than that of local interacting score and global interacting score, which shows that by combining local interacting score and global interacting score, we can obtain both higher sensitivity and higher precision than using them alone.

Note that here we regard only those interactions in the DIP core dataset as true interactions. However, some interactions not in the core dataset may be true interactions, so using the core dataset as the golden standard may underrate the performance of the methods. The actual performance of the methods tested should be better than what reported here.

4. Discussion and Conclusion

In this paper, we have proposed a computational approach to assessing and predicting protein interactions. The proposed method uses both local topological information of protein pairs and global topological structures discovered from the whole network to calculate interacting scores of protein pairs, and it outperforms existing methods, especially for predicting new interactions. We used an iterative approach to calculate local interacting scores. We have tried to apply this iterative approach to FSweight, and we also observed a significant improvement on the performance of FSweight. Here we uses a simple method to combine the local interacting score and global interacting score of a protein pair. It is possible to use a more sophisticated method to achieve better results.

In this paper, we use only the network topology to assess and predict interactions. It is complementary to those methods using other information for assessing and predicting protein interactions. The performance of our method, and other methods using solely interaction network topology, is limited by the availability and quality of existing interaction data. Chua et al. [7] proposed a framework for integrating multiple information sources. We can use their method or other methods to integrate other information sources into our approach, or integrate our method with other methods to obtain better results.

Acknowledgments

This research was supported in part by Singapore MOE Tier 1 grant R-252-000-274-112 (Liu, Wong) and in part by NTU Tier 1 grant RG66/07 (Li).

References

[1] Bock JR, and Gough DA, Predicting protein-protein interactions from primary structure. *Bioinformatics*, 17(5):455–460, 2001.

[2] Brun C, Chevenet F, Martin D, Wojcik J, Guenoche A, and Jacq B, Functional classification of proteins for the prediction of cellular function from a protein-protein interaction network. *Genome Biology*, 5(1):R6, 2003.

[3] Chen J, Chua HN, Hsu W, Lee ML, Ng SK, Saito R, Sung WK, and Wong L, Increasing confidence of protein-protein inteactomes. In *Proc. of 17th International Conference on Genome Informatics*, pp. 284–297, 2006.

[4] Chen J, Hsu W, Lee ML, and Ng SK, Discovering reliable protein interactions from high-throughput experimental data using network topology. *Artificial Intelligence in Medicine*, 35(1-2):37–47, 2005.

[5] Chen J, Hsu W, Lee ML, and Ng SK, Increasing confidence of protein interactomes using network topological metrics. *Bioinformatics*, 22(16):1998–2004, 2006.

[6] Chua HN., Sung WK., and Wong L., Exploiting indirect neighbours and topological weight to predict protein function from protein-protein interactions. *Bioinformatics*, 22(13):1623-30, 2006.

[7] Chua, HN., Sung, WK., and Wong L., An efficient strategy for extensive integration of diverse biological data for protein function prediction. *Bioinformatics*, 3(24): 3364-3373, 2007.

[8] Dandekar, T., Snel, B., Huynen, M., and Bork, P., Conservation of gene order: a fingerprint of proteins that physically interact. *Trends in Biochemical Sciences*, 23(9):324-8, 1998.

[9] Deane CM., Salwinski L., Xenarios I., and Eisenberg D., Protein interactions: two methods for assessment of the reliability of high throughput observations. *Mol Cell Proteomics.*, 1(5):349-56, 2002.

[10] Edwards AM, Kus B, Jansen R, Greenbaum D, Greenblatt J, and Gerstein M, Bridg-

ing structural biology and genomics: assessing protein interaction data with known complexes. *Trends in Genetics*, 18(10):529–536, 2002.

[11] Goh CS, Bogan AA, Joachimiak M, Walther D, and Cohen FE., Co-evolution of proteins with their interaction partners. *Journal of Molecular Biology*, 299(2):283-93, 2000.

[12] Gomez SM, and Rzhetsky A, Towards the prediction of complete protein-protein interaction networks. In *Pacific Symposium on Biocomputing*, pp. 413–424, 2002.

[13] Han D, Kim HS, Seo J, and Jang W, A domain combination based probabilistic framework for protein-protein interaction prediction. *Genome Informatics Series: Workshop on Genome Informatics*, 14:250–259, 2003.

[14] Kim WK, Park J, and Suh JK, Large scale statistical prediction of protein-protein interaction by potentially interacting domain (pid) pair. *Genome Informatics Series: Workshop on Genome Informatics*, 13:42–50, 2002.

[15] Legrain P, Wojcik J, and Gauthier JM, Protein-protein interaction maps: a lead towards cellular functions. *Trends in genetics*, 17(6):346–352, 2001.

[16] Li H, Li J, and Wong L, Discovery motif pairs at interaction sites from protein sequences on a proteome-wide scale. *Bioinformatics*, 22(8):989–996, 2006.

[17] Liu G., Lu H., Lou W., Xu Y., Yu X. J., Efficient Mining of Frequent Patterns Using Ascending Frequency Ordered Prefix-Tree. *Data Mining and Knowledge Discovery*, 9(3): 249-274, 2004.

[18] Marcotte EM., Pellegrini M., Ng HL., Rice DW., Yeates TO., and Eisenberg D., Detecting protein function and protein-protein interactions from genome sequences. *Science*, 285(5428):751-3m, 1999.

[19] Morrison JL, Breitling R, Higham DJ, and Gilbert DR, A lock-and-key model for protein-protein interactions. *Bioinformatics*, 22(16):2012–2019, 2006.

[20] Ng SK, Zhang Z, and Tan SH, Integrative approach for computationally inferring protein domain interactions. *Bioinformatics*, 19(8):923–929, 2003.

[21] Oliver S, Proteomics: guilt-by-association goes global. *Nature*, 403:601–603, 2000.

[22] Pazos F, and Valencia A., Similarity of phylogenetic trees as indicator of protein-protein interaction. *Protein Engineering*, 14(9):609-14, 2001.

[23] Ramani AK., Bunescu RC., Mooney RJ., and Marcotte EM., Consolidating the set of known human protein-protein interactions in preparation for large-scale mapping of the human interactome. *Genome Biology*, 6(5):R40, 2005.

[24] Saito R, Suzuki H, and Hayashizaki Y, Interaction generality, a measurement to assess the reliability of a protein-protein interaction. *Nucleic Acids Research*, 30(5):1163–1168, 2002.

[25] Saito R, Suzuki H, and Hayashizaki Y, Construction of reliable protein-protein interaction networks with a new interaction generality measure. *Bioinformatics*, 19(6):756–763, 2002.

[26] Sprinzak E, Sattath S, and Margalit H, How reliable are experimental protein-protein interaction data? *Journal of Molecular Biology*, 327(5):919–923, 2003.

[27] Tan SH., Hugo W., Sung WK., and Ng SK., A correlated motif approach for finding short linear motifs from protein interaction networks. *BMC Bioinformatics*,7:502, 2006.

[28] von Mering C, Krause R, Snel B, Cornell M, Oliver SG, Fields S, and Bork P, Comparative assessment of large-scale data sets of protein-protein interactions. *Nature*, 417:399–403, 2002.

[29] Yu H, Paccanaro A, Trifonov V, and Gerstein M, Predicting interactions in protein networks by completing defective cliques. *Bioinformatics*, 22(7):823–829, 2006.

Modelling the evolution of protein coding sequences sampled from Measurably Evolving Populations

Matthew Goode[1] Stéphane Guindon[1,2]

Allen Rodrigo[1]
a.rodrigo@auckland.ac.nz

[1] *The Bioinformatics Institute New Zealand and the Allan Wilson Centre for Molecular Ecology and Evolution, University of Auckland, Private Bag 92019, Auckland, New Zealand*
[2] *Department of Statistics, University of Auckland, Private Bag 92019, Auckland, New Zealand*

Models of nucleotide or amino acid sequence evolution that implement homogeneous and stationary Markov processes of substitutions are mathematically convenient but are unlikely to represent the true complexity of evolution. With the large amounts of data that next generation sequencing promises, appropriate models of evolution are important, particularly when data are collected from ancient and sub-fossil remains, where changes in evolutionary parameters are the norm and not the exception. In this paper, we describe a new codon-based model of evolution that applies to Measurably Evolving Populations (MEPs). A MEP is defined as a population from which it is possible to detect a statistically significant accumulation of substitutions when sequences are obtained at different times. The new model of codon evolution permits changes to the substitution process, including changes to the intensity of selection and the proportions of sites undergoing different selective pressures. In our serial model of codon evolution, changes in the selective regime occur simultaneously across all lineages. Different regions of the protein may also evolve under distinct selective patterns. We illustrate the application of the new model to a dataset of HIV-1 sequences obtained from an infected individual before and after the commencement of antiretroviral therapy.

Keywords: Measurably Evolving Populations; ancient DNA; HIV; serial samples; codon models of evolution; positive selection; likelihood.

1. Introduction

Next generation sequencing is capable of sequencing a bacterial genome in 24 hours and a human genome in 2 months. These times and the costs of sequencing will decrease in a few years. The sensitivity of our methods to amplify and sequence DNA also means that we now have the ability to obtain sequences from sub-fossil remains. There have been several high-profile studies with ancient DNA including one where portions of the genome of the mammoth were obtained using short-read sequencing [23]. We and others have been involved in the development of evolutionary methods to model the evolution of sequences obtained over time [8, 19].

If homologous gene sequences are sampled from a population at different times, and one is able to detect a statistically significant accumulation of genetic substitutions between sequences sampled earlier and those sampled later, the population is called a Measurably Evolving Population [8, MEP]. Rapidly evolving pathogens, for example, the RNA viruses [16], are MEPs, because the rates of substitution are sufficiently high that over the course of an infection within a host or a population of hosts, significant numbers of changes can be detected in the viral genomes. The definition of a MEP applies equally well to extant eukaryotic populations that have left sub-fossil traces of their ancient past and from which DNA may still be extracted. Here, the long intervals between samples of sequences – thousands to tens of thousands of years – mean that the accumulation of substitutions is statistically detectable, despite substitution rates that are several orders of magnitude lower than those of rapidly evolving pathogens.

Serial samples of sequences obtained from MEPs provide the opportunity to estimate substitution rates directly. The "calibration points" which are required for molecular dating correspond here to actual sequences. Indeed, with serial samples, each sequence has an associated "time stamp" that corresponds to an actual chronological date specifying when that sequence was collected. Since time is measured absolutely, the differences between sampling intervals is measured in units of chronological time and we are able to obtain direct estimates of substitution rates by estimating the expected number of substitutions that accrue over a sampling interval, and dividing this estimate by the interval length. Since we can estimate substitution rates directly, serial samples also allow us to estimate population genetic parameters, for example, effective population size, migration rates and recombination rates, that would otherwise be estimated as composite parameters when all that is available are sequences collected at the same time (i.e., isochronously). Additionally, serially-sampled sequences allow us to estimate changes in the values of these parameters over time. Frequently, a change in the environment – one that is accompanied by a tectonic shift of the adaptive landscape, for instance – has evolutionary consequences for all individuals in the population simultaneously. From a phylogenetic perspective, these changes in the evolutionary process cut across all lineages at the same time. Analyses of serial samples are particularly suitable for modelling these types of changes [6, 10]. Given the rich complexity of models that MEPs permit, it is not surprising, then, that the evolutionary analysis of MEPs challenges standard population genetic and phylogenetic methods, which are not conditioned to take account of sequences sampled at different times (i.e., heterochronously).

Over the last few years, methods have been developed to reconstruct phylogenies of serially-sampled sequences under the constraint of a molecular clock [5], and to estimate one (or more) substitution rate(s) spanning the sampling interval(s) [6, 25]. Additionally, population genetic methods have also been extended to include the development of the serial coalescent (or s-coalescent; [26]), a version of the standard Kingman n-coalescent [17, 18] suitable for heterochronous sequences. The s-coalescent has, in turn, been used to estimate mutation rates, effective population size (and the rate of change of population size) [7] and migration rates across several demes [9].

In this paper, we develop a new method that focuses on the estimation of selection intensity in protein-coding gene sequences, using phylogenies reconstructed under the criterion of maximum-likelihood [11] incorporating codon models of evolution that permit changes of selection regimes over time. Models of codon evolution, and details of how they can be used in likelihood-based phylogenetic reconstruction, have been available since the early 1990s [14, 20]. However, more recent models allow for greater flexibility, and perhaps the most widely used models are those that have been described by Nielsen and Yang [22]. Codon models typically include a parameter, ω, that measures the ratio of the rate of nonsynonymous substitutions (i.e., those substitutions at a given codon that result in a change of amino acid) to the rate of synonymous substitutions. In more sophisticated codon models, different classes of sites are permitted. For instance, we may posit the existence of the following three classes of codons: (1) a class of sites at which amino-acid changes decrease the fitness of the organism and, consequently, at which the rate of nonsynonymous substitution per codon is lower than the rate of synonymous substitution (= negatively selected sites with $\omega < 1$); (2) a class of sites where any change is equally tolerated, regardless of whether the substitution is nonsynonymous or synonymous (neutrally evolving sites with $\omega = 1$); and (3) a class where a nonsynonymous substitution confers a selective advantage to the organism (positively selected sites with $\omega > 1$). Nielsen and Yang [22] designated this Model 2, and, although it is one of a host of other models, it is perhaps the most widely used. We will refer to this model as M2. The model we have developed applies to serially sampled sequences and is an extension of M2, although, as we note in the discussion, it is relatively straightforward to see how it may apply to other models of codon substitution. However, our model also allows a change in the value of ω at some pre-specified time along the sampling interval between the earliest and most recent samples (Fig. 1). We will refer to this codon model as *serial*-Model 2 or, simply, sM2. This paper is organised as follows. First, we will briefly describe M2, then sM2, and its variants. We will illustrate the application of sM2 to a phylogeny of HIV-1 partial envelope (*env*) gene sequences sampled from a single infected individual before and after anti-HIV therapy begins.

2. The M2 codon substitution model

Consider an alignment of homologous protein-coding nucleotide sequences. If the sequences are genes encoded with the universal genetic code, there are 61 possible non-terminating codon types, each specified by a particular triplet of nucleotides. The M2 model defines a first-order continuous time Markov process over these 61 discrete states. This process acts identically and independently at each codon site along the alignment. The instantaneous rate of substitution between codons i and j ($i, j \in \{1 \ldots 61\}$) is given by :

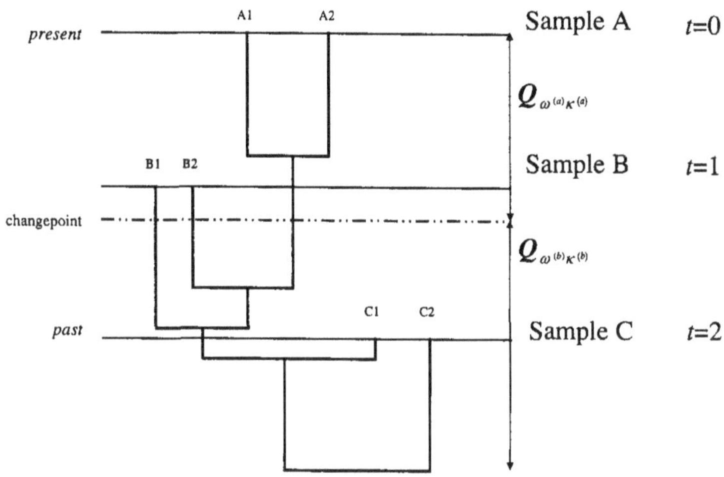

Fig. 1. An example of a phylogeny of six serially sampled sequences. Two sequences have been obtained at each of three sampling occasions. The number of nucleotide substitutions per codon site, μ is constant across the sampling intervals and can be estimated directly. By convention, sampling time is labelled from present to past. There is a change of the instantaneous rate matrix from $\mathbf{Q}_{\omega^{(a)}\kappa^{(a)}}$ to $\mathbf{Q}_{\omega^{(b)}\kappa^{(b)}}$ at the "changepoint". The parameters $\omega^{(x)}$ and $\kappa^{(x)}$ (where $x \in \{a, b\}$) are, respectively, ratios of nonsynonymous to synonymous rates and transitions to transversions, after (a) or before (b) the changepoint.

$$Q_x(ij) = \begin{cases} 0, & \text{if } i \text{ and } j \text{ differ at more} \\ & \text{than one position,} \\ \pi_j, & \text{for synonymous transversions,} \\ \kappa\pi_j, & \text{for synonymous transitions,} \\ \omega_x\pi_j, & \text{for nonsynonymous transversions,} \\ \kappa\omega_x\pi_j, & \text{for nonsynonymous transitions,} \end{cases} \tag{1}$$

where π_j is the equilibrium frequency of the j-th codon type, κ is the transition to transversion rate ratio, ω_x is the nonsynonymous to synonymous substitution rate ratio. $Q_x(ii) = -\sum_{j \neq i} Q_x(ij)$ for all $i \in \{1 \dots 61\}$ and the matrix $\mathbf{Q}_x = (Q_x(ij))$ is the infinitesimal generator of the Markov chain. M2 relies on the assumption that the values of π_j's, ω_x and κ are constant during the evolution. Hence, the substitution process is homogeneous, stationary and time-reversible [33]. $\mathbf{Q}_x$ is also normalised such that $-\sum_i \pi_i Q_x(ii) = 1$. Therefore, the expected number of codon substitutions occuring at a given site during a time t is simply μt, where μ is the rate at which substitutions occur.

Let $\mathbf{\Pi} = (\pi_j)$ be the vector of codon stationary frequencies. The probability, $P_t(j|i, \omega_x, \kappa, \mathbf{\Pi}, \mu)$, that codon i will be substituted by codon j after a period of time t, for any i and j, conditioned on ω_x, κ, $\mathbf{\Pi}$, and μ, is retrievable from the matrix :

$$\mathbf{P}_x(t) = \exp(\mathbf{Q}_x \mu t), \tag{2}$$

which can be calculated using the eigen decomposition of matrix $\mathbf{Q}_x$.

Now, assume that the substitution model is a mixture of three classes of sites along the alignment, each evolving under a different selective pressure: a proportion, p_0 of sites evolve under strong negative selection ($x = 0$ with $\omega_0 = 0$), a proportion, p_1, evolve neutrally ($x = 1$ with $\omega_1 = 1$) and a proportion ($p_2 = 1 - p_0 - p_1$), evolve under positive selection ($x = 2$ with $\omega_2 > 1$). *A priori*, for any given site, we have no knowledge of the class to which it belongs; consequently, the proportions p_0, p_1 and p_2 represent the prior probabilities of class membership. Let $\Xi = \{\omega_0, \omega_1, \omega_2, p_0, p_1, p_2\}$ denote the mixture of selection regimes. If ω_2, κ, p_0, p_1 and μ are known, we may write the marginal probability (summed over all possible selection classes) that codon i will be substituted by j, over a period of time, t, as :

$$\begin{aligned} P_t(j|i, \Xi, \kappa, \mathbf{\Pi}, \mu) =& p_0 P_t(j|i, \omega_0, \kappa, \mathbf{\Pi}, \mu) + \\ & p_1 P_t(j|i, \omega_1, \kappa, \mathbf{\Pi}, \mu) + \\ & p_2 P_t(j|i, \omega_2, \kappa, \mathbf{\Pi}, \mu), \end{aligned} \tag{3}$$

where $P_t(j|i, \omega_x, \kappa, \mathbf{\Pi}, \mu)$ is obtained from $\exp(\mathbf{Q}_x \mu t)$.

3. The serial codon substitution model, sM2

The most general extension to M2 permits changes to the values of ω_2, κ, p_0 and p_1 over a period of time. Formally, let $t^* = t(a) + t(b)$, and let $\Xi^{(a)}$, $\kappa^{(a)}$ designate the parameters associated with $t(a)$, with an equivalent notation for $t(b)$. Note that under this formulation, not only do we allow the nonsynonymous/synonymous and the transition/transversion rates to change over t^*, we also allow the prior probability of a site belonging to a particular selection class to change. Therefore, the probability that codon i will be substituted by codon j over the period t^* is equal to the product of the probability of going from codon i to codon k over period $t(a)$, and the probability of going from codon k to codon j over period $t(b)$, summed over all possible values of k:

$$P_{t^*}(j|i, \Xi^{(a)}, \kappa^{(a)}, \Xi^{(b)}, \kappa^{(b)}, \mathbf{\Pi}, \mu) = \sum_{k \in C} \sum_{x=0}^{2} p_x^{(a)} P_{t(a)}(k|i, \omega_x^{(a)}, \kappa^{(a)}, \mathbf{\Pi}, \mu) \sum_{y=0}^{2} p_y^{(b)} P_{t(b)}(j|k, \omega_y^{(b)}, \kappa^{(b)}, \mathbf{\Pi}, \mu), \tag{4}$$

where C is the set of all possible codons. Equation 4 can be expressed as :

$$P_{t^*}(j|i, \Xi^{(a)}, \kappa^{(a)}, \Xi^{(b)}, \kappa^{(b)}) = \sum_{x=0}^{2} \sum_{y=0}^{2} p_x^{(a)} p_y^{(b)} \sum_{k \in C} P_{t(a)}(k|i, \omega_x^{(a)}, \kappa^{(a)}, \mathbf{\Pi}, \mu) P_{t(b)}(j|k, \omega_y^{(b)}, \kappa^{(b)}, \mathbf{\Pi}, \mu), \tag{5}$$

where the innermost sum corresponds to the probability of moving from codon i to codon j, over the interval t^*, for a given combination of $\omega_x^{(a)}$, $\kappa^{(a)}$, $\omega_y^{(b)}$ and $\kappa^{(b)}$. There is, of course, one striking difference between this transition probability and those used in Equation 3: the probability given in Equation 5 is defined by a Markov process that is not homogeneous over the interval t^* (although the Markov processes over intervals $t(a)$ and $t(b)$ are homogeneous). This non-homogeneity imposes a time-irreversible process that can be accommodated by serial-sample phylogenies, as we will see later.

4. Variants of sM2

Equation 5 is, in fact, equivalent to Equation 3 except that instead of a mixture made of three categories of sites, we now have nine categories, each with its own prior probability. It is convenient to write these probabilities as a matrix :

$$\mathbf{p} = \begin{bmatrix} p_{00} & p_{01} & p_{02} \\ p_{10} & p_{11} & p_{12} \\ p_{20} & p_{21} & p_{22} \end{bmatrix},$$

where x and y in p_{xy} designate the selection classes before and after the change-point, respectively. Note that $\sum_x \sum_y p_{xy} = 1$. In Equation 5, $p_{xy} = p_x^{(a)} p_y^{(b)}$ but it is not necessary to constrain p_{xy} in this way. In fact, we can assign to eight of the nine probabilities in $\mathbf{p}$ a value in the interval $[0, 1)$ (the ninth is equal to 1 minus the sum of the other eight). This is the most general characterisation of sM2, and borrowing terminology from standard contingency table and logistic regression analysis, we refer to this variant as an *interaction-effects sM2*. In contrast, a *main-effects sM2* retains the constraint $p_{xy} = p_x^{(a)} p_y^{(b)}$ but permits $p_x^{(a)} \neq p_x^{(b)}$, $\omega_2^{(a)} \neq \omega_2^{(b)}$, and $\kappa^{(a)} \neq \kappa^{(b)}$. Whereas the main-effects model posits differences in the prior probabilities associated with selection classes before and after the changepoint, the probability that a site evolves under a given selective regime after is independent of its selective category before. This differs from the interaction-effects sM2, where – perhaps because of the functional and structural constraints under which a given site evolves – the selective pressure acting on it before the changepoint conditions the selective regime after.

A *mean-effects sM2* also keeps $p_{xy} = p_x^{(a)} p_y^{(b)}$ but, additionally, constrains $p_x^{(a)} = p_x^{(b)}$, $\omega_x^{(a)} = \omega_x^{(b)}$, and $\kappa^{(a)} = \kappa^{(b)}$. With the mean-effects sM2, the values of ω_x, κ and the prior probabilities associated with different selection classes remain the same across the entire interval t^*. At first glance, it may appear that this is simply the reduction of sM2 to M2. In fact, it is not, because the mean-effects sM2 permits sites to move between selection classes along t^*, whereas M2 does not. To obtain M2 from sM2, we need to impose even stricter constraints: we set $\omega_2^{(a)} = \omega_2^{(b)}$, $\kappa^{(a)} = \kappa^{(b)}$, and $p_{xy} = 0$ when $x \neq y$; that is, M2 sets all off-diagonal entries of matrix $\mathbf{p}$ to 0.

There are several other ways we can think of constraining the selection class probabilities and the values of ω in our model to test a variety of different evolutionary hypotheses. Regardless of which variant of sM2 is used, the row and column sums of $\mathbf{p}$ give the marginal probabilities of the different selection classes before and after the change in the substitution process, respectively.

5. Phylogenetic estimation of substitution parameters

Frequently, of course, we have no knowledge of some or all the parameters in our substitution models. The tree topology, $\mathcal{T}$, is usually known *a priori*, while the vector of codon stationary frequencies, $\mathbf{\Pi}$, is deduced from the observed nucleotide frequencies. Given $\mathcal{T}$ and $\mathbf{\Pi}$, homologous coding sequences are then used to estimate the values of Ξ's, κ's, μ, and the vector of internal node time-stamps, noted as $\mathbf{\Theta}$. Maximum likelihood estimates (MLE's) of these numerical parameters are obtained by maximizing the probability of the data given the model. We have :

$$\{\widehat{\Xi}, \widehat{\kappa}, \widehat{\mathbf{\Theta}}, \widehat{\mu}\} = \underset{\{\Xi,\kappa,\mathbf{\Theta},\mu\}}{\operatorname{argmax}} \left(P(D|\mathcal{T}, \Xi, \kappa, \mathbf{\Theta}, \mu, \mathbf{\Pi})\right) \tag{6}$$

where D is the sequence alignment [36]. The value of $P(D|\mathcal{T}, \Xi, \kappa, \mathbf{\Theta}, \mu, \mathbf{\Pi})$ is computed using the standard pruning algorithm developed by Felsenstein [11]. The algorithm sums over all the possible codon states and selection classes at ancestral nodes. The calculation of the likelihood under a serial codon substitution model is performed using a slight modification of the standard pruning algorithm. Since changepoints are effectively nodes that occur along one or more branches at which selection regimes can change, the pruning algorithm for sM2 is applied at both ancestral nodes and changepoints.

As we have noted above, phylogenies of serially sampled sequences permit the estimation of substitution rate, μ (measured as the expected number of substitutions per codon site per unit of chronological time), directly [8]. The serial-Model 2 that we have proposed permits changes to the homogeneous substitution process defined by the instantaneous rate matrix specified in Equation 1. As a non-homogenous, time-irreversible Markov model of evolution, it is one of a class of models that have been available for some time [3, 12, 35]. However, unlike other phylogenetic applications of irreversible models of evolution, where changes in the substitution process can occur independently along each lineage [12], serial phylogenies encounter these changes at a specified point in (chronological) time across all lineages simultaneously (Fig. 1). The fact that serial-sample phylogenies are suspended in chronological time also means that they need to respect a molecular clock, such that sequences with the same time-stamp terminate at equal distances from the root (Fig. 1). Since, in our application of sM2, we posit changes across all lineages at the same point in time, the molecular clock constraint is a necessary one because it allows a time-stamp to identify the node on each affected lineage that corresponds to the change in selection regime.

6. Likelihood Ratio Tests (LRTs)

The three variants of sM2 we have described in this paper have nested parametric constraints. It is therefore possible to constrain the parameters of the interaction-effects sM2 so that we obtain the main-effects sM2, and it is possible to constrain parameters of the main-effects sM2 to obtain the mean-effects sM2. To compare any two nested models, we use the likelihood ratio statistic given by :

$$\Delta = 2\ln\frac{P(D|\mathcal{T},\boldsymbol{\Xi}_1,\kappa_1,\boldsymbol{\Theta}_1,\mu_1,\boldsymbol{\Pi})}{P(D|\mathcal{T},\boldsymbol{\Xi}_0,\kappa_0,\boldsymbol{\Theta}_0,\mu_0,\boldsymbol{\Pi})} \tag{7}$$

where $\{\mathcal{T},\boldsymbol{\Xi}_0,\kappa_0,\boldsymbol{\Theta}_0,\mu_0,\boldsymbol{\Pi}\}$ is the set of parameters of the constrained model, with fewer free parameters than the model defined by $\{\mathcal{T},\boldsymbol{\Xi}_1,\kappa_1,\boldsymbol{\Theta}_1,\mu_1,\boldsymbol{\Pi}\}$. Asymptotically, Δ is distributed as χ^2 with degrees of freedom, *df*, equal to the difference in the number of free parameters between the two models being compared.

For the test between the interaction-effects sM2 and the main-effects sM2, $df = 4$ (since the interaction-effects sM2 has 8 free p's, 2 free ω_2's and 2 free κ's, whereas the main-effects sM2 has 2 free p's for each period – giving a total of 4 – and 2 free ω_2's and 2 free κ's). Similarly, for the test between the main-effects sM2 and the mean-effects sM2, $df = 4$. Interestingly, M2 can only be compared to the interaction-effects sM2 because there is no way to parameterise the main-effects sM2 or mean-effects sM2 to obtain M2. For the test between the interaction-effects sM2 and M2, $df = 8$.

7. Empirical Bayes' classification of sites

Nielsen and Yang [22] demonstrated how one may assign sites to different selection categories, using an empirical Bayes' procedure. The principle is simple: calculate the posterior probability of a selection category for each site, and choose the category with the highest posterior probability. The posterior probability that site D_s belongs to selection class x ($x \in \{0, 1, 2\}$) along an alignment is :

$$P(\omega_x | D_s, \mathcal{T}, \Xi, \kappa, \boldsymbol{\Theta}, \mu, \boldsymbol{\Pi}) = \frac{p_x . P(D_s | \omega_x, \mathcal{T}, \kappa, \boldsymbol{\Theta}, \mu, \boldsymbol{\Pi})}{\sum_{y=0}^{2} p_y . P(D_s | \omega_y, \mathcal{T}, \kappa, \boldsymbol{\Theta}, \boldsymbol{\Pi}, \mu)}, \tag{8}$$

where $P(D_s | \omega_x, \mathcal{T}, \kappa, \boldsymbol{\Theta}, \boldsymbol{\Pi}, \mu)$ is the component of the likelihood at site D_s that corresponds to class x of the mixture Ξ. Of course, the posterior probability of each selection class x is estimated using the MLE of each model parameter.

Extending Equation 8 to sM2 is trivial. In this case, the posterior probability that site D_s belongs to selection class x before the split, and selection class y after the split is :

$$P(\omega_x, \omega_y | D_s, \mathcal{T}, \Xi, \kappa, \boldsymbol{\Theta}, \mu, \boldsymbol{\Pi}) = \frac{p_{xy} . P(D_s | \omega_x, \omega_y, \mathcal{T}, \kappa, \boldsymbol{\Theta}, \mu, \boldsymbol{\Pi})}{\sum_{v=0}^{2} \sum_{w=0}^{2} p_{vw} . P(D_s | \omega_v, \omega_w, \mathcal{T}, \kappa, \boldsymbol{\Theta}, \mu, \boldsymbol{\Pi})}. \tag{9}$$

Equation 9 returns the posterior probabilities for every pair of selection classes that a given site might move between as it evolves before and after the changepoint.

8. An example

Here, we illustrate the application of sM2 in a phylogenetic analysis of HIV-1 sequences. There are several caveats that apply to this analysis (we discuss these later), and within the context of this paper, our aim is simply to provide an example of how sM2 may be used, and the types of inferences that may be made.

HIV-1 partial *env* sequences spanning the variable regions V3 – v5 were obtained from the viral population from a single HIV infected patient over a period of three years [27]. Five samples were obtained, with the last four taken at 7, 22, 23, and 34 months after the first sample. 30 non-identical sequences – a subset of the 60 sequences reported in [27] – were used in this analysis. Ambiguously aligned sites were removed from the alignment, leaving 211 codon sites.

The patient began a course of zidovudine, a nucleoside reverse transcriptase inhibitor, 13 months after the study began, and continued until at least the end of the sampling period. There is evidence that selective pressures on the HIV *env* gene may be different before and after antiretroviral therapy begins [1]. The *env* gene, the most variable region of the HIV genome, contains epitopes recognised by both the humoral and cellular arms of the immune system, but plays no known role in antiretroviral resistance. Positive selection in *env* is typically associated with immune evasion [29]. However, after the commencement of therapy, the substitutions associated with large increases in viral fitness are those that confer drug resistance and, with zidovudine, these occur in the *pol* gene. Consequently, one expects to see a change in the selective regime operating on the *env* gene after therapy begins. With the sequences we use, therefore, it is reasonable to ask if there is any evidence of a change in the pattern of selection across all lineages after the start of therapy.

Codon equilibrium frequencies (i.e., the values of π in Equation 1) were estimated using the $F1 \times 4$ method implemented in PAML [36]. A phylogenetic tree estimated using Neighbour-Joining [30] and rooted with two basal sequences obtained with sUPGMA [5], was constructed with a HIV-1-specific GTR model reported by Anderson et al. [2]. The interaction-effects sM2 and the sM2 constrained to M2 (as described above) were fitted to the tree, and for each model both the substitution parameters and the branch-lengths were re-estimated, subject to the constraints of a molecular clock. For each model, a mutation rate, μ, constant across all intervals, was also estimated.

Table 1. Results obtained by fitting M2 under the assumption of a molecular clock

ω_2	7.321
κ	2.946
μ	3.4×10^{-5}
p_0	0.477
p_1	0.411
p_2	0.112

Table 2. Results obtained by fitting a saturated sM2, under the assumption of a molecular clock

	$\omega_0^{(a)} = 0$	$\omega_1^{(a)} = 1$	$\omega_2^{(a)} = \infty$	Totals
$\omega_0^{(b)} = 0$	0.413	0.062	0.002	0.477
$\omega_1^{(b)} = 1$	0.388	0.002	0.002	0.392
$\omega_2^{(b)} = 8.241$	0.126	0.002	0.002	0.130
Totals	0.927	0.066	0.006	1.00

$\kappa^{(b)}$	$\kappa^{(a)}$	μ
3.212	0.580	6.4×10^{-5}

Results for M2 and sM2 are shown in Tables 1 and 2, respectively. The maximum log-likelihood of M2 is -2776.468, and that of sM2 is -2761.350; applying the LRT described above, sM2 provides a statistically better fit to the data than M2 ($\Delta =$ 30.2; critical $\chi^2_{8,0.01} = 20.9; p < 0.01$). The estimated substitution rate is 6.4×10^{-5} nucleotide substitutions per codon site per day, or 2.3×10^{-2} nucleotide substitutions per codon site per year.

Interestingly, the marginal probabilities of the three selection classes before and after therapy are different and they accord with what we expect may be the case with the HIV-1 *env* gene. Whereas before therapy, quite a high proportion of sites (13.3%) are under positive selection, after therapy, less than 1% are in that class ($\omega_2^{(a)} = \infty$ because it reached the upper bound set in our computer program). Since the *env* gene does not have mutations that confer resistance to zidovudine – resistance mutations are found on the *pol* gene – our results are consistent with the hypothesis that after therapy, viral fitness is strongly associated with drug resistance, and not with immune evasion [1].

It is also interesting to compare how M2 and sM2 classify sites into different selection classes using the posterior probabilities calculated with Equations 8 and 9, respectively. In Table 3, where this comparison is summarised, there are a few points of interest. First, all sites identified by M2 to be under purifying selection ($\omega = 0$) are inferred by sM2 to also be purifying both before and after antiretroviral therapy. This is not surprising – by definition, if a site evolves under purifying selection (i.e., $\omega = 0$) we expect to see no nonsynonymous substitutions at that site, over the entire tree. In contrast, a site inferred by M2 to be under neutral or positive selection across the entire tree, may fall into different selection classes before and after therapy, when classified using sM2. In fact, of the 211 codon sites analysed, 71 encounter different selective pressures before and after treatment, providing additional support for a model that permits changes to the selective environment, over one which assumes that no changes have occurred. Of even greater interest, only 10 of the 53 sites classified as neutral under M2, were classified as neutral both before and after therapy by sM2 – the remaining 43 sites were found to be under some form of selection over the period of evolution. As others have pointed out [4], the assignment of neutrality under a model that does not permit changes in the selective regime at a given site (such as M2) may simply be a consequence of an "averaging" of nonsynonymous and synonymous substitution rates across the entire tree. Our results indicate that M2 may underestimate the proportion of sites that are impacted by natural selection, either positively or negatively. As noted above, our principle aim in this section has been to illustrate how the sM2 codon substitution model may be applied. A complete and comprehensive analysis of these data would constitute a second paper. Similarly, the biological interpretation of the results deserves more than the cursory treatment we give it here. For instance, most sites classed as neutral or under positive selection before therapy appear to be under strong negative selection after therapy (Table 3). For these sites, then, neutrality is context-dependent, but how so? How does therapy influence the fitness advantage

Table 3. Comparison of the number of sites assigned to different selection categories under M2 and sM2

		M2			
		$\omega = 0$	$\omega = 1$	$\omega > 1$	Totals
	$\omega^{(b)} = 0, \omega^{(a)} = 0$	129	0	0	129
	$\omega^{(b)} = 0, \omega^{(a)} = 1$	0	10	0	10
	$\omega^{(b)} = 0, \omega^{(a)} > 1$	0	0	0	0
	$\omega^{(b)} = 1, \omega^{(a)} = 0$	0	30	0	30
sM2	$\omega^{(b)} = 1, \omega^{(a)} = 1$	0	10	1	11
	$\omega^{(b)} = 1, \omega^{(a)} > 1$	0	0	0	0
	$\omega^{(b)} > 1, \omega^{(a)} = 0$	0	3	21	24
	$\omega^{(b)} > 1, \omega^{(a)} = 1$	0	0	6	6
	$\omega^{(b)} > 1, \omega^{(a)} > 1$	0	0	1	1
Totals	129	53	29	211	

of a site on the *env* gene that evolves neutrally when therapy is absent? These and other interesting questions are beyond the scope of the present paper.

9. Conclusions

The sM2 model described here permits changes to the codon substitution process over time. The methods we have used to derive the sM2 model can also be applied to other models of codon evolution, although some practical restrictions apply. Nielsen and Yang [22] and later, Yang et al. [13] defined 14 models in total, each differing in how selection classes and ω is modelled. With Models 7 and 8, for instance, ω is distributed according to a β distribution, that is discretized into rate classes for computational ease. In these Models, we may hypothesize different parameterisations of the distribution function before and after a split.

One other class of models developed by Yang, Nielsen and colleagues needs to be mentioned – the branch-sites models [4, 34] permit different branches (or sets of branches) to evolve under different selective constraints. With these models, different sets of branches may have a different values of ω. There are three main differences between sM2 and the branch site models. First, with the sM2 it is possible to change the model of selection at some internal point along one or more branches (Fig. 1). Second, in branch-sites models the prior probabilities of selection classes do not change. Finally, there is no equivalent description of an interaction-effects branch sites model available.

There are other methods that permit changes in the evolutionary parameters associated with phylogenetic reconstruction. As we have noted above, non-homogenous models of evolution have been around for a while, and permit changes in nucleotide composition [12]. Other methods model changes along the phylogeny stochastically – covarion models allow changes to nucleotide composition and relative rates of substitutions [32] or selection classes [15] by permitting sites to move probabilistically between sets of evolutionary parameters. Finally, there are methods that permit changes to mutation rates along a clock-constrained phylogeny [24]. In all these cases, however, changes in the evolutionary process (and the parameters that describe this process) occur independently along the different lineages of a phylogeny. Arguably, these approaches are ideal if we are dealing with phylogenies of species or higher taxa, but may not be so useful when we deal with intraspecific phylogenies. Within a species or a population, it seems reasonable to suggest that changes to the environment can have a selective effect on all individuals simultaneously. If we are fortunate enough to have samples of fossils from which we can obtain DNA, we may be able to study the selective impact of large-scale environmental changes on macrobiota. The methods we have described in this paper are particularly amenable to this, because they have been developed to model changes in the selective environment and the evolutionary processes that shape genetic diversity.

Finally, we note that the model we have proposed here is appropriate when there is a stepwise change to the evolutionary dynamics. Recently, mathematical models for continuous change in evolutionary parameters have been developed [28]. It is possible to apply these developments to codon models of evolution, although it is not clear how computationally intensive this will be.

Acknowledgments

We thank Greg Ewing and Alexei Drummond for helpful discussions about the computational problems of serial sample analysis. We also thank Maria Anisimova, Nicolas Galtier, and Nicolas Lartillot for helpful comments on the manuscript, and Peter Tsai for assistance in preparing the final copy. This research was supported by funding from the Allan Wilson Centre for Molecular Ecology and Evolution. AG completed this research while on sabbatical at Olivier Gascuel's laboratory in the Laboratoire d'Informatique, de Microélectronique et de Robotique de Montpellier.

References

[1] S. Almodovar, I. M. Maldonado, S. Gonzalez, S. E. Costa, M. D. Hill, R. Mendoza, G. Sepulveda, R. Yanagihara, V. Nerurkar, R. Kumar, Y. Yamamura, W. A. Scott, A. Kumar, E. Lorenzo, and M. C.Colon. (2004) Influence of cd4+ t cell counts on viral evolution in hiv-infected individuals undergoing suppressive haart. *Virology* **330**, 116–126.

[2] J. P. Anderson, A. G. Rodrigo, G. H. Learn, Y. Wang, H. Weinstock, M. L. Kalish, K. E. Robbins, L. Hood, and J. I. Mullins. (2001) Substitution model of sequence evolution for the human immunodeficiency virus type 1 subtype b gp120 gene over the c2-v5 region. *Journal of Molecular Evolution* **53**,55–62.

[3] D. Barry and J. A. Hartigan. (1987) Statitstical analysis of hominoid molecular evolution. *Statistical Science* **2**,191–207.

[4] J. P. Bielawski and Z. Yang. (2003) Maximum likelihood methods for detecting adaptive evolution after gene duplicaiton. *Journal of Structural and Functional Genomics* **3**,201–212, 2003.

[5] A. Drummond and A. Rodrigo. (2000) Reconstruction genealogies of serial samples under the assumption of a molecular clock using serial-sample UPGMA. *Molecular Biology and Evolution* **17**, 1807–1815.
[6] A. Drummond, R. Forsberg, and A. G. Rodrigo. (2001) The inference of stepwise changes in substitution rates using serial sequence samples. *Molecular Biology and Evolution* **18**,1365–1371.
[7] A. J. Drummond, G. K. Nicholls, A. G. Rodrigo, and W. Solomon. (2002) Estimating mutation parameters, population history and genealogy simultaneously from temporally spaced sequence data. *Genetics* **161**, 1307–1320.
[8] A. J. Drummond, O. G. Pybus, A. Rambaut, R. Forsberg, and A. G. Rodrigo. (2003) Measurably evolving populations. *Trends in Ecology and Evolution* **18**, 481–488.
[9] G. B. Ewing, G. K. Nicholls, and A. G. Rodrigo. (2004) Using temporally spaced sequences to simultaneously estimate migration rates, mutation rate and population sizes in measurably evolving populations. *Genetics* **168**, 2407–2420.
[10] G. B. Ewing and A. G. Rodrigo. (2006) Coalescent-based estimation of population parameters when the number of demes changes over time. *Molecular Biology and Evolution* **23**, 988 – 996.
[11] J. Felsenstein. (1981) Evolutionary trees from DNA sequences: a maximum likelihood approach. *Journal of Molecular Evolution* **17**, 368 –376.
[12] N. Galtier and M. Guoy. (1998) Inferring pattern and process: Maximum-likelihood implementation of a nonhomogeneous model of DNA sequence evolution for phylogenetic analysis. *Molecular Biology and Evolution* **15**, 871–879.
[13] N. Goldman, A. M. Pedersen, Z. Yang, and R. Nielsen. (2000) Codon-substitution models for heterogeneous selection pressure at amino acid sites. *Genetics* **155**, 431–439.
[14] N. Goldman and Z. Yang. (1994) A codon based model of nucleotide substitution for protein-coding dna sequences. *Molecular Biology and Evolution* **11**, 725–736.
[15] S. Guindon, A. G. Rodrigo, K. A. Dyer, and J. P. Huelsenbeck. (2004) Modeling the site-specific variation of selection patterns along lineages. *Proceedings of the National Academy of Sciences, USA* **101**, 12957–12962.
[16] G. M. Jenkins, A. Rambaut, O. G. Pybus, and E. C. Holmes. (2002) Rates of molecular evolution in rna viruses: a quantitative phylogenetic analysis. *Journal of Molecular Evolution* **54**, 156-165.
[17] J. F. C. Kingman. (1982a) The coalescent. *Stochastic Processes and their Applications* **13**, 235–248.
[18] J. F. C. Kingman. (1982b) On the genealogy of large populations. *Journal of Applied Probability* **19A**, 27–43.
[19] X. Liu, and Y. X. Fu. (2007). Test of Genetical Isochronism for Longitudinal Samples of DNA Sequences. *Genetics* **176**, 327–342.
[20] S. V. Muse and B. S. Gaut. (1994) A likelihood approach for comparing synonymous and nonsynonymous nucleotide substitution rates, with applications to the chloroplast genome. *Molecular Biology and Evolution* **11**, 715–724.
[21] S. Nadarajah and S. Kotz. (2005) Some bivariate beta distributions. *Journal of Theoretical and Applied Statistics* **39**, 457–466.
[22] R. Nielsen and Z. Yang. (1998) Likelihood models for detecting positively selected amino acid sites and applications to the HIV-1 envelope gene. *Genetics* **148**, 929–936.
[23] N. N Poinar, C. Schwarz, J. Qi, B. Shapiro, R. D. E. MacPhee, B. Buigues, A. Tikhonov, D. H. Huson, L. P. Tomsho, A. Auch, M. Rampp, W. Miller, S. C. Schuster (2006). Metagenomics to Paleogenomics: Large-Scale Sequencing of Mammoth DNA. *Science* **311**, 392-394.

[24] O. G. Pybus. (2006) Model selection and the molecular clock. *PLoS Biology* **4**, e151.

[25] A. Rambaut. (2000) Estimating the rate of molecular evolution: incorporating non-contemporaneous sequences into maximum likelihood phylogenies. *Bioinformatics*, **16**, 395–399.

[26] A. G. Rodrigo and J. Felsenstein. (1998) Coalescent approaches to HIV-1 population genetics. In K. A. Crandall, editor, *The Evolution of HIV*, pages 233–272. Johns Hopkins University Press, Baltimore, USA.

[27] A. G. Rodrigo, E. G. Shpaer, E. L. Delwart, A. K. Iversen, and M. V. Gallo et al. (1999) Coalescent estimates of HIV-1 generation time in vivo. *Proceedings of the National Academy of Sciences, USA* **96**, 2187–2191.

[28] A. G. Rodrigo, F. Bertels, J. Heled, R. Noder, H. Shearman, P. Tsai. (In press) The Perils of Plenty:What are we going to do with all these genes? *Philosophical Transactions of the Royal Society, Series B*

[29] H. A. Ross and A. G. Rodrigo. (2002) Immune-mediated positive selection drives human immunodeficiency virus type 1 molecular variation and predicts disease duration. *Journal of Virology* **76**, 11715–11720.

[30] N. Saitou and M. Nei. (1987) The neighbor-joining method: a new method for reconstructing phylogenetic trees. *Molecular Biology and Evolution* **4**,406–25.

[31] R. Shankarappa, J. B. Margolick, S. J. Gange, A. G. Rodrigo, D. Upchurch, H. Farzadegan, P. Gupta, G. H. Learn, C. R. Rinaldo, X. He, X.-L. Huang, and J. I. Mullins. (1999). Consistent viral evolutionary changes associated with the progression of HIV-1 infection. *Journal of Virology* **78**, 10489–10502.

[32] C. Tuffley and M. A. Steel.(1998) Modelling the covarion hypothesis of nucleotide substitution. *Mathematical Biosciences* **147**, 63–91.

[33] Z. Yang and R. Nielsen. (2000) Estimating synonymous and nonsynonymous substitution rates under realistic evolutionary models. *Molecular Biology and Evolution* **17**,32–43.

[34] Z. Yang and R. Nielsen. (2002) Codon substitution models for detecting molecular adaptation at individual sites along specific lineages. *Molecular Biology and Evolution* **19**, 908–917.

[35] Z. Yang and D. Roberts. (1995) On the use of nucleic acid sequences to infer early branchings in the tree of life. *Molecular Biology and Evolution* **12**, 451–458.

[36] Z. Yang. (1997) PAML: a program package for phylogenetic analysis by maximum likelihood. *CABIOS* **13**, 555-556.

[37] A. Zharkikh. (1994) Estimation of evolutionary distances between nucleotide sequences. *Journal of Molecular Evolution* **39**, 315–329.

A PHYLOGENOMIC APPROACH FOR STUDYING PLASTID ENDOSYMBIOSIS

AHMED MOUSTAFA[1]* CHEONG XIN CHAN[2]*
ahmed-moustafa@uiowa.edu cx-chan@uiowa.edu

MEGAN DANFORTH[2] DAVID ZEAR[2] HIBA AHMED[2]
megan-danforth@uiowa.edu drzear@gmail.com hiba-ahmed@uiowa.edu

NAGNATH JADHAV[2] TREVOR SAVAGE[2] DEBASHISH BHATTACHARYA[1,2]
n-jadhav@uiowa.edu trevor-savage@uiowa.edu debashi-bhattacharya@uiowa.edu

*These authors contributed equally to this work.
[1]Interdisciplinary Genetics Program, University of Iowa, Iowa City, IA 52242, U.S.A.
[2]Department of Biology and Roy J. Carter Center for Comparative Genomics, University of Iowa, Iowa City, IA 52242, U.S.A.

Gene transfer is a major contributing factor to functional innovation in genomes. Endosymbiotic gene transfer (EGT) is a specific instance of lateral gene transfer (LGT) in which genetic materials are acquired by the host genome from an endosymbiont that has been engulfed and retained in the cytoplasm. Here we present a comprehensive approach for detecting gene transfer within a phylogenetic framework. We applied the approach to examine EGT of red algal genes into *Thalassiosira pseudonana*, a free-living diatom for which a complete genome sequence has recently been determined. Out of 11,390 predicted protein-coding sequences from the genome of *T. pseudonana*, 124 (1.1%, clustered into 80 gene families) are inferred to be of red algal origin (bootstrap support $\geq$ 75%). Of these 80 gene families, 22 (27.5%) encode novel, unknown functions. We found 21.3% of the gene families to putatively encode non-plastid-targeted proteins. Our results suggest that EGT of red algal genes provides a relatively minor contribution to the nuclear genome of the diatom, but the transferred genes have functions that extend beyond photosynthesis. This assertion awaits experimental validation. Whereas the current study is focused within the context of secondary endosymbiosis, our approach can be applied to large-scale detection of gene transfer in any system.

Keywords: phylogenomics; endosymbiotic gene transfer; lateral gene transfer; plastid; chromalveolates.

1. Introduction

Lateral gene transfer (LGT) is a phenomenon in which genetic materials are transmitted between non-lineal individuals (*e.g.*, between two different strains or species). This phenomenon is one of the major mechanisms for functional innovation in the genomes of prokaryotes [1, 2] and eukaryotes [3, 4], as well as for the acquisition of new virulence genes in pathogens [5]. Therefore, the elucidation of gene transfer events will enhance our understanding of how genomes evolve. Here we present a systematic approach for detecting LGT within the context of plastid endosymbiosis.

1.1. *Plastid endosymbiosis and gene transfer*

The origin and establishment of the photosynthetic organelle (plastid) in algae and plants are important for understanding biotic evolution because these taxa form the primary food source for all life on earth. The endosymbiosis hypothesis postulates that the plastid originated from the ancient engulfment and retainment of a free-living cyanobacterium (the endosymbiont) by a heterotrophic, unicellular protist. This ancestral photosynthetic eukaryote diversified into the red, green, and glaucophyte algae [6, 7]. Subsequent to this, a secondary endosymbiosis occurred, in which a red alga, that had gained its photosynthetic capability from primary endosymbiosis, was itself engulfed by a non-photosynthetic protist, giving rise to the progenitor of the eukaryote supergroup Chromalveolata [7, 8]. The process of endosymbiosis and the origin of plastid are detailed in [9–11] and Figure 1 in [6]. The phenomenon of endosymbiosis led to the transfer of genetic material from the endosymbiont to the host nuclear genome *via* endosymbiotic gene transfer (EGT), which is a specific case of LGT.

Chromalveolata is one of the six major "supergroups" of eukaryotes. This lineage consists of a taxonomically diverse group of species that are of high ecological and economic importance, including diatoms, seaweeds, dinoflagellates, and the malaria parasite *Plasmodium*. Our group has previously demonstrated EGT (and LGT) in chromalveolate genomes [3, 12–14], but the extent of EGT from red algae into chromalveolates, *vis-à-vis* secondary endosymbiosis, has not been studied in a rigorous manner.

Among the chromalveolates, diatoms are unicellular eukaryotes and one of the primary contributors to the marine food chain. The diatoms are estimated to generate $\approx$ 40% of the organic carbon produced annually in the sea [15]. These taxa affect the flux of atmospheric carbon dioxide into the oceans, which in turn has effects on global climate [16]. Recently, the genome of the free-living diatom *Thalassiosira pseudonana* was sequenced to completion [17]. Using the available genomic sequences, here we present a rigorous, phylogenomic pipeline to examine the extent of EGT of red algal genes in *T. pseudonana*, and investigate if these transferred genes are restricted to photosynthesis-related functions.

2. A phylogenomic approach for inferring phylogenies

With the increasing amount of available genome data, phylogenomics, the intersection of evolutionary and genomic approaches [18], has become a key instrument in studying genomes on a gene-by-gene basis. This is done primarily by the automated generation and inspection of phylogenetic trees. In many recent studies, phylogenomics has been employed to answer various questions including, *e.g.*, prediction of biochemical gene functions [19], evolution of gene functions [20], detection of gene transfer events [1, 3], and resolution of complex taxonomic relationships [13].

Our phylogenomic pipeline consists of four basic steps as shown in Figure 1. First, homologous genes for the target sequences are identified (step 1) using WU-

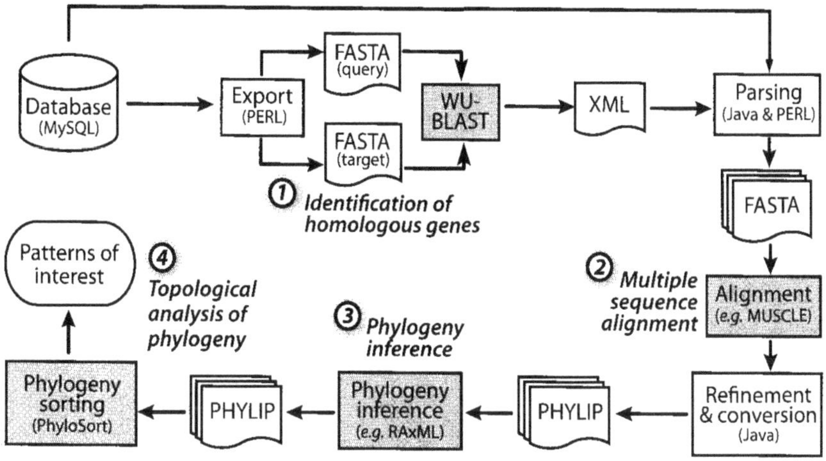

Fig. 1. A schematic diagram of the phylogenomic pipeline: functional components and data flow.

BLAST (http://blast.wustl.edu/) searches against a database containing sequences collected from public resources, *e.g.* NCBI (http://www.ncbi.nlm.nih.gov/) and JGI (http://www.jgi.doe.gov/). We used WU-BLAST because this program shows higher time-efficiency than the original BLAST algorithm [21]. Following this, multiple sequence alignment (step 2) is performed for each homologous gene family prior to phylogeny inference (step 3). We used MUSCLE [22] to align the sequences, and both neighbor-joining (NJ) [23] and maximum likelihood (ML) [24] to reconstruct the phylogenies, because these yield high accuracy in a reasonably short period of time [22, 24]. However, other approaches for sequence alignment and phylogeny inference can easily be incorporated into our pipeline. Finally, once the phylogeny for each gene family is obtained, these can be searched for topological patterns of interest (step 4). In the current study, we used PhyloSort [25] to sort and examine monophyletic relationships between chromalveolates and other taxa of interest.

2.1. *Analysis of EGT in Thalassiosira pseudonana*

We obtained all 11,390 predicted protein-coding sequences from the complete *Thalassiosira pseudonana* genome from JGI (http://www.jgi.gov/). We performed a preliminary screening using BLAST (at e-value ≤ 0.001) for sequences that are highly similar to and thus possibly share a common ancestry (*i.e.*, homologous) with the genes in red algae. Using 5,014 protein sequences from the complete genome of the red alga *Cyanidioschyzon merolae* [26], we found 4,894 (43.0% of 11,390) protein-coding sequences in *T. pseudonana* to have homologs in *C. merolae*.

These protein-coding sequences were used as input in our phylogenomic pipeline that utilizes our local database, which consists of 2,555,575 sequences from 62 eukaryote genomes, inclusive of complete and partial expressed sequence tag (EST) sequences spanning Plantae, chromalveolates, Rhizaria, excavates, animals, fungi, and Amoebozoa, and 500 complete bacterial genomes. Initially, the phylogenetic trees were constructed using NJ with a Poisson-distance correction and 100 replicates for the bootstrap analysis. By searching for the monophyly of cyanobacteria and chromalveolates, with or without Plantae, we identified and removed 1,907 chromalveolate genes with a potential cyanobacterial origin. This step was designed to exclude genes that were introduced *via* EGT into the red algal nucleus as a result of primary endosymbiosis. For the remaining 2,987 trees, we searched for the monophyly of red algae and chromalveolates, with or without green and glaucophyte algae ($\geq$ 75% bootstrap support). We identified 288 protein-coding sequences in *T. pseudonana* with potential red algal origin through EGT (as a result of secondary endosymbiosis).

Following this, we inferred ML phylogenies for each of the 288 genes using RAxML [24] (WAG model [27]; 100 bootstrap replicates). Using the same approach for detecting secondary EGT (described above), we identified 124 genes in chromalveolates with a putative red algal origin, and clustered these into 80 distinct families. We manually annotated the functions of these gene families. Blast2GO [28] was used to annotate each family based on significant matches (e-value $\leq 10^{-5}$) in the Gene Ontology (GO) database (http://geneontology.org/), for the three GO classes: molecular function, biological processes, and cellular components. The GO protein target prediction was complemented with PSORT [29] and Predotar [30]. Plastid-targeting localization was inferred when two out of the three prediction methods yielded positive results.

To examine the significance of the observed monophyly between chromalveolates and Plantae, we repeated the phylogenomic analysis using a dataset that excluded

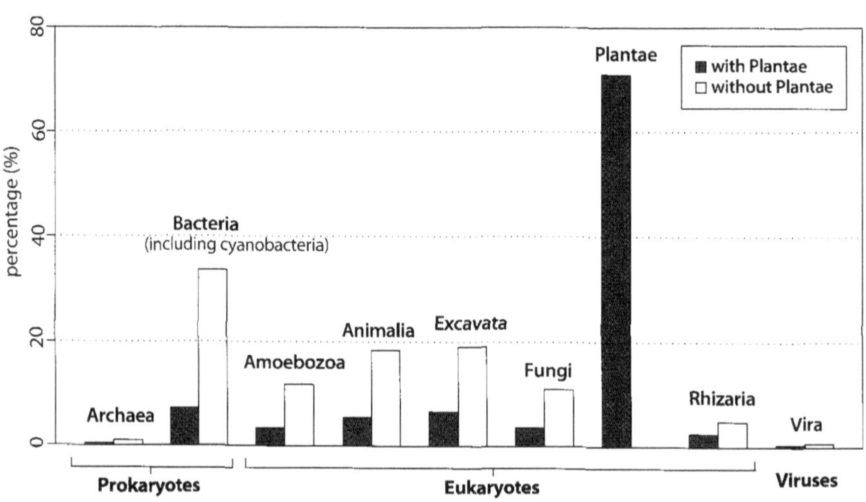

Fig. 2. Distribution of monophyly between chromalveolates and different lineages, for *Thalassiosira pseudonana* genes that showed a potential algal ancestry. The Y-axis represents the percentage of monophyletic relationships recovered, the X-axis represents the different lineages of prokaryotes, eukaryotes, and viruses. The blue and red bars represent the distributions across the dataset inclusive and exclusive of Plantae genomes, respectively.

Plantae genomes (glaucophytes, red, and green algae), and compared the observed monophyly between chromalveolates and the other lineages, with the existing results (dataset inclusive of Plantae genomes). As shown in Figure 2, the distributions of the observed monophyly between chromalveolates and non-Plantae are not significantly different between the two instances, *i.e.*, when Plantae genomes are included or not (Kolmogorov-Smirnov test [31], p-value > 0.05). This finding suggests that the observed monophyletic relationship between chromalveolates and Plantae is non-random, and not biased by a secondary or tertiary association between chromalveolates and the other lineages. The strong association between chromalveolates and Bacteria (33.6%) in the dataset that excluded Plantae genomes can be explained by the presence of cyanobacterial genes, which have originated *via* primary EGT (most of which are of plastid function). The (cyano)bacterial association with diatom genes can therefore be explained by endosymbiosis and not by other scenarios that involve LGT from prokaryotes.

3. EGT of red algal genes in *Thalassiosira pseudonana*

We observe 124 (1.1% of the total 11,390) protein-coding sequences from the genome of *T. pseudonana* to have a red algal origin. The phylogenetic trees built with each of these genes and their respective homologs show monophyly of the red algae and chromalveolates with bootstrap support $\geq 75\%$. The genes are clustered into 80 putative families (Table 1). Among these gene families, 40 (50.0%) are well-annotated with gene ontologies (complete annotation for $\geq 90\%$ of the sequences in each family), whereas 18 (22.5%) are partially annotated (complete annotation for $< 90\%$ of the sequences in each family). The remaining 22 (27.5%) are either incompletely annotated or have no significant match in the gene ontology database. We consider these 22 gene families to encode novel, unknown functions in the diatom.

The majority of genes from *T. pseudonana* in each of these families is primarily represented by single-copy sequences (58, 72.5%), with some containing two (14, 17.5%) or three (6, 7.5%) gene copies. There are two families in which the gene is highly duplicated within the genome of *T. pseudonana*. These are the ABC-1 domain protein (7 copies) and light-harvesting protein (13 copies). As shown in the last column of Table 1, 23 (28.8%) of the 80 gene families putatively code for proteins targeted to the plastid, 21 (26.3%) putatively code for proteins targeted to multiple organelles with the majority going to the plastid, 19 (23.8%) of the proteins are potentially targeted to multiple organelles with the minority being the plastid, whereas the remainder (17, 21.3%) putatively code for proteins that are not targeted to the plastid. In parallel with gene ontology analysis, we do not observe a N-terminal extension in the bacterial homologs of these 17 eukaryotic gene families, suggesting that these genes are not targeted to membrane-bounded organelles. The families in which the gene copy is highly duplicated in *T. pseudonana* are found to be targeted to multiple organelles in the cell (including the mitochondrion and nucleus) and are not restricted to the plastid.

Table 1: Gene families showing a red algal origin in *T. pseudonana*. The number of genes from the species in each family is shown. Indication whether a family encodes for a putative plastid-targeted proteins is shown in the last column, based on GO annotations of cellular components for each family: completely plastid-targeted (+++), targeted to multiple membrane-bounded organelles with majority to plastid (++), targeted to multiple membrane-bounded organelles with minority being plastid (+), and not targeted in plastid at all (-).

No.	ID	Description	No. of genes in *T. pseudonana*	Plastid-targeted (+/-)
1	49	bile acid:sodium symporter	3	+++
2	33	sodium hydrogen exchanger	3	+++
3	15	ATP-dependent CLP protease proteolytic subunit	2	+++
4	21	HAD-superfamily hydrolase subfamily variant 3	2	+++
5	12	protease Do	2	+++
6	24	unknown protein	2	+++
7	63	2-c-methyl-d-erythritol 4-phosphate cytidylyltransferase	1	+++
8	17	3-dehydroquinate synthase	1	+++
9	50	aspartate aminotransferase	1	+++
10	34	aspartate kinase	1	+++
11	31	carboxyl-terminal protease	1	+++
12	57	fkbp-type peptidyl-prolyl cis-transisomerase	1	+++
13	67	glycosyl transferase group 1	1	+++
14	54	GTP pyrophosphokinase	1	+++
15	39	monogalactosyldiacylglycerol synthase	1	+++
16	52	serine acetyltransferase	1	+++
17	56	small drug exporter protein	1	+++
18	45	sulfolipid (UDP-sulfoquinovose) biosynthesis protein	1	+++
19	41	tRNA pseudouridine synthase a	1	+++
20	44	unknown protein	1	+++
21	53	unknown protein	1	+++
22	78	unknown protein	1	+++
23	81	unknown protein	1	+++
24	4	light-harvesting protein	13	++
25	8	ABC-1 domain protein	7	++
26	27	phosphoglycolate phosphatase precursor	2	++
27	5	trehalose-6-phosphate synthase	2	++
28	3	ABC family transporter	1	++
29	7	ATP-dependent RNA helicase	1	++
30	32	cysteinyl-tRNA synthase	1	++
31	61	cytochrome C peroxidase	1	++
32	48	dihydrodipicolinate reductase	1	++
33	69	methionyl aminopeptidase	1	++
34	64	peptidyl-prolyl cis-transcyclophilin type	1	++
35	66	RNA polymerase sigma factor	1	++
36	72	thioredoxin-1	1	++

continued on next page...

Table 1 – Continued

No.	ID	Description	No. of genes in *T. pseudonana*	Plastid-targeted (+/-)
37	28	translation elongation factor g	1	++
38	14	unknown protein	1	++
39	18	unknown protein	1	++
40	22	unknown protein	1	++
41	26	unknown protein	1	++
42	42	unknown protein	1	++
43	76	unknown protein	1	++
44	75	valyl-tRNA synthetase	1	++
45	55	peroxisomal membrane protein	3	+
46	23	unknown protein	3	+
47	16	zinc finger protein	3	+
48	62	histone deacetylase family protein	2	+
49	11	hypothetical protein	2	+
50	51	phosphate phosphoenolpyruvate translocator precursor	2	+
51	43	protein phosphatase 2c related protein	2	+
52	9	ABC transporter related protein	1	+
53	46	cell division protein	1	+
54	74	DNA topoisomerase VI subunit a	1	+
55	73	elongation factor 1 alpha	1	+
56	60	GTP binding protein	1	+
57	30	HAD superfamily (subfamily ig) 5-nucleotidase	1	+
58	20	heat shock protein 90	1	+
59	37	homogentisate solanesyltransferase	1	+
60	80	NADH dehydrogenase	1	+
61	68	ribosomal protein s7	1	+
62	19	unknown protein	1	+
63	79	unknown protein	1	+
64	35	p-ATPase family transporter: cation	3	-
65	10	anion exchange family protein	2	-
66	40	prolyl-tRNA synthase	2	-
67	2	unknown protein	2	-
68	6	unknown protein	2	-
69	38	amine oxidase	1	-
70	59	chromodomain helicase DNA binding protein	1	-
71	71	DNA topoisomerase VI subunit b	1	-
72	36	glucose-6-phosphate isomerase	1	-
73	70	glycerol-3-phosphate dehydrogenase (NAD+)	1	-
74	65	HSP associated protein like	1	-
75	47	s-adenosyl-l-homocysteine hydrolase	1	-
76	1	unknown protein	1	-
77	25	unknown protein	1	-
78	29	unknown protein	1	-
79	58	unknown protein	1	-
80	77	unknown protein	1	-

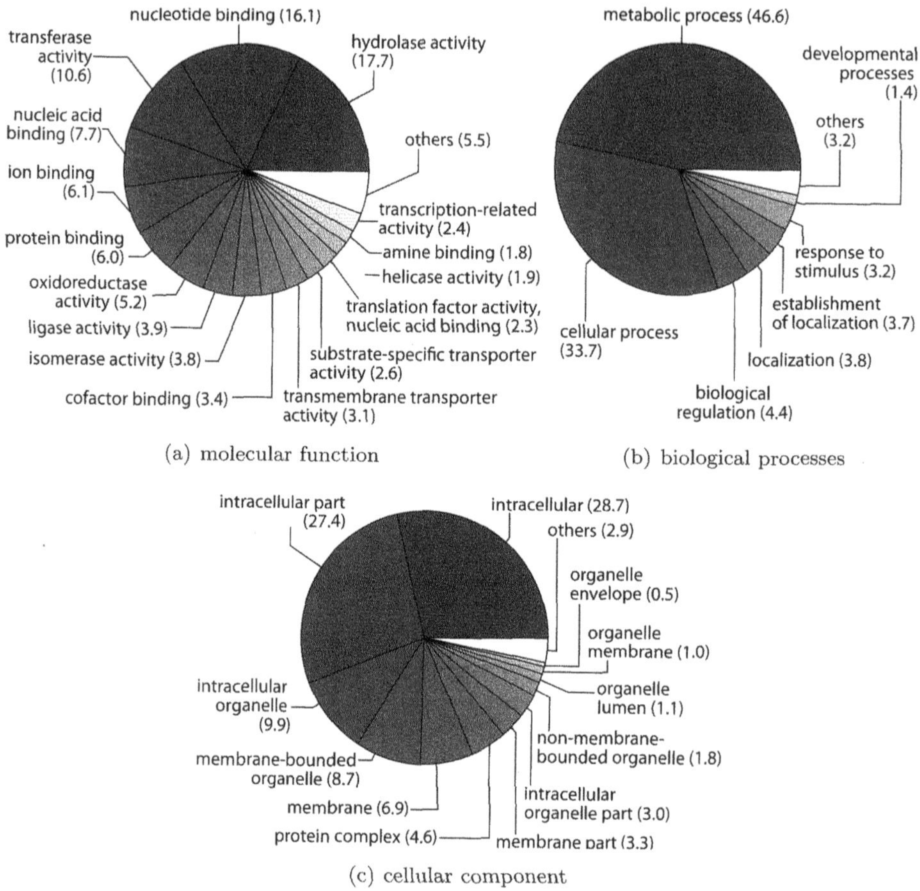

Fig. 3. Gene ontology (GO) annotations of all homologous sequences in the 80 gene families that show support for red algal origin in *T. pseudonana*. Annotations is shown for the classes (a) molecular function at GO level 3; (b) biological process at GO level 2; (c) cellular component at GO level 3. The numbers shown are in percentage.

Figure 3 shows the gene ontology annotations for all homologous sequences from the 80 gene families, for each class of (a) molecular function, (b) biological process and, (c) cellular component. As shown in the panels (a) through (c), the families are of diverse functions that are involved in a variety of biological processes and the encoded proteins are targeted to various compartments within the cell. The gene functions range from biomolecule-binding, transporters, to catalytic activities. Most of these genes are annotated to engage in metabolic processes, whereas some are related to cellular, regulatory, and localization processes.

3.1. *Examples of EGT in chromalveolates*

Figure 4 and Figure 5 shows three examples of EGT of red algal genes into the nucleus of chromalveolates.

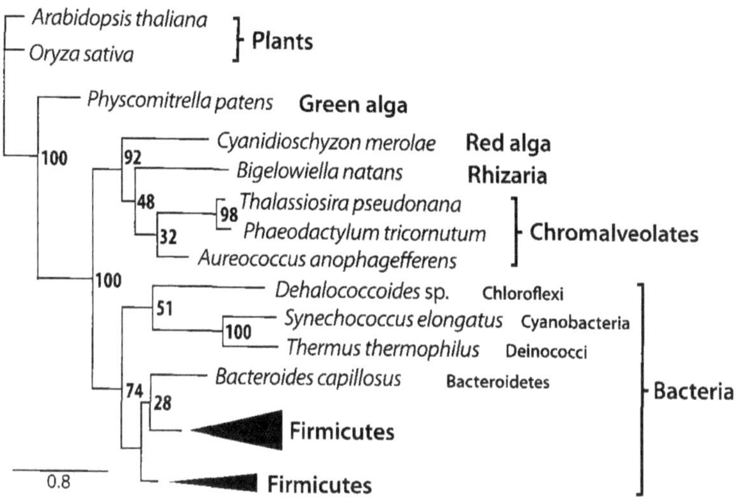

Fig. 4. A maximum likelihood phylogeny showing an example of EGT of an annotated plastid-targeted protein from red algae to *T. pseudonana* (monophyly support for chromalveolates and red algae). Numbers shown are bootstrap support values for each node. The scale bar is shown in unit of substitution per site.

Figure 4 is the phylogeny of a gene family that putatively encodes plastid-targeted small drug exporter proteins, showing strong bootstrap support (92%) for monophyly of an RRC group: a red alga, *Cyanidioschyzon merolae*, a Rhizaria, *Bigelowiella natans*, and three species of chromalveolates, including *T. pseudonana*. In the absence of genetic transfer, the red algae and Rhizaria would be sister taxa to the green algae. This phylogeny implies EGT between the ancestral lineage of the red algae to the ancestral lineage of chromalveolates. In addition, the RRC grouping also forms a monophyletic relationship with all gene copies present in bacteria (bootstrap support 100%), suggesting that the transferred gene is of an ancient bacterial origin. The observation supports the notion of plastid endosymbiosis that plastids in chromalveolates originated from red algae, which in turn are of a cyanobacterial origin.

In contrast, Figure 5 shows the phylogenies of (a) a plastid-targeted gene family and (b) a non-plastid-targeted gene famaily of unknown (and likely novel) functions. In the gene phylogeny shown in Figure 5(a), three species of red algae form the sister taxa with three species of chromalveolates rather than with the green algae. The monophyly of red algae and chromalveolates is strongly supported at bootstrap support 100%. Although the gene function is unknown, this family putatively encodes proteins targeted only to plastids and might therefore play roles in the process of photosynthesis. For the gene phylogeny shown in 5(b), homologous sequences are absent in a large number of lineages. A non-EGT explanation would involve many gene loss events along a large number of lineages. The most parsimonious explanation for such a gene phylogeny is an EGT event from the ancestral lineage of the red alga *Cyanidioschyzon merolae* to the ancestral lineage of the chromalveolates.

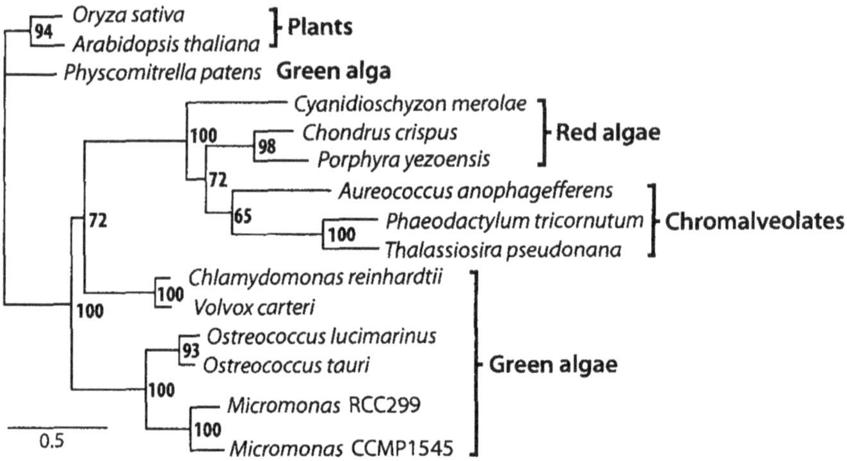

(a) Gene family ID 81, plastid-targeted

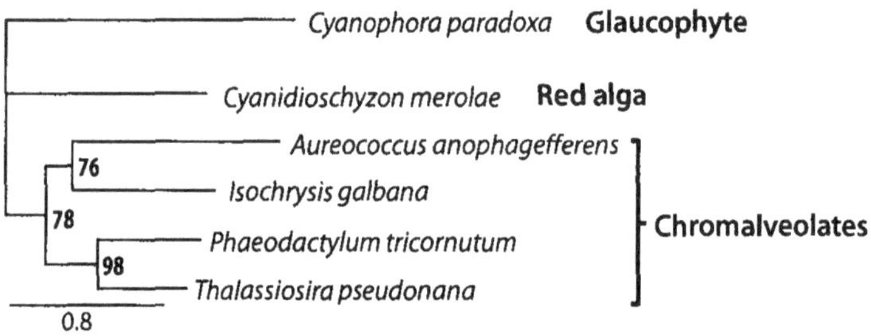

(b) Gene family ID 58, non-plastid-targeted

Fig. 5. Two maximum likelihood phylogenies showing EGT of red algal genes in *T. pseudonana* (monophyly support for chromalveolates and red algae). The genes are of unknown function for (a) a plastid-targeted gene family and (b) a non-plastid-targeted gene family. Numbers shown are bootstrap support values for each node. The scale bars are shown in unit of substitution per site.

4. Performance and limitations

We have demonstrated the use of a rigorous, computational phylogenomic approach to infer the events of gene transfer within the context of plastid endosymbiosis. Our approach is based on the implicit assumption that genes are transferred as a whole. The transfer of genes in smaller fragments, which introduces within-gene discrepancies of phylogenetic signal, might not be fully recovered using this approach. In addition, the efficiency of detecting phylogenetic signal can also be compromised by sequence divergence, presence or absence of informative and/or invariant sites. Therefore, the extent of genetic transfer inferred in this study is a conservative estimate.

In the current study, our approach shows a low false positive discovery rate of 1.23% (*e.g.*, trees that return the incorrect monophyly of chromalveolates and animals). In a preliminary study, we generated simulated eight-taxon protein sets (sample size = 100, sequence length = 1000 amino acids) that are evolved homogeneously at various degrees of sequence conservation. Our phylogenomic approach yielded 0% false positive in recovering the target monophyletic relationships (data not shown), with 0.17% false negative rate in cases where sequences are highly divergent (average substitution per site = 2). Under a more-realistic evolutionary regime, *e.g.*, heterogeneous evolution with varied substitution rates along the same or different lineages, the false positive and negative rates are expected to be higher.

Based on bioinformatic predictions and analysis at a high statistical (bootstrap) confidence, our findings suggest that genes that show a history of EGT from red algae into *T. pseudonana* extend beyond plastid-related (*e.g.*, photosynthetic) functions, and thus these transferred genes might make a much greater impact in genome innovation of *T. pseudonana* than previously thought. Nevertheless, the extent of such an impact in plastid endosymbiosis remains to be verified by experimental approaches. The current approach is suitable for an high-throughput detection of whole-gene transfer within broader biological contexts at a multi-genome scale.

5. Authors' contributions

AM designed and implemented the phylogenomic pipeline, conducted the phylogenomic analysis and contributed to the preparation of the manuscript draft. CXC conducted downstream functional analysis of the gene families, wrote and prepared the table, figures, and the manuscript draft. Both AM and CXC contributed to the analysis of the results. MD, DZ, HA, NJ and TS conducted gene-by-gene phylogenetic analysis to validate the results from the pipeline. DB conceived of and supervised this study. AM, CXC and DB conceived, edited and approved the final manuscript.

6. Acknowledgments

This work was supported by a grant from the National Institutes of Health (R01ES013679) awarded to DB. We acknowledge the intellectual input of Adrián Reyes-Prieto and Valérie Reeb (University of Iowa) in this project.

References

[1] R. G. Beiko, T. J. Harlow and M. A. Ragan, *Proc. Natl. Acad. Sci. U.S.A.* **102**, 14332 (2005).
[2] E. Lerat, V. Daubin, H. Ochman and N. A. Moran, *PLoS Biology* **3**, Art. e130 (2005).
[3] T. Nosenko and D. Bhattacharya, *BMC Evol. Biol.* **7**, Art. 173 (2007).
[4] D. Bhattacharya and T. Nosenko, *J. Phycol.* **44**, 7 (2008).
[5] V. M. D'Costa, K. M. McGrann, D. W. Hughes and G. D. Wright, *Science* **311**, 374 (2006).
[6] D. Bhattacharya, H. S. Yoon and J. D. Hackett, *Bioessays* **26**, 50 (2004).
[7] G. I. McFadden, *J. Phycol.* **37**, 951 (2001).

[8] T. Cavalier-Smith, *J. Eukaryot. Microbiol.* **46**, 347 (1999).
[9] A. Reyes-Prieto, A. P. M. Weber and D. Bhattacharya, *Ann. Rev. Genet.* **41**, 147 (2007).
[10] D. Bhattacharya, J. M. Archibald, A. P. M. Weber and A. Reyes-Prieto, *Bioessays* **29**, 1239 (2007).
[11] S. B. Gould, R. F. Waller and G. I. McFadden, *Annu Rev Plant Biol* **59**, 491 (2008).
[12] J. D. Hackett, H. S. Yoon, M. B. Soares, M. F. Bonaldo, T. L. Casavant, T. E. Scheetz, T. Nosenko and D. Bhattacharya, *Curr. Biol.* **14**, 213 (2004).
[13] J. D. Hackett, H. S. Yoon, S. Li, A. Reyes-Prieto, S. E. Rummele and D. Bhattacharya, *Mol. Biol. Evol.* **24**, 1702 (2007).
[14] A. Reyes-Prieto, A. Moustafa and D. Bhattacharya, *Curr Biol* **18**, 956 (2008).
[15] D. M. Nelson, P. Tréguer, M. A. Brzezinski, A. Leynaert and B. Quéguiner, *Global Biogeochem. Cycl.* **9**, 359 (1995).
[16] M. A. Brzezinski, C. J. Pride, V. M. Franck, D. M. Sigman, J. L. Sarmiento, K. Matsumoto, N. Gruber, G. H. Rau and K. H. Coale, *Geophys. Res. Lett.* **29**, 1564 (2002).
[17] E. V. Armbrust, J. A. Berges, C. Bowler, B. R. Green, D. Martinez, N. H. Putnam, S. G. Zhou, A. E. Allen, K. E. Apt, M. Bechner, M. A. Brzezinski, B. K. Chaal, A. Chiovitti, A. K. Davis, M. S. Demarest, J. C. Detter, T. Glavina, D. Goodstein, M. Z. Hadi, U. Hellsten, M. Hildebrand, B. D. Jenkins, J. Jurka, V. V. Kapitonov, N. Kröger, W. W. Y. Lau, T. W. Lane, F. W. Larimer, J. C. Lippmeier, S. Lucas, M. Medina, A. Montsant, M. Obornik, M. S. Parker, B. Palenik, G. J. Pazour, P. M. Richardson, T. A. Rynearson, M. A. Saito, D. C. Schwartz, K. Thamatrakoln, K. Valentin, A. Vardi, F. P. Wilkerson and D. S. Rokhsar, *Science* **306**, 79 (2004).
[18] J. A. Eisen and C. M. Fraser, *Science* **300**, 1706 (2003).
[19] J. Huang, G. S. V. Aller, A. N. Taylor, J. J. Kerrigan, W. S. Liu, J. M. Trulli, Z. Lai, D. Holmes, K. M. Aubart, J. R. Brown and M. Zalacain, *J. Bacteriol.* **188**, 5249 (2006).
[20] U. John, B. Beszteri, E. Derelle, Y. V. de Peer, B. Read, H. Moreau and A. Cembella, *Protist* **159**, 21 (2008).
[21] S. F. Altschul, W. Gish, W. Miller, E. W. Myers and D. J. Lipman, *J Mol Biol* **215**, 403 (1990).
[22] R. C. Edgar, *Nucl. Acids Res.* **32**, 1792 (2004).
[23] N. Saitou and M. Nei, *Mol. Biol. Evol.* **4**, 406 (1987).
[24] A. Stamatakis, *Bioinformatics* **22**, 2688 (2006).
[25] A. Moustafa and D. Bhattacharya, *BMC Evol. Biol.* **8**, Art. 6 (2008).
[26] M. Matsuzaki, O. Misumi, I. T. Shin, S. Maruyama, M. Takahara, S. Y. Miyagishima, T. Mori, K. Nishida, F. Yagisawa, Y. Yoshida, Y. Nishimura, S. Nakao, T. Kobayashi, Y. Momoyama, T. Higashiyama, A. Minoda, M. Sano, H. Nomoto, K. Oishi, H. Hayashi, F. Ohta, S. Nishizaka, S. Haga, S. Miura, T. Morishita, Y. Kabeya, K. Terasawa, Y. Suzuki, Y. Ishii, S. Asakawa, H. Takano, N. Ohta, H. Kuroiwa, K. Tanaka, N. Shimizu, S. Sugano, N. Sato, H. Nozaki, N. Ogasawara, Y. Kohara and T. Kuroiwa, *Nature* **428**, 653 (2004).
[27] S. Whelan and N. Goldman, *Mol. Biol. Evol.* **18**, 691 (2001).
[28] A. Conesa, S. Götz, J. M. García-Gómez, J. Terol, M. Talón and M. Robles, *Bioinformatics* **21**, 3674 (2005).
[29] P. Horton, K. J. Park, T. Obayashi, N. Fujita, H. Harada, C. Adams-Collier and K. Nakai, *Nucl. Acids Res.* **35**, W585 (2007).
[30] I. Small, N. Peeters, F. Legeai and C. Lurin, *Proteomics* **4**, 1581 (2004).
[31] F. J. Massey, *J. Am. Stat. Assoc.* **46**, 68 (1951).

CIS-REGULATORY ELEMENT BASED GENE FINDING: AN APPLICATION IN ARABIDOPSIS THALIANA

YONG LI[a,1] YANMING ZHU[a,1*] YANG LIU[2]
yong@neau.edu.cn ymzhu2001@neau.edu.cn liuyang@cuhk.edu.hk

YONGJUN SHU[1] FANJIANG MENG[3] YANMIN LU[3]
syjun@neau.edu.cn fjmeng@neau.edu.cn luyanmin@neau.edu.cn

BEI LIU[3] XI BAI[1] DIANJING GUO[2*]
liubei@neau.edu.cn maixi@neau.edu.cn djguo@cuhk.edu.hk

[1] *Plant Bioengineering Laboratory, Northeast Agricultural University, Harbin, China*
[2] *State Key Lab for Agrobiotechnology and Department of Biology, The Chinese University of Hong Kong, Shatin, N.T., Hong Kong*
[3] *Department of Computer Science, Northeast Agricultural University, Harbin, China*

[a] *These authors contributed equally to this work*

[*] *Corresponding author*

Abstract

Using cis-regulatory motifs known to regulate plant osmotic stress response, an artificial neural network model was built to identify other functionally releted genes involved in the same process. The rationale behind our approach is that gene expression is largely controlled at the transcriptional level through the interactions between transcription factors and cis-regulatory elements. Gene Ontology enrichment analysis on the 500 top-scoring predictions showed that, 60% of the enriched GO classification was related to stress response. RT-PCR analysis showed that nearly 70% of the top-scoring predictions exhibited altered expression under various stress treatments. We expect that similar approach is widely applicable to infer gene function in various cellular processes in different species.

Keywords: Artificial Neural Network; Gene Expression; Gene Finding; Cis-regulatory element; *Arabidopsis thaliana*

1. Introduction

Gene expression is largely controlled at the transcriptional level, where the interactions between transcription factors (TFs) and cis-regulatory elements in the promoter region of a gene play crucial roles [6]. Previous research suggests that functional related genes tend to be co-regulated by similar sets of transcription factors. Therefore, using cis-regulatory motifs are known to regulate gene expression in certain cellular process, one can identify other functionally relevant genes involved in the same process. When combined with experimental verification, this has been proved to be an effective approach to genome-wide targeted gene identification [28].

Drought, high salinity, and low temperature are three major osmotic stresses that adversely affect plant growth, development, or productivity. Osmotic stress elicits a dehydration response in plants that shares many common elements and interacting signaling pathways [5, 6, 28], which have been suggested to be Abscisic Acid (ABA) dependent [20]. Subsequent analysis of the ABA-regulated gene promoter region has led to the identification of several ABA-responsive elements (ABREs) [7, 12]. Zhang et al. [28] reported a computational approach to identifying putative ABA responsive genes using conserved ABA-responsive element (ABRE) and its coupling element (CE). Using similar cis-element based approach, promoters that contains known binding motifs were used for targeted gene finding in *Drosophila melanogaster* [13] and *C. elegans* [24]. Despite the proved success, the previous researchers all used one or two specifically defined motifs for gene screening. In fact, a growing body of evidence suggests that functional related genes tend to be regulated by a common set of regulatory proteins to form namely transcription regulatory modules, in order to respond to internal and external signals. By organizing the genome into such modules, a living cell can coordinate the activities of many genes and carry out complex functions [25]. For gene function inference in complex cellular process such as stress response, more sophisticated approaches are required.

Identification of genes that specifically respond to internal and external cues remains one of the most compelling yet elusive areas in computational genomics. Currently the commonly used gene finding approach is consensus-based comparative analysis that relies on sequence homology among genes in closely related species [27]. Such method has limited application because a large portion of those sequenced genomes still remain uncharacterized. Furthermore, such consensus-based method may not be efficient for identification of genes that are induced under specific environmental stimuli. In this study, we applied an Artificial Neural Network (ANN) modeling approach [8, 12, 16, and 17] to plant functional genomics and identified genes respond to osmotic stress in *A. thaliana*. We demonstrate its efficacy by Gene Ontology enrichment analysis as well as by RT-PCR analysis.

2. Materials and Methods

2.1. *Stress Response Genes and Cis-regulatory Elements*

Cis-regulatory elements in the promoter region of drought, salinity, and/or cold stress responsive genes were collected from public database PLACE [9, 29], PlantCARE [18, 32], and DoOP [2]. Other motifs were collected through literature-mining approach. The redundant motifs were eliminated and in total 55 cis-acting elements were collected for further analysis. A bioperl module was used to search for significant motifs occurred in the promoter region. P-value was calculated to confirm the significance of motif detection (Poisson distribution [19]).

2.2. *Promoter Sequences*

Arabidopsis genome sequences were downloaded from TAIR [33]. Transcription start site (TSS) was predicted using TSSP-TCM software from Shahmuradov's group [35].

When multiple TSSs were predicted, the one closest to the ORF was chosen. For each given TSS, we retrieved a segment from 500 bases upstream to 20 bases downstream of the TSS for motif analysis. In total, the TSSs of 18061 ORFs were retrieved.

2.3. *Scoring algorithms*

A Bioperl module was used to search for significant motifs occurred in the promoter region of reported stress responsive genes. P-value was calculated to confirm the significance of motif detection. The ANN toolkit in Matlab was used to establish a feed-forward cascade neural network model. For network training and simulation, we retrieved the promoter region of 362 genes annotated as "response to drought, high salinity, or cold stress" according to Gene Ontology terminology [30, 31] and used these as positive dataset. The promoter sequences of a randomly selected 1086 ORFs (3 fold of positive dataset) from the rest of the gene pool (not annotated as "response to stress or ABA treatment" according to GO) were used as negative dataset. The number of times each cis-regulatory element appears in the promoter region and the ratio of cis-element length to promoter length (we defined it as coverage) were taken as inputs for the network training. Principle component analysis was conducted to eliminate the input node with least effect.

2.4. *Gene Expression Data Analysis and GO Enrichment*

Microarray gene expression data was collected from AtGenExpress [32]. The dataset include global Arabidopsis transcriptome profile change over UV-B light, high salinity, drought and cold stress responses. The raw data was normalized using RMAExpress [32, 33] and differentially expressed genes were detected using BRB ArrayTools [34] ($p<0.05$). The subset of differentially expressed genes contains 3276 gene transcripts identifiers with significantly higher abundance at least once per treatment and per time-point. We performed GO term enrichment analysis using software suite DAVID 2007 (The **D**atabase for **A**nnotation, **V**isualization and **I**ntegrated **D**iscovery 2007) [35]. The enrichment analysis is indeed to compare the annotation composition in the analyzed gene list to that of population background genes. DAVID default population background, which is the corresponding genome-wide genes with at least one annotation in the analyzing categories, was used in enrichment calculation.

2.5. *Plant Materials, stress treatment, and RT-PCR analysis*

Four-week-old seedlings grown in a controlled environment growth chamber under a 16 hr light/8 hr dark period, a photo fluency rate of 3000 lux, and a temperature of 22 °C. For salinity stress treatment, seedlings were subjected to 250 mM NaCl and time series samples were collected at 4hr, 12hr, and 24hr respectively. For cold stress treatment, seedlings were incubated at 4°C under darkness condition for 24hr, 48hr, and 72hr respectively. For drought treatment, seedlings were subjected to drought for 24hr, 48hr, and 72hr. Total RNA was extracted from whole seedlings of the control plants and stressed plants using RNeasy Plant Mini Kit (Qiagen,Germany). cDNA was synthesized using super script II Kit (Invitrogen) following the manufacturer's instruction. PCR was performed using 0.5–2 μl of the cDNA in a total of 25 μl reaction volume and carried out at 94◦C for 2 min, 29 cycles of 94◦C for 1 min, 58◦C for 1 min and 72◦C for 1 min, and then 72◦C for 5 min. Expression analysis of each gene was confirmed in at least two independent RT-reactions using forward and reverse primers.

3. Results and Discussions

3.1. *In silico Gene Identification and Gene Ontology Analysis*

Using cis-regulatory motifs known to regulate osmotic stress response, an artificial neural network model was built to identify other relevant genes involved in the same process. The trained model was able to distinguish between genes that do and do not respond to stress, based on the motif patterns of gene promoters in the training dataset. We then applied the model to the candidate dataset to infer the function of the unknown genes.

According to network theory [1, 23], genes within co-expression context often share conserved biological functions. To investigate the significant functional annotation of our predictions, we selected the 500 top ranking genes predicted by ANN model and performed Gene Ontology (GO) enrichment analysis (Table 1). GO provides a controlled vocabulary for describing genes and gene products in living organism [19]. We used terms from "Biological Process" (GOBP), which is one of the three broad GO categories (the other two being "Molecular Function" and "Cellular Component"), to represent gene function. GOBP terms are organized into a directed acyclic graph (DAG) to reflect the hierarchical relationships between the terms. Parent GOBP terms are subdivided into increasingly specific child GOBP terms. This GOBP term: "reponse to stress" has 19 child terms, such as *GO:0009409 [response to cold]*, *GO:0009408 [response to heat]*, and *GO:0009414 [response to water deprivation]*, etc. The GO enrichment analysis is indeed to compare the annotation composition in the analyzed gene list to that of population background genes. We used the DAVID default population background in enrichment calculation, which is the corresponding genome-wide genes with at least one annotation in the analyzing categories. The default background is a good choice for the studies in genome-wide scope or close to genome-wide scope.

According to GO enrichment analysis, except for the un-annotated ORFs (~40%), about 60 %of the significantly enriched GO classification was related to stress response or ABA response. In fact, ABA plays a protective role in plant response to osmotic stress [15, 21, 26] and a large number of genes respond to abiotic stress are also inducible by ABA treatment [15]. The GO enrichment analysis clearly demonstrated that our regulatory motif based computational model is a reliable means for gene function inference at the genome level.

Table 1. GOBP description of top predictions using ANN model

GOBP term	No. of Genes	P-Value
photosynthesis	9	1.32E-04
cold acclimation	5	2.49E-04
response to abscisic acid stimulus	10	6.45E-04
response to water	7	0.006
response to temperature stimulus	10	0.007
response to cold	7	0.008
response to water deprivation	6	0.013
development	28	0.015
response to hormone stimulus	18	0.015
response to abiotic stimulus	34	0.02
response to salt stress	5	0.034
seed development	7	0.035
response to chemical stimulus	25	0.035
reproductive structure development	7	0.039
actin cytoskeleton organization	4	0.041
response to stimulus	44	0.042
embryonic development	6	0.047
response to stress	24	0.054
cytoskeleton organization	7	0.057
response to endogenous stimulus	20	0.057
response to osmotic stress	5	0.072
reproduction	8	0.078
carbohydrate biosynthesis	8	0.084
cellular response to water	2	0.09

3.2. *Cross Validation Using Gene Expression Profiling Data*

Since many functionally related genes display coordinated transcriptional regulation, large-scale gene expression measurements can therefore serve as a check point for our *in silico* prediction. Comparison of our prediction with stress microarray data from AtGenExpress revealed that about 30% of the top-scoring gene transcripts were

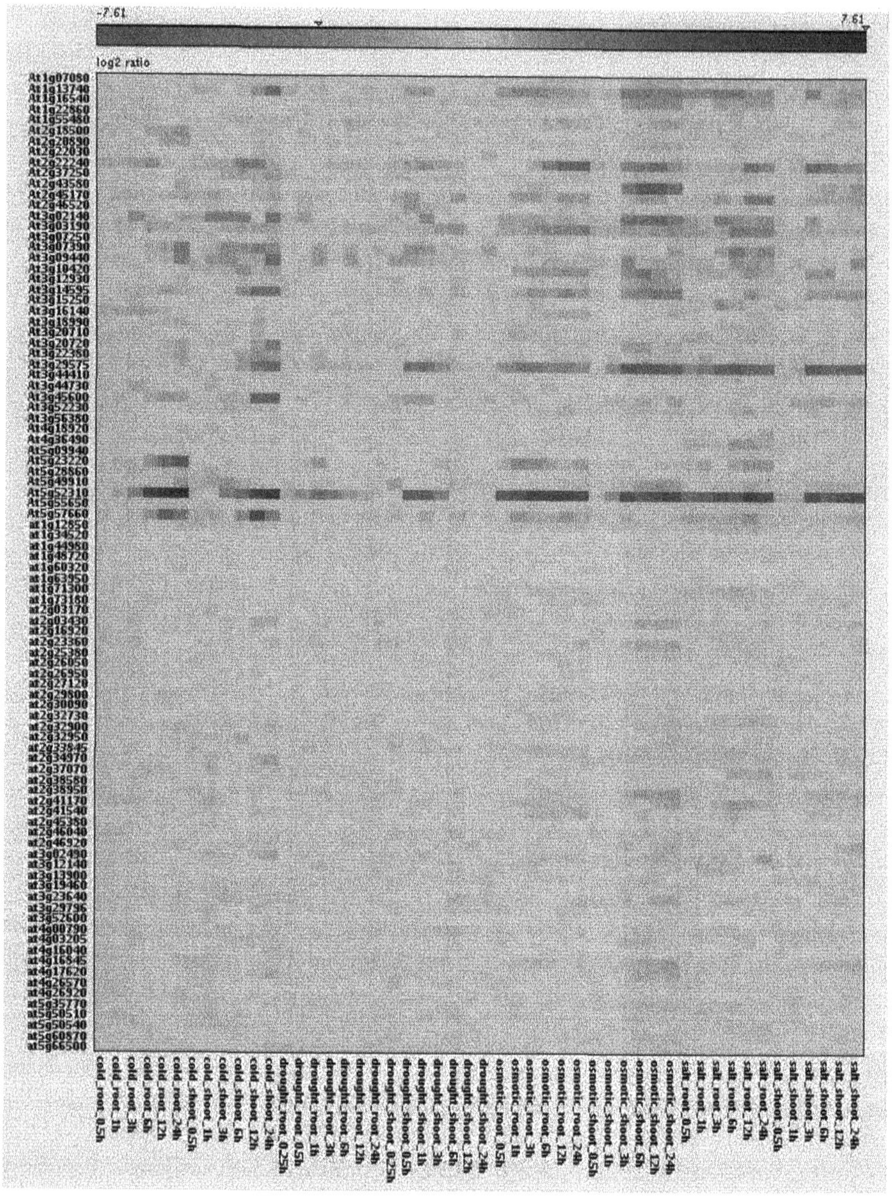

Figure 1. Comparative E-northern analysis of top scoring genes (upper part of the image) vs. randomly selected genes (lower part of the image). High scoring genes and randomly selected genes can be distinguished by different IDs ("At-----" for high scoring genes and "at------" for randomly selected gene). Each row in the heat map is a gene, and each column is an experimental condition, such as high salinity, cold, or drought stress etc. The color at a point represents the log2 of the ratio of the average of replicate treatments relative to the average of corresponding controls.

significantly changed upon stress treatment (p-value ≈ 0). It is well-known that microarray gene expression data, although powerful in providing global transcriptome information, is highly noisy and discontinuous. Some genes that are rapidly and transiently induced by stress may not be detected by microarray analysis. At1g16540,which encodes molybdenum cofactor sulfurase, was predicted as stress response gene by our method. Although direct measurement of its transcript abundance is lacking, previous research indicate that this protein may play important role in regulating many stress relevant genes including *RD29A, COR15, COR47, RD22*, and *P5CS* [20, 21]. However, this gene was not differentially expressed in published Arabidopsis stress microarray data. Similarly, another top-scoring gene At2g23430 (*GO: 0009737*), which encodes a cyclin-dependent kinase inhibitor known to respond to ABA stimulus [22], was also not detected in the stress microarray data. Furthermore, gene expression at the mRNA level does not always reflect the protein function due to post-transcriptional and post-translational regulation. Our method may serve as a complimentary approach to gene expression analysis in stress response gene identification.

To provide visual evidence supporting our computational predictions, we generated a E-Northern heat map (Fig. 1) for 41 top scoring genes (upper part of the image) and 41 randomly selected genes beyond the prediction list (lower part of the image) using the Expression Browser tool of the Botany Array Resource (BAR) [22] and the AtGenExpress Stress data. From E-Northern analysis, we observed significant up-regulation and down-regulation of many of our predictions. The up- and down-regulation were also observed among the genes in the training data (data not shown). In contrast, randomly selected genes show much less change upon stress treatment. Since a gene that shows significant alteration at transcript level under stress conditions is likely to be involved in response to stress, we conclude that regulatory motif in the promoter region is highly indicative of gene function and can therefore be used for *in silico* gene identification. The rapid development of cis elements databases, such as TRANSFAC [14], PLACE [9, 29] and PlantCARE [18, 32], provides valuable sources that can be used to identify more cis-regulatory motifs relevant to cellular response to various internal and external stimuli.

3.3 *Experimental Validation and Comparison with Other Methods*

To further assess our method, the Arabidopsis seedlings were subjected to various stressful conditions (method) and the transcript abundance of 41 top-scoring predictions were monitored by RT-PCR analysis (Fig. 2). Gene-specific primers were designed based on the cDNA full-length sequence. Overall, 27 of 41 tested genes showed altered level of expression compared to control, giving a prediction accuracy of 65.8% for the top scoring predictions. The full list of the RT–PCR results is provided in supplementary file. Some of the tested genes exhibited varied expression patterns in different types of stress treatment, suggesting possible synergistic effects of multiple transcription factors involved in various type of osmotic stress. We believe that identification of distinctive class of transcription-factor binding motifs under specific stress condition will facilitate more efficient computational discovery of stress relevant genes. Such knowledge is also important in understanding the cross-talks between distinctive signaling pathways and the underlying regulatory machinery of cellular stress response.

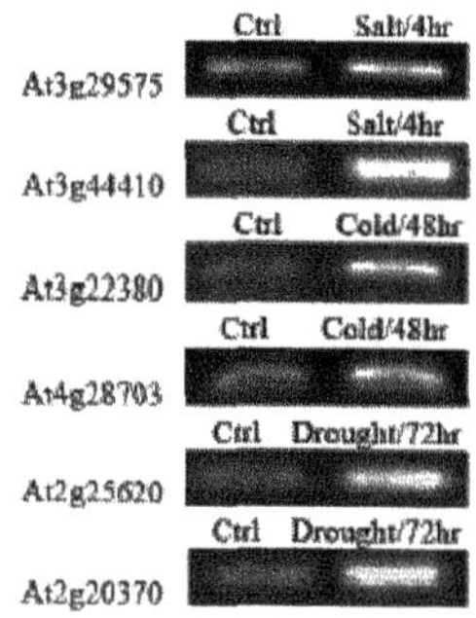

Figure2. RT-PCR analysis of selected predictions. For salinity stress treatment, 4-week old Arabidopsis seedlings were subjected to 250 mM NaCl and samples were collected at 4hr time point. For cold stress and drought stress, seedlings were incubated at 4°C and 22 °C respectively in the dark and samples were collected at 48hr, and 72hr repectively. Samples were also collected from control plants grown under the same condition for parallel comparison. Actin was used as loading control and loading was estimated by staining the gel with ethidium bromide. Expression analysis of each gene was confirmed in at least two independent RT-reactions using forward and reverse primers.

Cis-regulatory element based gene identification has been reported previously [13, 20, 24]. These researchers utilized one or two well defined cis-regulatory motifs to search for functional related genes and achieved varied prediction accuracy (ranging from 34% to 72%). As a comparison, we expand the motif list and used a set of diversed regulatory motifs (55 in total) identified from promoters of both experimentally validated and computational predicted stress responsive genes to train an ANN model. The learned model achieved a comparable prediction accuracy of 67.8% for the top scoring predictions in our study. Our results, together with those from other groups, further demonstrate that cis-regulatory motifs are highly indicative of gene function. With more information of transcription factors and their DNA binding information becoming freely available, we anticipate this computational approach can be widely used for gene function inference in different organisms.

4. Conclusions

In this study, we present a cis-regulatory element based ANN modeling method for genome wide gene function prediction in Arabidopsis thaliana. By explicitly utilizing the information of transcription regulation of known gene and cis-regulatory motifs, our method gives reliable result with high prediction accuracy. We demonstrate this is a practical and compelling means for genome wide gene identification. We anticipate that identification of more condition-specific cis-regulatory motifs and further understanding of the synergistic effects of different regulatory motifs will facilitate more efficient computational discovery of stress relevant genes. One promising aspect of the approach as applied in this study is in its potential use for gene finding in various cellular processes in different organisms. The software codes are available upon request.

Acknowledgments

This project was supported by grant from National Natural Science Foundation of China (30570990), and in part by grants from the Science and Technology Department of Heilongjiang province and Hong Kong UGC/AoE Plant & Agricultural Biotechnology Project AoE-B-07/09.

References

1 Barabasi, A.L. and Oltvai Z.N., Network biology: understanding the cell's functional organization, *Nat Rev Genet*, 5(2):101-113, 2004.

2 Barta, E., Sebestyén, E., Pálfy, T.B., Tóth, G., Ortutay, C.P. and Patthy L., DoOP: Databases of Orthologous Promoters, collections of clusters of orthologous upstream sequences from chordates and plants, *Nucl. Acids Res.*, 33:D86-D90 2005.

3 Bittner, F., Oreb, M. and Ralf, Mendel, R.R., ABA3 Is a Molybdenum Cofactor Sulfurase Required for Activation of Aldehyde Oxidase and Xanthine Dehydrogenase in *Arabidopsis thaliana*, *The Journal of Biological Chemistry*, 276(44):40381-40384, 2001.

4 Bolstad, B.M., Irizarry, R.A., Astrand, M. and Speed, T.P., A Comparison of Normalization Methods for High Density Oligonucleotide Array Data Based on Bias and Variance, *Bioinformatics*, 19(2):185-193, 2003.

5 Bray, E.A., Molecular responses to water deficit, *Plant Physiol.*, 103:1035–1040, 1993.

6 Brivanlou, A. and Darnell, J., Signal transduction and the control of gene expression. *Science*, 295:813-818, 2002.

7 Busk, P.K. and Pages, M., Regulation of abscisic acid-induced transcription, *Plant Mol. Biol.*, 37:425-435, 1998.

8 Demeler, B. and Zhou, G., Neural Network Optimization for *E. coli* Promoter Prediction, *Nucleic Acids Research*, 19:1593-1599, 1991.

9 Higo, K., Ugawa, Y., Iwamoto, M. and Korenaga, T., Plant cis-acting regulatory DNA elements (PLACE) database, *Nucleic Acids Research*, 27:297-300, 1999.

10 Irizarry, R.A., Bolstad, B.M., Collin, F., Cope, L.M., Hobbs, B. and Speed, T.P., Summaries of Affymetrix GeneChip probe level data, *Nucleic Acids Research*, 31(4):e15, 2003.

11 Kilian, J., Whitehead, D., Horak, J., Wanke, D., Weinl, S., Batistic, O., D'Angel, C., Bauer, E.B., Kudla, J. and Harter K., The AtGenExpress global stress

expression dataset: protocols, evaluation and model data analysis of UV-B light, drought and cold stress responses, *The Plant Journal*, 50(2):347-363, 2007.

12 Mahadevan, I. and Ghosh, I., Analysis of *E. coli* Promoter Structures Using Neural Networks, *Nucleic Acids Research*, 22:2158-2165, 1994.

13 Markstein, M. et al., Genome-wide analysis of clustered Dorsal binding sitesidentifies putative target genes in the Drosophila embryo, *Proc. Natl Acad. Sci. USA*, 22:763–768. 2002.

14 Matys, V., TRANSFAC: transcriptional regulation, from patterns to profiles, *Nucleic Acids Res.*, 31:374-378, 2003.

15 Narusaka, Y., Nakashima, K., Shinwari, Z.K., Sakuma, Y., Furihata, T., Abe, H., Narusaka, M., Shinozaki, K. and Yamaguchi-Shinozaki, K., Interaction between two cis-acting elements, ABRE and DRE, in ABA-dependent expression of Arabidopsis rd29A gene in response to dehydration and high-salinity stresses, *Plant J*, 34:137-148, 2003.

16 Pedersen, A. G. and Nielsen H., Neural network prediction of translation initiation sites in eukaryotes, *ISMB*, 5:226-233, 1997.

17 Reese, M.G., Application of a time-delay neural network to promoter annotation in the Drosophila melanogaster genome, *Comput Che.*, 26(1):51-56. 2001.

18 Rombauts, S., Déhais, P., Van Montagu, M. and Rouz,é P., PlantCARE, a plant cis-acting regulatory element database, *Nucleic Acids Ressearch*, 27(1):295-296, 1999.

19 Shah, N.H., King, D.C., Shah, P.N. and Fedoroff N.V., A tool-kit for cDNA microarray and promoter analysis, *Bioinformatics*, 19(14):1846-1848, 2003.

20 Shinozaki, K., and Yamaguchi-Shinozaki, K., Molecular responses to dehydration and low temperature: difference and cross-talk between two stress signaling pathways, *Current Opinion in Plant Biology*, 3:217-223, 2000.

21 Shinozaki, K., Yamaguchi-Shinozaki, K., and Seki, M., Regulatory network of gene expression in the drought and cold stress responses, *Current Opinionin Plant Biology*, 6:410-417, 2003.

22 Toufighi, K., Brady, S.M., Austin, R., Ly, E. and Provart, N.J., The Botany Array Resource: e-Northerns, Expression Angling, and promoter analyses, *Plant J.*, 43(1):153-63, 2005.

23 Ubramanian, A., Tamayo, P., Mootha, V.K., Mukherjee, S., Ebert, B.L., Gillette, M.A., Paulovich, A., Pomeroy, S.L., Golub, T.R., Lander, E.S. and Mesirov, J.P., From the Cover: Gene set enrichment analysis: A knowledge- based approach for interpreting genome-wide expression profiles, *Proceedings of the National Academy of Sciences*, 102(43):15545-15550, 2005.

24 Wenick, A. and Hobert, O., Genomic cis-regulatory architecture and trans-acting regulators of a single interneuron-specific gene battery in *C.elegans.*, *Cell*, Dev. 6, 2004.

25 Wu, W.S., Li, W.H. and Chen, B.S., Computational reconstruction of transcriptional regulatory modules of the yeast cell cycle, BMC *Bioinformatics*, 7:421, 2006.

26 Xiong, L. and Zhu, J.K., Molecular and genetic aspects of plant responses to osmotic stress, *Plant, Cell and Environment*, 25:131-139, 2002.

27 Zhang, M., Computational prediction of eukaryotic protein-coding genes, *Nat. Rev. Genet.*, 3:698-709, 2002.
28 Zhang, W.X., Ruan, J.H., Ho, T.H., You, Y.S., Yu, T.T. and Quatrano, R.S., Cis-regulatory element based targeted gene finding: genome-wide identification of abscisicacid-and abiotic stress-responsive genes in *Arabidopsis thaliana*, *Bioinformatics*, 21(14):3074-3081, 2005.
29 A Database of Plant Cis-acting Regulatory DNA Elements: http://www.dna.affrc.go.jp/PLACE/.
30 BRB ArrayTools: http://linus.nci.nih.gov/BRB-ArrayTools.html.
31 Gene Ontology: http://www.geneontology.org/.
32 PlantCARE: http://bioinformatics.psb.ugent.be/webtools/plantcare/html/.
33 The Arabidopsis Information Resource (TAIR): http://www.arabidopsis.org.
34 The Gene Ontology Consortium. Gene Ontology: tool for the unification of biology, *Nature Genet.*, 25:25-29, 2000.
35 TSSP-TCM software: http://mendel.cs.rhul.ac.uk/mendel.php?topic=genom
36 DAVID: http://david.abcc.ncifcrf.gov/
37 Wang H, Qi Q, Schorr P, Cutler AJ, Crosby W.L., and Fowke LC., ICK1, a cyclin-dependent protein kinase inhibitor from Arabidopsis thaliana interacts with both Cdc2a and CycD3, and its expression is induced by abscisic acid. *Plant Journal*, 15(4):501-10, 1998

USING SIMPLE RULES ON PRESENCE AND POSITIONING OF MOTIFS FOR PROMOTER STRUCTURE MODELING AND TISSUE-SPECIFIC EXPRESSION PREDICTION

ALEXIS VANDENBON[1]
alexvdb@hgc.jp

KENTA NAKAI[1,2,3]
knakai@ims.u-tokyo.ac.jp

[1] *Department of Medical Genome Sciences, Graduate School of Frontier Sciences, University of Tokyo, 4-6-1 Shirokanedai, Minato-ku, Tokyo 108-8639, Japan*
[2] *Institute of Medical Science, University of Tokyo, 4-6-1 Shirokanedai, Minato-ku, Tokyo 108-8639, Japan*
[3] *Institute for Bioinformatics Research and Development (BIRD), Japan Science and Technology Agency, 5-3 Yonbancho, Chiyoda-ku, Tokyo 102-0081, Japan*

Regulation of transcription is controlled by sets of transcription factors binding specific sites in the regulatory regions of genes. It is therefore believed that regulatory regions driving similar expression profiles share some common structural features. We here introduce a computational approach for finding a small set of rules describing the presence and positioning of motifs in a set of promoter sequences. This rule set is subsequently used for finding promoters that drive similar expression profiles from a genomic set of sequences. We applied our approach on muscle-expressed genes in *Caenorhabditis elegans*. We obtained a high average performance, and in the best case we found that almost 50% of true positive test genes scored higher than 90% of the true negative test genes. High scoring non-training sequences were enriched for muscle-expressed genes, and predicted motifs fitting the rules showed a significant tendency to be present in experimentally verified regulatory regions. Our model is more general than existing *cis*-regulatory module models, as rules selected by our model contain a variety of information, including not only proximal but also distal positioning of pairs of motifs, positioning with regard to the translation start site, and simply presences of motifs. We believe our model can help to increase our understanding about transcription factor cooperation and transcription initiation.

Keywords: regulation of transcription, promoter modeling, *C. elegans*, muscle tissue.

1. Introduction

Initiation of transcription in eukaryotes is regulated by the binding of proteins to *cis*-regulatory motifs present in the regulatory sequences of genes. As a first step in the regulation of gene expression, regulation of transcription is of major importance in determining when, where (e.g. in which tissues), and under what conditions a gene is expressed. Since this process is controlled by transcription factors (TFs) binding motifs in the regulatory sequences, we can make the assumption that regulatory regions containing similar sets of motifs are bound by similar sets of these TFs, and will drive similar expression profiles. Thus, the study of regulatory regions is of key importance in the elucidation of the process of transcription initiation.

One difficulty lies within the nature of the transcription factor binding sites (TFBSs) themselves: in general they are short (typically 6 to 15 bps), degenerate motifs, which causes biologically meaningless matches to occur frequently just by chance. Several approaches for detection and *de novo* prediction of TFBSs have been introduced [23]. In order to improve the specificity of the prediction of functional *cis*-regulatory elements, a number of restrictions can be considered. An often used restriction is to limit the sequence search space to parts that are evolutionary conserved. The idea behind this is that biologically meaningful patterns in the regulatory regions are subjected to a more stringent evolutionary conservation than the surrounding non functional parts. A second popular restriction is to focus attention to local clusters of motif sites. In this case the idea is that proximally located clusters of sites might be bound by sets of cooperating proteins. The local clusters of sites are referred to as *cis*-regulatory modules (CRMs), and are thought to be typically a few hundred to 1000 bps in length, containing 2 to 15 binding sites for about 1 to 8 different TFs per CRM [1]. Several successful applications have made use of these restrictions [6, 15].

However useful the above restrictions might be, they are not absolute requirements for a predicted site to be functional. A TF does not check whether a site is preserved in other species before deciding whether or not to bind it. In addition, thanks to the bendability of the DNA, even pairs of sites located very distantly in the sequence can come into close proximity of each other and can be bound by cooperating TFs. The use of the restrictions mentioned above might thus result in the exclusion of functional sites.

Recently, some studies have shown great progress in describing and modeling the structures of regulatory regions [8, 10]. A number of studies have addressed the problem of modeling the structures of promoters that drive a specific expression pattern. Segal *et al.* used a thermodynamic model to predict expression patterns as a function of TFBSs and expression of their corresponding TFs in *Drosophila* [20]. Vandenbon *et al.* used a probabilistic model taking into account positioning, orientation, and order of motifs, and their predictions in *Ciona intestinalis* muscle tissues were experimentally shown to be partially correct [24]. The method proposed by Beer and Tavazoie uses a set of constraints on *cis*-regulatory element features incorporated in a Bayesian Network (BN), which are subsequently successfully used to predict expression profiles in yeast [3].

In this study we have followed a similar approach for unraveling the structure of muscle-specific promoters in *C. elegans*. We introduce an approach for modeling promoter architectures from a set of input promoter sequences, and for subsequent prediction of regulatory regions with a similar structure and tissue expression using the trained model. Our model takes into account presence or absence of motifs, positional bias with regard to the translation (or transcription) start sites, and positional preferences between pairs of sites. Not only the presence or absence of each of these patterns is considered, but also their number of presences. We applied our models on a dataset of muscle-expressed sequences that has been used successfully before in the study of *C. elegans* muscle-specific gene regulation [26]. *C. elegans* muscle tissue has been extensively studied, and a number of muscle-specific genes have been identified. Analysis of their regulatory regions has lead to the identification of important regulatory regions and candidate TF binding motifs [12, 13, 26]. Extensive tissue expression data is available for this nematode, making it a useful organism for the study of tissue-specific transcriptional regulation. After training the model on muscle-specific promoter sequences, we scored the genomic set of promoters. Using a set of true positive (TP) and true negative (TN) test sequences we constructed an ROC curve, and obtained AUC values of up to 0.76, with almost 50% of the TP test sequences scoring higher than 90% of the TN test sequences, which outperforms a previous study on the same dataset [26]. We showed that the high scoring non-input promoters are enriched for promoters driving expression in muscle tissues. Finally, we found that motifs fitting the rules in our model show a significant tendency to be present in regions that have previously been shown to be of importance in driving expression in muscle tissue.

2. Methods

2.1. *Sequences and Expression Data*

In this study a number of sequence datasets were used for different purposes. First of all, as a set of true positive (TP) genes we used 121 sequences that have been used before in the study of *C. elegans* muscle tissue transcriptional control [26]. For 78 of these genes, orthologs have been reported in *C. briggsae*. These 78 genes were used as positive training set in our approach. The remaining 43 genes were used as a true positive test set. A test set of true negative (TN) genes was made by removing all genes reported to be expressed in muscle tissues from the expression data present in WormBase (WS188; http://www.wormbase.org/), which gave us a set of 2955 genes. As negative training set we used randomly selected sets of genes from the genomic set, excluding any true positive genes and test genes. As sequence data we used the 2000 bps upstream of the translation start site for each gene. Repeats were masked using RepeatMasker (version 3.1.9; http://www.repeatmasker.org). The positive training dataset of 78 genes was further divided into 2 sets; one for motif prediction and generation of rules, and one for rule selection (see Supplemental material Fig. 1 and further explanation below). This division was done randomly.

2.2. *Motif Prediction and Selection*

In the motif prediction and rule generation sequence dataset we predicted motifs in conserved parts of the promoter sequences using a number of motif prediction programs. Predictions were done both on the entire masked sequences, and on sub-regions of different lengths. All predicted motifs were collected, converted to PWMs, and their Over-Representation Index was calculated as described by Bajic *et al.* [2]. PWMs with an ORI value less than 2 were discarded and redundancy in the remaining motif set was removed (See Supplemental material for a more detailed description). Finally, the genomic set of promoter sequences was scanned for sites of the selected set of PWMs.

2.3. *Rule Types and the Scoring Procedure*

In a next step we identified patterns concerning the presence and positioning of motifs in the input sequences that might be useful for distinguishing muscle-expressed genes from genes not expressed in muscle. In our approach we considered three types of rules. A first type of rules concerns the positioning of TFBSs relative to the translation start site ("absolute positioning rules"). A second type concerns the positioning of pairs of TFBSs relative to each other ("relative positioning rules"). Finally, we considered that some TFs might not be restricted by positioning relative to the translation start site or to other TFs, and that thus just the presence of their binding site wherever inside the promoter region is meaningful. Therefore we included a third type of rules, the "presence rule" type. Figure 2 shows some examples of rules.

Hits of rules were defined as follows: for "absolute positioning rules" concerning a certain motif, each occurrence of the motif in question within the promoter region described in the rule is a hit to the rule. For "relative positioning rules", each pair of sites present in the relative positioning described in the rule is a hit to the rule. In the case of the "presence rule" type, each motif occurrence of the motif in question is a hit to the rule. From the number of hits for each rule the final score of a gene was calculated. We tried four different scoring functions (See Supplemental material). In this paper we will focus on the standard scoring function, in which the score of a sequence to a set of rules is equal to the sum of the hits to each individual rule.

2.4. *Rule Generation*

Keeping the above in mind, the construction of "presence rules" is trivial: we simply add the presence rule for each motif in our motif set to the rule set. However, the "absolute positioning rule" and "relative positioning rule" types are more difficult to train. Below we focus on the training of the "absolute positioning rules".

To evaluate whether a motif *m* shows a significant preference for a certain region in the input sequence we scanned the promoter regions with windows of different window sizes, at different positions. Each region defined by a window at a certain position corresponds to a rule describing the positioning of the sites of motif *m* relative to the translation start site. Over-representation of motif *m* in this region is calculated using an adapted version of the ORI value we used to measure over-representation of PWMs. The strong point of this measure for over-representation is that it not only takes into account the number of hits, but also the fraction of sequences containing at least one hit. Equation 1 shows the formula we used to calculate the ORI value of rules:

$$ORI_i = \frac{density_{TP,i}}{density_{genomic,i}} \times \frac{proportion_{TP,i}}{proportion_{genomic,i}} \quad (1)$$

where i is the index for the rule being considered, *density* stands for the number of hits for a rule per unit of sequence length, *proportion* stands for the ratio of sequences in a sequence set having at least one hit for a rule, and *TP* and *genomic* indicate the TP training set and the genomic training set used as negatives, respectively.

The significance of each ORI_i value was evaluated using the distribution of one million ORI values obtained using "shuffled" training sets. These sets are constructed by shuffling the positions of the sites for all motifs randomly over a set of sequences with the same length and number as the training sequence set. Rules with significant ORI values were retained. We tried the following p-value threshold values: 0.001, 0.005, 0.01, and 0.05. For each motif m typically several regions passed the p-value thresholds. In a second step, to avoid unnecessarily wide regions to be selected we introduced a threshold on the ORI value level. Rules with an ORI value lower than $ORI_{threshold,i} = ORI_{max,i} \times 0.80$ were discarded. For each motif m the regions that passed the above restrictions are merged together to form one, non-redundant rule concerning the positioning of the motif (See Supplemental material for more details).

The "relative positioning rules" were trained in a similar way (See Supplemental material for more details).

2.5. *Rule Selection*

Despite the thresholds we introduced to increase the quality of the generated rules, we still end up with a fairly high number of rules (depending on the p-value threshold from on average about 20 for $p = 0.001$ to on average about 70 for $p = 0.05$), a considerable part of which might still be biologically meaningless. We used the remaining true positive training set to select from this larger set a subset of biologically meaningful rules, which can subsequently be used for detecting new muscle-expressed genes.

We used a Genetic Algorithm (GA) approach in which set of individuals are generated with in their "chromosomes" a set of "genes". Each rule is represented by one "gene" and the GA algorithm subsequently proceeds to find the "fittest" individual (the best set of rules), over a number of generations using mutations and crossovers. As a measure for fitness the Area Under the Curve (AUC) of Receiver Operator Characteristic (ROC) curves was used. For each set of rules encountered during the search, the ROC curve can be easily constructed based on the scores of the genes in the true positive rule selection training set and a set of genomic sequences as negatives. The AUC value of the curve is calculated and used as a measure of the performance of the rule subset. The final set of selected rules (typically 5 to 15 rules) represents our promoter structure model.

2.6. *Performance Evaluation*

As an evaluation of the performance the genomic set of promoters was scored using all rules in our final selection, and an ROC curve was constructed using the TP and TN test sets. The AUC of this curve is one measure of performance. A second measure is the percentage of TP test sequences scoring higher than 90% or 95% of the TN test sequences. Other indications of the relevance of our selected rules are the enrichment of muscle-expressed genes among the top scoring genes, and the overlap of predicted motif sites fitting the rules with known regions of importance in regulation of expression in muscle tissues.

Table 1. An overview of the performance of the model is shown. For 4 different rule generation p-value s the average AUC, median AUC, and the average percentage of TP test sequences scoring higher than 95% and 90% of TN test sequences is shown. These results are based on 10 bootstrap runs.

Rule generation p-value	Average AUC	Median AUC	Average percentage of TP test sequences scoring higher than 95% of TNs	Average percentage of TP test sequences scoring higher than 90% of TNs
0.05	0.7086	0.7085	22.8	36.4
0.01	0.7175	0.7230	27.2	36.5
0.005	0.7197	0.7156	27.9	36.3
0.001	0.7210	0.7218	25.7	35.9

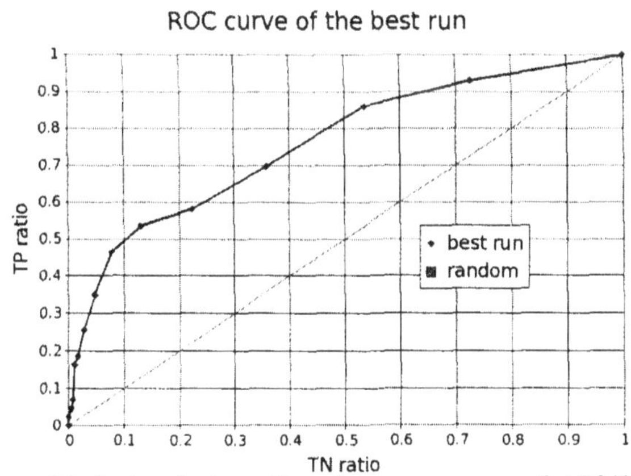

Figure 1. The ROC curve of the best run is shown. From the curve we van see that 36.1% of TP test genes score higher than 95% of TN test genes. 49.5% of TPs score better than 90% of TN sequences.

3. Results and Discussion

3.1. Overall Performance

Table 1 shows an overview of the results. As can be seen from the table, our model has a high performance, with average AUC values in the range of 0.70 to 0.73. The top scoring genes are highly enriched for TP test genes. For rule generation p-value threshold 0.01, the rule sets contain on average 9.2 rules (range 5 to 14). On average 27.2% of TP test sequences (range 18,1% to 36,1%) score higher than 95%, and 36.5% (range 20,0% to 49,5%) higher than 90% of the TN test sequences. These results were obtained using 50 training sequences for motif prediction and rule generation, and 28 training sequences for rule selection, and the standard scoring procedure (see Methods section). Using different allocations for both training tasks, and using different scoring schemes did not improve results (Supplemental material). In the further discussion we will focus on the run with the highest AUC value among the runs with p-value 0.01.

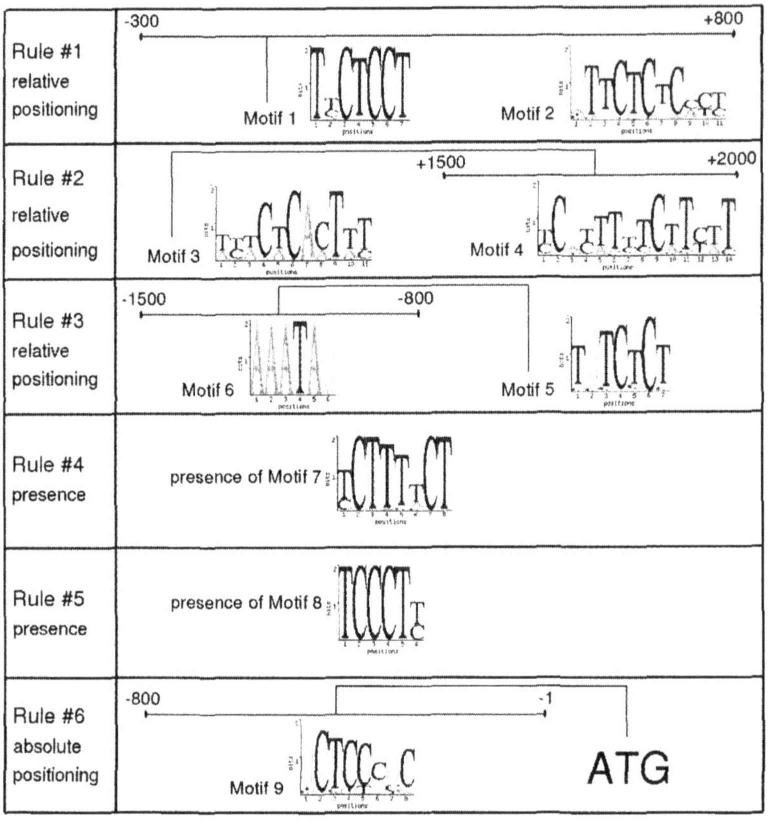

Figure 2. A visual representation of the 6 selected rules in the best run is shown. For each rule the motif logo of the motif(s) and the nature of the rule is shown.

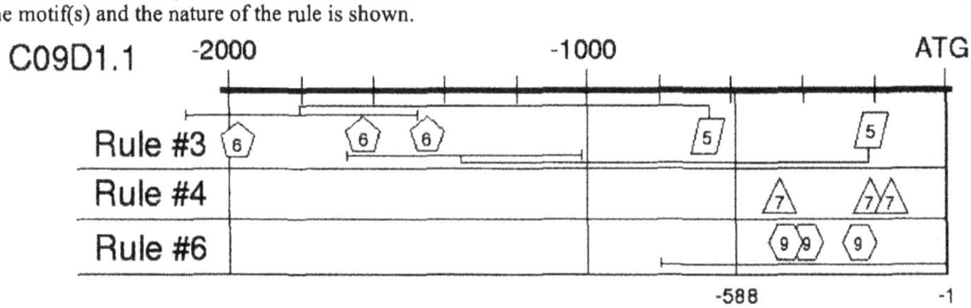

Figure 3. An example of a comparison between the predicted motifs fitting the rules and experimentally verified regulatory regions is shown. The 2000 bps upstream of the translation start site (indicated at the right side) are shown. For gene C09D1.1 the verified region is from -1 to -588 [13], as indicated in grey. Numbered symbols indicate predicted sites fitting the rule restrictions, the number indicating the motif number as shown in Fig. 2. The presence of distal Motif 6 sites outside the verified region might indicate an additional regulatory region.

3.2. Focus on the Best Run

In the best run an AUC value of 0.7601 was obtained, with about 36.1% of TP test sequences score higher than 95% of the TN test sequences. As much as 49.5% of the TP test sequences score higher than 90% of the TN test sequences (see Fig. 1). Given our definition of the TN test set, it is possible that some of the high scoring TN test sequences are in fact false negatives in the sense that their expression in muscle tissue might not be known yet. Figure 2 shows the 6 rules that were selected in the best run. Note that 3 rules are relative positioning rules, 2 are presence rules, and 1 is a rule on absolute positioning. Two of the relative positioning rules describe relatively distal positioning, one proximal positioning of pairs of motifs. In this way, a small set of rules can contain various kinds of structural information, and achieve high performance.

When we compared the positions of motif sites fitting the selected rules with the regulatory regions that have been experimentally verified to be of biological importance for muscle tissue expression in 16 upstream sequences, we found that sites fitting the rule descriptions are preferentially present in experimentally confirmed regulatory regions (*p*-value: 0.0017). Figure 3 shows an example of a comparison between predicted motif sites fitting the rules and experimentally verified regulatory regions driving expression in muscle tissue (see Supplemental material for more figures).

Figure 4 shows the motif logos of 4 motifs that were present in the rule set, as well as those of known PWMs that show a certain degree of similarity. Motifs 2 and 5 are similar to a motif that has been previously reported in a study on the regulation of *C. elegans* muscle genes [12, 13, 26]. In our study Motif 5 is found together with Motif 6 in a relative positioning rule. Motif 6 shows similarity to myocyte enhancer factor 2 (MEF2). MEF2 has been shown to play a role in muscle development in *Drosophila* and vertebrates, in cooperation with myogenic proteins [5, 17]. Its role in *C.elegans* is however less clear and it has been hypothesized that CeMEF2 might have adopted a divergent role in *C. elegans* [9, 11]. Nevertheless, MEF2 expression is reported in neurons and muscle cells. Motif 8 shows similarities to the EBF and TFII-I motifs. TFII-I factors have been reported to be regulators of muscle genes in human [18], however, it is unclear whether or not *C. elegans* possesses a homolog of this protein. EBF on the other hand has a known homolog in *C. elegans*, named unc-3. This gene plays a role in the development of sensory neurons and is expressed in motor neurons, amphid neurons, ventral cord, and other neural tissues [14, 19]. In other sets similar motifs were found.

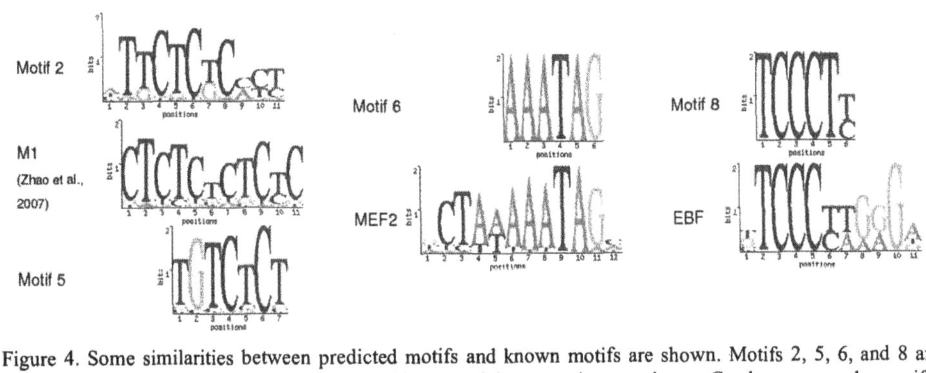

Figure 4. Some similarities between predicted motifs and known motifs are shown. Motifs 2, 5, 6, and 8 are motifs predicted in this study. M1 is a motif reported in a previous study on *C. elegans* muscle-specific expression regulation. The MEF2 and EBF motif logos were taken form the TRANSFAC database [16].

Since our model was trained on promoter sequences of muscle-expressed genes we expected the top scoring non-training sequences to be expressed in muscle tissues, too. Table 2 shows the over-represented tissues in the top scoring genes. We found a significant over-representation of genes expressed in muscle tissues like body wall musculature, vulval muscle, etc. On the other hand, expression in a number of neuronal tissues is also over-represented, especially nerve ring, ventral cord neuron, and a number of specific neurons (AVKL, AVKR, RMGL, and RMGR) or the QR neuroblast cell. It has been shown previously that muscle tissues and neuron tissues share a number of TFs [25], so the over-representation of genes expressed in neuronal tissues can be considered as a positive result. In addition to that, many TP training and test sequences are also reported to be expressed in neuronal tissues. The over-representation of genes expressed in gonad or seam cell might indicate that these tissues too share some regulatory mechanism with muscle tissue, or that a number of genes expressed in certain types of muscle tissue tend to be also expressed in gonad or seam cells.

Table 2. This table shows the tissues in which the top scoring non-training sequences are expressed. Since our scoring procedure returns integer values, here the 121st sequence has the same score as the 100th sequence. Therefore, the data shown is based on the top scoring 121 non-training genes. The Wormbase anatomy ID and term is shown, with its observed and expected counts of genes expressed in this tissue, and the corresponding *p*-value. Only terms with at least 10 genes in the total expression annotation data were considered.

Anatomy ID	Anatomy Term	Observed Count	Expected Count	*P*-value
WBbt:0006749	nerve ring	36	17.8	1.76e-5
WBbt:0005813	body wall musculature	47	28.0	7.68e-5
WBbt:0005175	gonad	17	6.4	2.35e-4
WBbt:0005821	vulval muscle	30	16.0	4.44e-4
WBbt:0005753	seam cell	23	11.0	5.19e-4
WBbt:0005300	ventral cord neuron	30	16.5	7.27e-4
WBbt:0003675	muscle cell	9	2.7	1.64e-3
WBbt:0003737	pharyngeal muscle 5	4	0.3	3.78e-3
WBbt:0005796	intestinal muscle	6	0.6	4.28e-3
WBbt:0003845	AVKL	4	1.7	6.99e-3
WBbt:0004054	QR	4	0.8	7.12e-3
WBbt:0003844	AVKR	4	0.8	8.29e-3
WBbt:0005017	RMGL	4	0.8	8.29e-3
WBbt:0005013	RMGR	4	0.1	8.47e-3

4. Concluding Remarks

We have introduced a method for modeling promoter architecture using a simple set of rules on motif presence and positioning, and used it to predict promoter sequences that drive a similar expression as the training promoters. The rules are generated by scanning for preferred positioning of motifs in a training set of sequences compared to both a background set of sequences and sets of "shuffled" training sequences. A small set of highly meaningful rules are selected by a GA, using a second set of training sequences.

We applied the model to muscle-expressed genes in *C. elegans* and obtained a high performance, both in terms of the AUC value of the ROC curve, as well as in the enrichment of TP test sequences in the top scoring sequences. Moreover, we found that high scoring non-training sequences were enriched for muscle-expressed genes. In addition, some interesting similarities between our predicted motifs and known TFBSs were observed, and we found that predicted sites fitting the restrictions described in the rules showed a significant tendency to be present in regions that have been experimentally shown to be of importance for expression in muscle tissues.

It has been suggested that it is not unusual to find 10 to 50 TFBSs for 5 to 15 different factors in a single promoter [1]. We therefore believe the use of AUC values of ROC curves in the selection of the rules to be of special importance, as it allows us to take into account regulatory regions having multiple hits to individual rules. In addition, we would like to mention that in a number of cases (such as the case we discussed in more detail in this paper) our approach was able to outperform previous analyses using only clustering of motifs [26]. Since our approach uses a small set of simple rules containing a variety of information, it might offer more insight into the complex cooperation between TFs and their binding sites, than just the clustering-based methods. In the example we introduced in this paper, too, the set of rules contained information on positioning relative to the translation start site, proximal as well as distal positioning of pairs of sites for motifs, and rules just stressing the presence of motifs. Despite the small number of rules, and their simplicity, we were able to get a high performance (AUC up to 0.75 and more, up to 50% of the TP test sequences in the top 10% scoring sequences). The success of the CRM-based models is thanks to the fact that it is unlikely to find a number of biologically meaningless sites clustered together. Basically, the rules on positioning work in a similar way, except that the positioning does not necessarily have to be proximal and that the translation (or transcription) start site is considered as a reference motif as well. We did not restrict our study to evolutionary conserved regions of the promoter sequences, except in the *de novo* motif prediction step.

Even though we constructed our rules in a way that they are as general as possible, we decided not to use the orientation and the strength of sites in the final application. Orientation of TFBSs has been shown to be of importance in determining the biological function of motif sites [4, 21]. However, the overall importance of orientation of motifs in transcription regulation is still unclear. Brown *et al.* showed how some functional TFBSs could be reversed and still retained functionality in muscle tissues of *Ciona intestinalis* [7]. The score of a sequence against a PWM has been shown to be correlated with the strength by which it is bound by the corresponding TF [22]. However, the complex cooperation between TFs makes it difficult to make a simple statement concerning the strengths of sites and their contribution to transcriptional regulation. In our study too, during preliminary study we did not find an improvement of performance when we included orientation or scores of sites in the model (data not shown). This might hint at a less important role of these two features, or it might indicate that the modeling of these features will require a better model or more training data.

In our model we did not use species specific or tissue-specific information, so we expect it to be universally applicable. Motifs included in the model can be *de novo* predicted motifs, as well as known motifs. Optionally, core promoter elements can be included although they might not necessarily be of importance in tissue-specific regulation. As for performance in *C. elegans*, taking into account alternative splicing, possible regulatory elements in introns, and operon structures is likely to improve accuracy of predictions in the future.

Acknowledgments

The authors would like to thank Dr. Nicolas Sierro and other members of Dr. Nakai Laboratory for helpful discussions and advice. Computation time was provided by the supercomputer system at the Human Genome Center, Institute of Medical Science, University of Tokyo. A.V. is supported by the Japanese Government Scholarship (Monbukagakusho; MEXT).

References

[1] Arnone, M. I. and Davidson, E. H., The hardwiring of development: organization and function of genomic regulatory systems, *Development*, 124:1851-64, 1997.

[2] Bajic, V. B., Choudhary, V. and Hock, C. K., Content analysis of the core promoter region of human genes, *In Silico Biol*, 4:109-25, 2004.

[3] Beer, M. A. and Tavazoie, S., Predicting gene expression from sequence, *Cell*, 117:185-98, 2004.

[4] Berendzen, K. W., Stuber, K., Harter, K. and Wanke, D., Cis-motifs upstream of the transcription and translation initiation sites are effectively revealed by their positional disequilibrium in eukaryote genomes using frequency distribution curves, *BMC Bioinformatics*, 7:522, 2006.

[5] Black, B. L. and Olson, E. N., Transcriptional control of muscle development by myocyte enhancer factor-2 (MEF2) proteins, *Annu Rev Cell Dev Biol*, 14:167-96, 1998.

[6] Blanchette, M., Bataille, A. R., Chen, X., et al., Genome-wide computational prediction of transcriptional regulatory modules reveals new insights into human gene expression, *Genome Res*, 16:656-68, 2006.

[7] Brown, C. D., Johnson, D. S. and Sidow, A., Functional architecture and evolution of transcriptional elements that drive gene coexpression, *Science*, 317:1557-60, 2007.

[8] Carninci, P., Sandelin, A., Lenhard, B., et al., Genome-wide analysis of mammalian promoter architecture and evolution, *Nat Genet*, 38:626-35, 2006.

[9] Dichoso, D., Brodigan, T., Chwoe, K. Y., et al., The MADS-Box factor CeMEF2 is not essential for Caenorhabditis elegans myogenesis and development, *Dev Biol*, 223:431-40, 2000.

[10] Elemento, O., Slonim, N. and Tavazoie, S., A universal framework for regulatory element discovery across all genomes and data types, *Mol Cell*, 28:337-50, 2007.

[11] Fukushige, T., Brodigan, T. M., Schriefer, L. A., Waterston, R. H. and Krause, M., Defining the transcriptional redundancy of early bodywall muscle development in C. elegans: evidence for a unified theory of animal muscle development, *Genes Dev*, 20:3395-406, 2006.

[12] Guhathakurta, D., Schriefer, L. A., Hresko, M. C., Waterston, R. H. and Stormo, G. D., Identifying muscle regulatory elements and genes in the nematode Caenorhabditis elegans, *Pac Symp Biocomput*, 425-36, 2002.

[13] GuhaThakurta, D., Schriefer, L. A., Waterston, R. H. and Stormo, G. D., Novel transcription regulatory elements in Caenorhabditis elegans muscle genes, *Genome Res*, 14:2457-68, 2004.

[14] Kim, K., Colosimo, M. E., Yeung, H. and Sengupta, P., The UNC-3 Olf/EBF protein represses alternate neuronal programs to specify chemosensory neuron identity, *Dev Biol*, 286:136-48, 2005.

[15] Li, L., Zhu, Q., He, X., Sinha, S. and Halfon, M. S., Large-scale analysis of transcriptional cis-regulatory modules reveals both common features and distinct subclasses, *Genome Biol*, 8:R101, 2007.

[16] Matys, V., Kel-Margoulis, O. V., Fricke, E., et al., TRANSFAC and its module TRANSCompel: transcriptional gene regulation in eukaryotes, *Nucleic Acids Res*, 34:D108-10, 2006.

[17] Molkentin, J. D., Black, B. L., Martin, J. F. and Olson, E. N., Cooperative activation of muscle gene expression by MEF2 and myogenic bHLH proteins, *Cell*, 83:1125-36, 1995.

[18] Polly, P., Haddadi, L. M., Issa, L. L., et al., hMusTRD1alpha1 represses MEF2 activation of the troponin I slow enhancer, *J Biol Chem*, 278:36603-10, 2003.

[19] Prasad, B. C., Ye, B., Zackhary, R., et al., unc-3, a gene required for axonal guidance in Caenorhabditis elegans, encodes a member of the O/E family of transcription factors, *Development*, 125:1561-8, 1998.

[20] Segal, E., Raveh-Sadka, T., Schroeder, M., Unnerstall, U. and Gaul, U., Predicting expression patterns from regulatory sequence in Drosophila segmentation, *Nature*, 451:535-40, 2008.

[21] Sharov, A. A., Dudekula, D. B. and Ko, M. S., CisView: a browser and database of cis-regulatory modules predicted in the mouse genome, *DNA Res*, 13:123-34, 2006.

[22] Stormo, G. D. and Fields, D. S., Specificity, free energy and information content in protein-DNA interactions, *Trends Biochem Sci*, 23:109-13, 1998.

[23] Tompa, M., Li, N., Bailey, T. L., et al., Assessing computational tools for the discovery of transcription factor binding sites, *Nat Biotechnol*, 23:137-44, 2005.

[24] Vandenbon, A., Miyamoto, Y., Takimoto, N., Kusakabe, T. and Nakai, K., Markov Chain-based Promoter Structure Modeling for Tissue-specific Expression Pattern Prediction, *DNA Res*, 15:3-11, 2008.

[25] Wasserman, W. W. and Fickett, J. W., Identification of regulatory regions which confer muscle-specific gene expression, *J Mol Biol*, 278:167-81, 1998.

[26] Zhao, G., Schriefer, L. A. and Stormo, G. D., Identification of muscle-specific regulatory modules in Caenorhabditis elegans, *Genome Res*, 17:348-57, 2007.

IMPROVING GENE EXPRESSION CANCER MOLECULAR PATTERN DISCOVERY USING NONNEGATIVE PRINCIPAL COMPONENT ANALYSIS

XIAOXU HAN
xiaoxu.han@emich.edu

Department of Mathematics and Bioinformatics Program, Eastern Michigan University, Ypsilanti MI, 48197 USA

Robust cancer molecular pattern identification from microarray data not only plays an essential role in modern clinic oncology, but also presents a challenge for statistical learning. Although principal component analysis (PCA) is a widely used feature selection algorithm in microarray analysis, its holistic mechanism prevents it from capturing the latent local data structure in the following cancer molecular pattern identification. In this study, we investigate the benefit of enforcing non-negativity constraints on principal component analysis (PCA) and propose a nonnegative principal component (NPCA) based classification algorithm in cancer molecular pattern analysis for gene expression data. This novel algorithm conducts classification by classifying meta-samples of input cancer data by support vector machines (SVM) or other classic supervised learning algorithms. The meta-samples are low-dimensional projections of original cancer samples in a purely additive meta-gene subspace generated from the NPCA-induced nonnegative matrix factorization (NMF). We report strongly leading classification results from NPCA-SVM algorithm in the cancer molecular pattern identification for five benchmark gene expression datasets under 100 trials of 50% hold-out cross validations and leave one out cross validations. We demonstrate superiority of NPCA-SVM algorithm by direct comparison with seven classification algorithms: SVM, PCA-SVM, KPCA-SVM, NMF-SVM, LLE-SVM, PCA-LDA and k-NN, for the five cancer datasets in classification rates, sensitivities and specificities. Our NPCA-SVM algorithm overcomes the over-fitting problem associative with SVM-based classifications for gene expression data under a Gaussian kernel. As a more robust high-performance classifier, NPCA-SVM can be used to replace the general SVM and k-NN classifiers in cancer biomarker discovery to capture more meaningful oncogenes.

Keywords: Nonnegative principal component analysis (NPCA)

1. Introduction

With the recent development of genomics and proteomics, molecular diagnostics has appeared as a novel tool to diagnose cancers. It picks a patient's tissue, serum or plasma samples and uses DNA chips or mass spectrometry (MS) based proteomics techniques to generate gene/protein expressions of these samples. The gene/protein expressions reflect gene/protein activity patterns in different types of cancerous or precancerous cells: i.e., they are molecular patterns or molecular signatures of cancers. Different cancers have different molecular patterns and the molecular patterns of a normal cell will be different from those of a cancer cell. In modern oncology, clinicians more and more rely on the robust classifications of gene/protein expression patterns to identify cancerous tissues and find their corresponding biomarkers. However, it is still a challenge for oncologists and computational biologists to robustly classify cancer molecular patterns because of the special characteristics of gene/protein expression data. In this study, we mainly focus on the cancer molecular pattern identification for gene expression data.

Gene expression data are characterized by high dimensionalities. It can be represented by a $n \times m$ matrix after preprocessing. Each row represents the expression levels of a gene across different biological samples; each column represents the gene expression levels of a genome under a sample. Usually $n >> m$, i.e., the number of variables (genes) is much greater than the number of biological samples. The number of samples is <200 and the number of genes > 5000 generally. These data are not noise-free because their raw data contain a large amount of systematic noise and preprocessing algorithms can not remove them completely. Although there are a large amount of variables in these data, only a small set of variables have meaningful contributions to data variations.

Many feature selection algorithms are employed to reduce gene expression data dimensions before further classification/clustering analysis [1,2,3,4]. Principal component analysis (PCA) may be the mostly used approach among them. It projects data in an orthogonal subspace generated by the eigenvectors of a data covariance or correlation matrix. The data representation in the subspace is uncorrelated and the maximum variance directions based subspace spanning guarantees the least information loss in feature selection. However, as a holistic feature selection algorithm, PCA can only capture the features contributing to the global characteristics of data and miss the features contributing to the local characteristics of data. This holistic feature selection mechanism not only leads to the hard time to interpret each principal component (PC) intuitively, but also hurdles the subtle data local latent structure discovery in the following clustering/ classification, because each PC only contains some levels of global characteristics of data.

One important reason for the holistic mechanism in PCA is that data representation in classic PCA is not 'purely additive', i.e., linear combinations in the PCA mix with both positive and negative weights and each PC consists of both negative and positive entries. The positive and negative weights are likely to cancel each other partially in data representation. Actually, it is more likely that weights contributing to local features are cancelled out than weights contributing to global features in linear combinations. This mainly leads to the data locality loss and holistic feature selection characteristics in PCA.

Imposing nonnegative constraints on PCA, i.e., restricting the signs of all entries of the PC matrix U to be nonnegative, can remove the likelihood of partial-cancellations and make data representation consisting of only additive components. It also contributes to the intuitive interpretation and sparse representation of each PC. In the context of feature selection, adding nonnegative constraints on PCA can improve data locality in feature selection and make data latent structure explicitly. Actually, adding non-negativity on PCA is actually also motivated by cancer molecular pattern discovery itself. Gene expression data generally are represented as positive or nonnegative matrices naturally or after simple processing. It is reasonable to require their corresponding dimension reduction data to be positive or at least nonnegative to maintain data locality in feature selection and for the sake of following clustering /classification.

2. Nonnegative Principal Component Analysis (NPCA)

Nonnegative PCA can be viewed as an extension of classic PCA by imposing PCA with non-negativity constraints in order to capture data locality in feature selection.

Let $X = (x_1, x_2, \cdots x_n)$, $x_i \in \Re^d$, be a zero mean dataset, i.e., $\sum_{i=1}^{n} x_i = 0$, a nonnegative PCA can be formulated as a constrained optimization problem to find maximum variance directions under nonnegative constraints. For instance, the first nonnegative maximum variance direction u i.e., 1st nonnegative PC can be found by solving the following optimization problem:

$$\begin{aligned} &\max\ u^T C u, \quad s.t. \\ &u^T u = 1, \ u \geq 0 \end{aligned} \tag{1}$$

Where $C = \frac{1}{n} X X^T$ is the covariance matrix of the input dataset X. If X is not a zero mean data, the data covariance matrix can be estimated by the equation: $C = \frac{1}{n}(XX^T - \frac{1}{n} X \vec{1}\vec{1}^T X^T)$, where $\vec{1} \in \Re^d$ with all entries are '1's. Similarly, all nonnegative maximum variance directions can be found by solving the optimization problem,

$$\begin{aligned} &\max\ J(U) = \frac{1}{2}\left\|U^T X\right\|_F^2, \quad s.t. \\ &U^T U = I, \ U \geq 0 \end{aligned} \tag{2}$$

Where $U = [u_1, u_2, \cdots u_k]$, $k \leq d$, is a set of nonnegative PCs. In fact, the rigorous orthonormal constraint under non-negativity is too strict for practical cancer molecular pattern analysis. Because it requires only one nonnegative entry in each column of U, i.e., each PC only contains one nonnegative entry. However, such a constraint is not congruent to the fact that the expressions of many key genes can be involved in developing the cancerous cells. This fact requires that each corresponding PC has more than one nonnegative entry. That is, the quadratic programming problem with the orthonormal-nonnegativity condition can be further relaxed as,

$$\max_{U \geq 0} J(U, \alpha) = \frac{1}{2}\left\|U^T X\right\|_F^2 - \alpha\left\|I - U^T U\right\|_F^2 \tag{3}$$

Where $\alpha \geq 0$ is a parameter to control the orthonormal degree of each column of U. After relaxation, matrix U is a near-orthonormal nonnegative matrix, i.e., $U^T U \sim I$. We give a gradient learning method for this nonlinear optimization problem as follows. Computing the gradient of the objective function with respective to U, we have

$$U(t+1) = U(t) - \eta(t) \nabla_U J(t), U \geq 0 \tag{4}$$

Where $\nabla_U J(U, \alpha) = (U^T X) X^T + 2\alpha(I - U^T U)U$ and $\eta(t)$ is the iteration step size in the t time level iteration. For convenience, we select the step size in the iteration as 1. This is equivalent to finding the local maximum of the following function $f(u_{sl})$ in the equation (5) under the conditions: $u_{sl} \geq 0$, where $s = 1, 2 \cdots d; l = 1, 2 \cdots n$ in the scalar level.

$$\max_{u_{sl} \geq 0} f(u_{sl}) = -\alpha u_{sl}^4 + c_2 u_{sl}^2 + c_1 u_{sl} + c_0 \tag{5}$$

Where c_2 and c_1 are the coefficients of u_{sl}^2 and u_{sl} ; c_0 is the sum of the constant items independent of u_{sl}. Computing stationary points for $f(u_{sl})$, we have a cubic function root finding problem. The final U matrix is a set of nonnegative roots of equation (6).

$$p(u_{sl}) = df(u_{sl}) / du_{sl} = -4\alpha u_{sl}^3 + 2c_2 u_{sl} + c_1 = 0 \tag{6}$$

By collecting the coefficients of u_{sl} and u_{sl}^2, we have

$$c_2 = \frac{1}{2}\sum_{i=1}^{n} x_{si}^2 - \alpha \sum_{j=1, j\neq l}^{k} u_{sj}^2 - 2\alpha \sum_{t=1, t\neq s}^{d} u_{tl}^2 + 2\alpha \tag{7}$$

$$c_1 = \sum_{i=1}^{n} \sum_{t=1, t\neq s}^{d} x_{si} u_{tl} x_{ti} - 2\alpha \sum_{j=1, j\neq l}^{k} \sum_{t=1, t\neq s}^{d} u_{sj} u_{tl} u_{tj} \tag{8}$$

Actually, the constant term $c_0 = -k\alpha$ does not have affects on the entries of U matrix. Only c_1 and c_2 are involved in the nonnegative root finding of equation (6). The complexity for the nonnegative principal component analysis algorithm (NPCA) is $O(dkn \times N)$, where N is the total iteration number in the algorithm to achieve the final termination threshold.

3. NPCA-based Cancer Molecular Pattern Classification

The nonnegative principal component analysis (NPCA) based cancer molecular pattern classification first employs NPCA to obtain the nonnegative representation of each biological sample in a purely additive low dimensional subspace spanned by meta-genes. A meta-gene is a linear combination of the expression levels of all genes in a cancer dataset. The nonnegative representation of a biological sample in NPCA is a meta-sample, which is the prototype of the sample with small dimensionalities. Then, a classification algorithm π_A, which can be any classification algorithm, is applied to the meta-samples to gain classification information. In this study, we choose support vector machines (SVM) as π_A to discriminate the meta-samples of cancer molecular patterns.

Theoretically, NPCA-based classification is rooted from a special nonnegative matrix factorization (NMF) that we propose in this study: the nonnegative principal component induced NMF. The principle of the NPCA-induced NMF can be briefed as follows. Let $X \in \Re^{d \times n}, d \ll n$, be a nonnegative matrix, which is a gene expression dataset with d number of samples for n number of genes in our context. Let $U \in \Re^{d \times d}$ be the corresponding nonnegative PC matrix for X, which is a near-orthogonal matrix before any further dimension selection. Projecting X^T into the column space generated by U, we obtain the nonnegative projection $X^T U = P$. Since U is a near-orthogonal matrix, we can view it as an orthogonal matrix to decompose the data matrix, i.e., $X^T \sim PU^T$, where P is equivalent to the basis matrix W and U^T is equivalent to the feature matrix H in the classic NMF $X^T \sim WH$ [5]. Unlike the general NMF, the basis matrix and the feature matrix in the NPCA-induced NMF both can be near-orthogonal.

The NPCA-based nonnegative matrix factorization can be explained alternatively. That is, each row of U is a corresponding meta-sample of each biological sample of X in the meta-gene space: $X_i^T \sim PU_i^T$. The meta-gene space $S = span(p_1, p_2 \cdots p_r)$, $p_i \geq 0$ is a column space of the nonnegative basis matrix P, where each basis is a meta-gene. It is a purely additive space, where each variable can be represented as nonnegative linear combinations of meta-genes: $X_i^T = \sum_{j=1}^{r} U_{ij}^T P_j \,, 1 \leq r \leq d$.

Since we use SVM as π_A in the NPCA-based classification, we brief NPCA-based SVM (NPCA-SVM) classification as follows. Considering gene expression data are naturally nonnegative data or can be converted to the corresponding nonnegative data easily, we conduct feature selection through nonnegative principal component analysis to obtain the low dimensional but data locality preserved meta-sample for each biological sample. Then, a SVM algorithm is employed to gain classification information from these meta-samples. To improve classification performance, we input the normalized meta-samples, i.e., $U = U / \|U\|_2$, to the following SVM classification.

Since different robust levels of the prior knowledge from different training sets affect classification results for a classification algorithm, we conducted NPCA-SVM classifications under two types of data cross validations. The first is the 50% holdout cross validation with N=100 times, i.e., 100 sets of training and test datasets generated by the 50% holdout cross validation for each dataset. The second is the leave one out cross validation (*LOOCV*). To improve computing efficiency, the matrix U is cached from previous trial and used as an initial point to compute the next trial principal component matrix in the classification.

4. Experimental Results

Table 1. Five Affymetrix oligonucleotide gene expression datasets

Dataset	#genes	#samples
colon	2,000	22 controls + 40 cancers
leukemia	5,000	27 ALL +11 AML
medulloblastoma	5,893	25 classic + 9 desmoplastic
hepatocellula carcinoma(HCC)	7129	20 early intrahepatic recurrence 40 non-early intrahepatic recurrence
glioma	12,625	28 glioblastomas + 22 anaplastic oligodendrogliomas

We applied our NPCA-SVM algorithm for five bench-mark Affymetrix oligonucleotide gene expression datasets: *colon*, *leukemia*, *medulloblastoma*, *hepatocellular carcinoma*, and *glioma* [6,7,8,9]. Table 1 presents detailed information of the datasets. Without loss of generality, we choose two mostly used kernels in our NPCA-SVM algorithm: a general linear kernel and Gaussian (*'rbf'*) kernel: $k(x,y) = (x \cdot y)$, $k(x,y) = \exp(-\|x-y\|^2/2)$

[10]. We compared classification results from NPCA-SVM algorithm under the orthonormal control $\alpha=100$ with those from PCA-SVM and SVM algorithm under linear and Gaussian kernels for each of the five microarray datasets under 100 times (trials) of 50% holdout cross validations. The average classification rates, sensitivities and specificities and their corresponding standard deviations from these three algorithms are given in Table 2.

Table 2. Average classification performances of three algorithms

Dataset	Average Classifying Rates (%)	Average Sensitivity (%)	Average Specificity (%)
Colon			
npca-svm-linear	89.77±4.79	95.24±4.89	81.76±11.24
npca-svm-rbf	88.90±5.33	93.90±5.62	81.10±11.60
svm-linear	46.10±11.25	39.80±29.62	55.14±31.59
svm-rbf	62.81±6.41	100.0±0.0	0.0±0.0
pca-svm-linear	75.90±7.96	85.24±9.69	61.73±16.80
pca-svm-rbf	62.81±6.41	100.0±0.0	0.0±0.0
Leukemia			
npca-svm-linear	96.11±5.57	94.64±14.53	99.16±2.69
npca-svm-rbf	91.84±8.51	86.96±20.81	94.25±7.04
svm-linear	44.21±22.08	66.00±47.61	34.00±47.61
svm-rbf	71.58±7.14	0.0±0.0	100.0±0.0
pca-svm-linear	95.32±6.97	87.40±20.34	99.26±2.46
pca-svm-rbf	71.58±7.14	0.0±0.0	100.0±0.0
Medulloblostoma			
npca-svm-linear	86.18±9.47	67.55±27.02	94.09±9.17
npca-svm-rbf	81.76±9.80	62.52±27.53	90.68±11.21
svm-linear	35.88±20.30	83.00±37.75	17.00±37.75
svm-rbf	73.47±7.55	0.0±0.0	100.0±0.0
pca-svm-linear	80.82±8.47	56.37±26.93	90.95±8.38
pca-svm-rbf	73.47±7.55	0.0±0.0	100.0±0.0
HCC			
npca-svm-linear	88.37±5.16	88.37±5.16	86.73±5.36
npca-svm-rbf	86.73±5.36	88.76±6.08	84.62±16.66
svm-linear	41.07±15.55	19.00±39.34	81.00±39.34
svm-rbf	66.87±5.93	0.0±0.0	100.0±0.0
pca-svm-linear	60.93±7.90	72.82±14.19	39.53±17.70
pca-svm-rbf	66.87±5.93	0.0±0.0	100.0±0.0
Glioma			
npca-svm-linear	91.24±5.11	91.11±8.99	91.83±6.68
npca-svm-rbf	90.20±5.40	90.24±9.90	90.66±7.69
svm-linear	49.56±8.22	53.18±27.70	45.72±28.72
svm-rbf	50.40±9.65	18.00±38.61	82.00±38.61
pca-svm-linear	72.68±6.75	69.30±14.19	76.17±11.64
pca-svm-rbf	50.40±9.65	18.00±38.61	82.00±38.61

We have following observations from these classification results. 1. It is clear that the PCA-SVM, SVM algorithms suffer from the over-fitting problem under a Gaussian ('*rbf*') kernel. This can be found through the complementary results of the sensitivities and specificities for the five gene expression datasets obviously. 2. There is no over-fitting problem, under a Gaussian ('*rbf*') kernel, for the NPCA-SVM algorithm; NPCA-SVM algorithm under a Gaussian kernel has the second best classification performance among all the results. 3. The classification results from our NPCA-SVM under a linear kernel have leading advantages over other two algorithms for all datasets.

Figure 1 shows the comparisons of the expectations of classification rates, sensitivities and specificities for the same 100 set of training and testing data for each gene expression dataset. Since PCA-SVM and SVM algorithms under a Gaussian kernel encountered the over-fitting problem, we did not include their sensitivities and specificities in the plot. It is obvious that NSPCA-SVM algorithm not only leads the PCA-SVM and SVM in the classification rates, sensitivities and specificities, but also it demonstrates the robust stability for the three measures. This can also be verified the relative small standard deviations for the three classifying performance measures of the NPCA-SVM classifications.

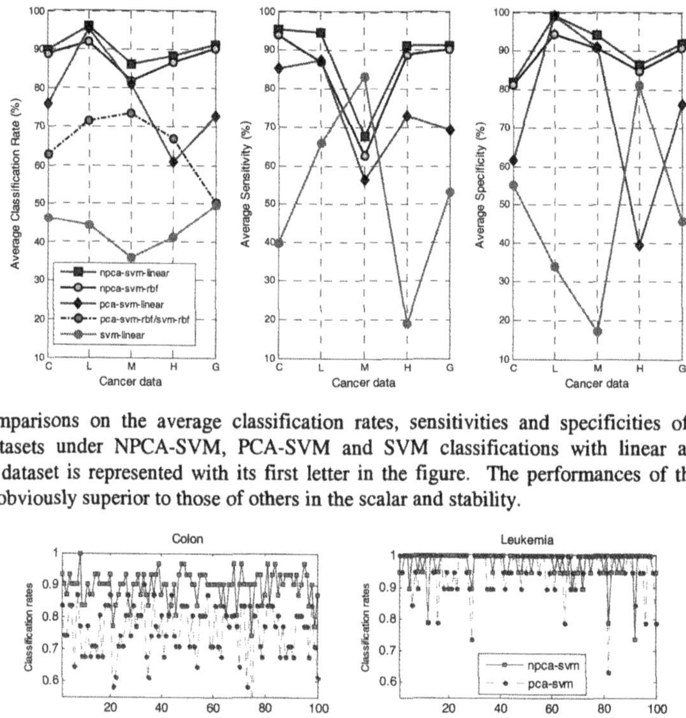

Figure 1. Comparisons on the average classification rates, sensitivities and specificities of the five gene expression datasets under NPCA-SVM, PCA-SVM and SVM classifications with linear and a Gaussian kernels. Each dataset is represented with its first letter in the figure. The performances of the NPCA-SVM algorithm are obviously superior to those of others in the scalar and stability.

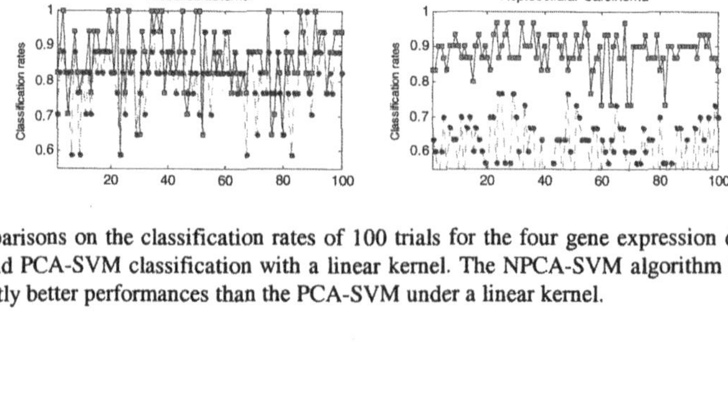

Figure 2. Comparisons on the classification rates of 100 trials for the four gene expression datasets under the NPCA-SVM and PCA-SVM classification with a linear kernel. The NPCA-SVM algorithm has the obviously leading or slightly better performances than the PCA-SVM under a linear kernel.

Since the PCA-SVM algorithm under a linear kernel has the best classification results among the PCA-SVM and SVM classifications, we compare its performances with those of NPCA-SVM under a same kernel in Figure 2 for the first four datasets. It is easy to find that our NPCA-SVM algorithm has achieved obviously leading performances for *colon, medulloblastoma* and *heptocellular carcinoma* datasets under a linear kernel, when compared with the PCA-SVM algorithm. Our NPCA-SVM algorithm also achieves slightly better results for the *leukemia* dataset than the PCA-SVM algorithm under a linear kernel. According to our experimental results, the NPCA-SVM classification results from *Glioma* dataset also strongly demonstrated its leading advantages over the PCA-SVM algorithm under a linear kernel.

4.1 Classification results comparisons with those of other algorithms

It is desirable to compare the nonnegative principal component analysis based SVM algorithm (NPCA-SVM) with other classification algorithms to further verify its superiority. In this section, we compare the classification performances of our NPCA-SVM algorithm with those of other five classification algorithms. These five algorithms include k-nearest neighbor (k-NN), linear discriminant analysis under principal component analysis (PCA-LDA) and three nonlinear feature selection based SVM classification algorithms: SVM classifications under kernel principal component analysis (KPCA-SVM); nonnegative matrix factorization based SVM (NMF-SVM) and SVM classifications under locally linear embedding (LLE-SVM). Details about these three feature selection algorithms can be found in [5, 11,12] .

The k-NN and PCA-LDA algorithms both are widely used algorithms in microarray data classifications. The k-NN is a simple Bayesian inference method. It determines the class type of a sample interested based on the class belongs of its nearest neighbors, which are measured by correlation, Euclidean or other distances. In PCA-LDA classifications, it conducts PCA processing for training samples and projects testing samples in the subspace spanned by the principal components of the training samples at first. Then, a linear discriminant analysis (LDA) is used to classify projections of the testing samples, which is equivalent to solving a generalized eigenvalue problem [2].

The three nonlinear feature selection based SVM classification algorithms conduct SVM classification for the meta-samples in the space generated by corresponding feature selection algorithms respectively. For instance, KPCA-SVM conducts classification for the projections of testing data in the space spanned by the PCs of training data, obtained by performing PCA in a kernel space; LLE-SVM conducts classification for the meta-samples of input biological samples, which are the low dimensional and neighborhood preserving embedding of the original high dimensional data. For convenience, we brief the NMF-SVM classification algorithm as follows.

The NMF-SVM algorithm is to decompose the nonnegative gene expression cancer data $X \in \Re^{n \times m}$ into the product of two nonnegative matrices: $X \sim WH$, under a rank r with the least reconstruction error. The matrix $W \in \Re^{n \times r}$ is called a basis matrix. Its column space sets up a new coordinate system for X ; the matrix $H \in \Re^{r \times m}$ is called a feature matrix. It stores the new coordinate values for each variable of X in the new space. Then, a SVM algorithm is used to classify the corresponding meta-sample of each sample in the gene expression matrix X , which is the corresponding column in the feature matrix H .

For each dataset, we still use previous 100 trial of training and testing data from 50% holdout cross validations in classifications. Table 3 shows the average sensitivities and specificities and their corresponding standard deviations for four algorithms.

Table 3. Average classification performances of NMF/LLE-SVM, PCA-LDA and k-NN

Dataset	Average Classifying Rates (%)	Average Sensitivity (%)	Average Specificity (%)
Colon			
Nmf-svm-linear	84.03±6.31	89.28±6.81	76.68±13.70
Nmf-svm-rbf	74.42±7.89	86.98±9.62	55.23±16.72
knn-euclidean	78.03±7.57	90.38±7.15	58.65±7.15
knn-correlation	80.52±8.78	93.31±4.36	61.61±18.96
pca-lda	86.39±5.64	89.37±6.74	81.65±10.91
lle-svm-linear	83.71±5.47	93.96±4.85	67.04±12.00
lle-svm-rbf	73.81±7.85	84.56±9.74	58.40±19.16
Leukemia			
Nmf-svm-linear	92.16±7.19	77.79±21.69	98.46±3.89
Nmf-svm-rbf	88.89±8.59	74.87±23.85	95.12±6.45
knn-euclidean	91.32±7.75	76.72±23.95	98.09±3.88
knn-correlation	93.05±5.93	83.42±17.08	97.79±4.36
pca-lda	94.21 ±6.79	81.98±19.62	100.00±0.0
lle-svm-linear	95.00±3.03	90.93±12.35	96.75±4.00
lle-svm-rbf	89.11±7.71	81.76±17.30	92.59±9.58
Medulloblostoma			
Nmf-svm-linear	81.76±9.70	64.95±25.56	88.22±10.87
Nmf-svm-rbf	82.18±8.98	58.98±27.03	91.98±10.87
knn-euclidean	76.59±10.51	24.22±29.61	98.66±4.79
knn-correlation	79.12±10.08	50.22±27.97	91.66±11.22
pca-lda	81.24±9.59	58.55±28.92	92.21±8.79
lle-svm-linear	76.00±10.47	53.34±29.50	85.60±12.36
lle-svm-rbf	73.71±7.95	3.17±9.82	99.57±1.71
HCC			
Nmf-svm-linear	61.30±8.91	71.17±13.47	43.47±16.67
Nmf-svm-rbf	58.67±8.49	79.52±9.21	22.83±13.14
knn-euclidean	61.83±7.92	79.54±14.73	25.84±19.58
knn-correlation	63.10±7.80	81.55±11.18	27.86±16.15
pca-lda	60.87±7.82	72.88±14.42	39.47±17.08
lle-svm-linear	62.77±7.36	91.78±13.87	5.81±12.93
lle-svm-rbf	66.83±5.87	99.96±0.42	0.0±0.0
Glioma			
Nmf-svm-linear	74.40±8.04	74.54±11.10	74.19±13.53
Nmf-svm-rbf	70.40±5.40	51.87±13.61	84.07±7.79
knn-euclidean	46.80±8.76	47.85±16.66	47.63±15.84
knn-correlation	74.56±7.66	74.14±13.12	76.24±11.90
pca-lda	73.44±6.93	70.82±13.91	76.41±12.47
lle-svm-linear	73.84±8.82	73.00±14.39	74.93±13.61
lle-svm-rbf	65.92±12.19	47.22±26.80	84.97±22.50

In the k-NN algorithm, the distance measure was chosen as the correlation or Euclidean distance and the number of nearest neighbors was selected from 2 to 7. In the LLE-SVM classification, we selected embedding dimensionalities from 2 to 20. In the NMF-SVM classification, the matrix decomposition rank in the NMF was selected from 2 to 18. The kernel function in SVM is still selected as a linear or Gaussian kernel. The final average classification rate for a dataset under each algorithm is selected as the best average classification rate among all possible cases.

In the KPCA-SVM algorithm, there are two kernel functions: one is kernel function $k_1(x, y)$ in kernel PCA and another is kernel $k_2(x, y)$ in the following SVM algorithm. When $k_1(x, y)$ is a linear kernel, then kernel PCA is just original PCA and KPCA-SVM has the same performance as the previous PCA-SVM algorithm; when $k_1(x, y)$ is a Gaussian kernel, we have found that KPCA-SVM encounters the over-fitting problem for all five datasets no matter $k_2(x, y)$ is a linear or Gaussian kernel.

From our experimental results, we have found that the NMF-SVM algorithm generally has better classification results under a linear kernel than a Gaussian kernel, although the NMF-SVM classification under a linear kernel has slightly better performance than that under a Gaussian kernel for the *medulloblastoma* dataset. NMF-SVM also overcomes the over-fitting problem under a Gaussian kernel. This is because that the meta-samples are from the space generated from nonnegative matrix factorization, is also a purely additive space. We have found LLE-SVM algorithm generally has better classification performances under a linear kernel than a Gaussian kernel. However, it still cannot avoid the over-fitting problem under a Gaussian kernel, because this manifold learning based algorithm encountered the over-fitting for *medulloblastoma* and *heptocellular carcinoma (HCC)* datasets. It is also easy to see that the k-NN classification under the correlation distance has the advantages over the Euclidean distance for the five cancer datasets. However, from these results, we can observe that the performances of all these four algorithms still can not compete with those of NPCA-SVM algorithm for the five microarray datasets, under the 100 trials of 50% holdout cross validations.

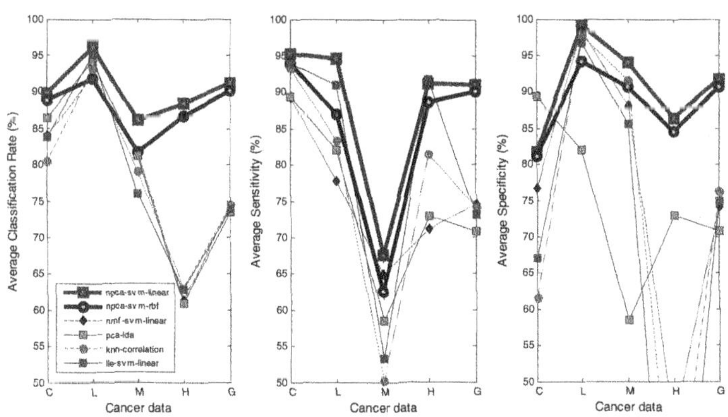

Figure 3. Comparisons on the average classification rates, sensitivities and specificities of the five gene expression datasets under the NPCA-SVM, NMF-SVM, LLE-SVM, PCA-LDA, KNN classifications. Each cancer dataset is represented with its first letter in Figure. The NPCA-SVM algorithm achieves the stably leading classification performances for the five datasets.

Figure 3 shows the comparisons on the average classification rates, sensitivities and specificities of the NPCA-SVM algorithm under the linear and Gaussian kernel with those of other four classification algorithms: *NMF-SVM, LLE-SVM, PCA-LDA, k-NN*, for the five gene expression datasets. We can observe that the average classification rates, sensitivities and specificities of our NPCA-SVM algorithms are clearly superior to those of others in a stable pattern. Alternatively, the curves of average classification rates, sensitivities and specificities from the other algorithms have relatively large oscillations, for the five microarray datasets.

We also compared our NPCA-SVM algorithm with six other algorithms: PCA-SVM, SVM, NMF-SVM, LLE-SVM, k-NN and PCA-LDA for the five datasets under the leave one out cross validations (*LOOCV*). The classification rate for each algorithm under *LOOCV* is the ratio between the total numbers of correctly classified samples over the number of total samples. Figure 4 demonstrates that the NPCA-SVM algorithm, under linear and Gaussian kernels, have strongly leading performances over the other six algorithms for the five cancer datasets under *LOOCV* on classification rates. In our plot, we selected the classification results from NMF-SVM, LLE-SVM algorithms under linear kernels for their better performances than Gaussian kernels. For the same reason, we selected k-NN classification results under correlation distances instead of Euclidean distances. We can observe that only the PCA-SVM algorithm under a linear kernel has achieved a comparable result at *leukemia* dataset, compared with those of our NPCA-SVM algorithm. For other four gene expression datasets, the classification results from the NPCA-SVM algorithm, under linear and Gaussian kernels, are obviously superior to those of the other six algorithms. Such a result is consistent with the previous results under the 50% holdout cross validations. Under a Gaussian kernel, algorithms PCA-SVM, SVM, LLE-SVM all suffer from the over-fitting problem. However, just as before, there is no over-fitting problem in the NPCA-SVM algorithm.

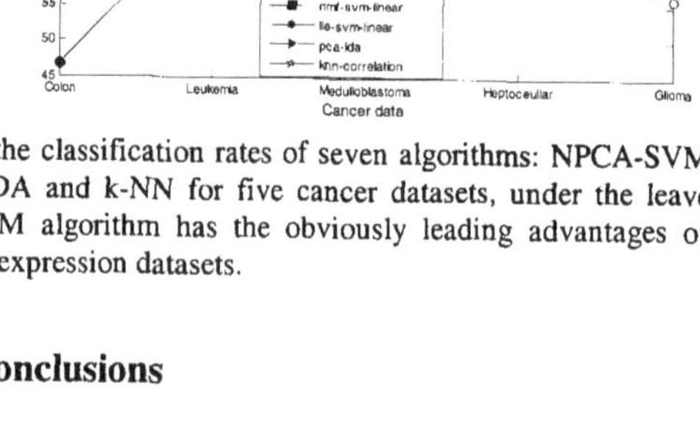

Figure 4. Comparisons of the classification rates of seven algorithms: NPCA-SVM, PCA-SVM, SVM, NMF-SVM, LLE-SVM, PCA-LDA and k-NN for five cancer datasets, under the leave one out cross validations (*LOOCV*). The NPCA-SVM algorithm has the obviously leading advantages over other six classification algorithm for the five gene expression datasets.

5. Discussions and Conclusions

In this study, we present a novel nonnegative principal component analysis (NPCA) algorithm and apply it in the gene expression data classification. We have demonstrated that NPCA-SVM algorithm has obviously leading advantages over other seven classification algorithms in the cancer pattern classification for five microarray datasets under the 50% hold-out and leave one out cross validations. The general over-fitting problem associative with SVM based classification in gene expression data under a Gaussian kernel is also overcome in our algorithm.

From the nonnegative principal component analysis, we can develop a family of NPCA-induced statistical learning algorithms by applying NPCA as a feature selection algorithm before a classification or clustering algorithm. For example, NPCA-based Fisher discriminant analysis (NPCA-FDA), etc. Alternatively, since NPCA-SVM is a more robust high-performance classifier than the general SVM and k-NN classifiers, it can replace the popular SVM and k-NN classifiers used in the cancer biomarker identification to capture oncology genes. In the following work, we plan to investigate applications of NPCA-based classification algorithms in the SNP array, exon-array and proteomics data and related biomarker discovery.

References

[1] Pochet N., De Smet F., Suykens J.A.K. and De Moor B.L.R., Systematic benchmarking of microarray data classification: assessing the role of non-linearity and dimensionality reduction, *Bioinformatics*, 20(17), 3185-3195, 2004.

[2] Lilien, R. and Farid, H., Probabilistic Disease Classification of Expression-dependent Proteomic Data from Mass Spectrometry of Human Serum, *Journal of Computational Biology*, 10(6), 925-946, 2003.

[3] Gao, Y. and Church, G., Improving molecular cancer class discovery through sparse nonnegative matrix factorization, *Bioinformatics*, 21(21), 3970–3975, 2005.

[4] Han, X., Cancer molecular pattern discovery by subspace consensus kernel classification, *Computational Systems Bioinformatics, Proceedings of the Conference CSB 2007*, 6:55-65, 2007.

[5] Daniel D. Lee and H. Sebastian Seung., Learning the parts of objects by non-negative matrix factorization. *Nature*, 401, 788–791, 1999.

[6] Alon,A.,et al., Broad patterns of gene expression revealed by clustering analysis of tumor and normal colon tissues probed by oligonucleotide arrays. *Proc. Natl Acad. Sci. USA*, 96, 6745–6750. 1999.

[7] Brunet, J., Tamayo, P., Golub, T. and Mesirov., J., molecular pattern discovery using matrix factorization. *Proc. Natl Acad. Sci. USA*, 101(12), 4164–4169, 2004.

[8] Iizuka,N., et al., Oligonucleotide microarray for prediction of early intrahepatic recurrence of hepatocellular carcinoma after curative resection, *The Lancet*, 361,923-929, 2003.

[9] Nutt,C.L., et al., Gene expression-based classification of malignant gliomas correlates better with survival than histological classification, *Cancer Research*, 63(7), 1602-1607, 2003.

[10] Vapnik,V.N., *Statistical Learning Theory*, John Wiley & Sons, New York,1998.

[11] Schölkopf, B., Smola, A. J., and Müller, K.-R., Nonlinear component analysis as a kernel eigenvalue problem, *Neural Computation*, 10, 1299-1319, 1998.

[12] Roweis, S. and Saul, L., Nonlinear dimensionality reduction by locally linear embedding, *Science*, v.290 no.5500, 2323-2326, 2000.

SIMULATION ANALYSIS FOR THE EFFECT OF LIGHT-DARK CYCLE ON THE ENTRAINMENT IN CIRCADIAN RHYTHM

NATSUMI MITOU[1]
natsumi.mitou@qdenbs.com
YUTO IKEGAMI[2]
ikegami@ib.sci.yamaguchi-u.ac.jp
HIROSHI MATSUNO[2]
matsuno@sci.yamaguchi-u.ac.jp
SATORU MIYANO[3]
miyano@ims.u-tokyo.ac.jp
SHIN-ICHI T. INOUYE[4]
inouye@yamaguchi-u.ac.jp

[1] *Kyuden Business Solution Co. Inc., 2-1-10, Watanabe-dori, Chuo-ku, Fukuoka 810-0004, Japan.*
[2] *Graduate School of Science and Engineering, Yamaguchi University, 1677-1 Yoshida, Yamaguchi 753-8512, Japan.*
[3] *Human Genome Center, Institute of Medical Science, University of Tokyo, 4-6-1 Shirokanedai, Minato-ku, Tokyo 108-8639.*
[4] *Research Institute for Time Studies, Yamaguchi University, 1677-1 Yoshida, Yamaguchi 753-8512, Japan.*

Circadian rhythms of the living organisms are 24hr oscillations found in behavior, biochemistry and physiology. Under constant conditions, the rhythms continue with their intrinsic period length, which are rarely exact 24hr. In this paper, we examine the effects of light on the phase of the gene expression rhythms derived from the interacting feedback network of a few clock genes, taking advantage of a computer simulation with Cell Illustrator. The simulation results suggested that the interacting circadian feedback network at the molecular level is essential for phase dependence of the light effects, observed in mammalian behavior. Furthermore, the simulation reproduced the biological observations that the range of entrainment to shorter or longer than 24hr light-dark cycles is limited, centering around 24hr. Application of our model to inter-time zone flight successfully demonstrated that 6 to 7 days are required to recover from jet lag when traveling from Tokyo to New York.

Keywords: circadian rhythm, light-dark cycle, entrainment, simulation, hybrid functional Petri net

1. Introduction

Circadian rhythms are endogenous oscillations with a period close to 24hr, found in most living organisms. They are driven by a central circadian clock located in the suprachiasmatic nuclei (SCN) of the hypothalamus. The genes involved in the regulation of circadian rhythms have largely been uncovered during the last decade in the organisms from cyanobacteria to plants, insects, and mammals. To understand the systematic behavior of circadian gene regulatory mechanism, now it is necessary to describe the consequence of dynamic individual interactions of the genes and these products that constitute the circadian clocks. Computer simulation is a powerful tool that enables us to predict complex behaviors along time axis over multilevel molecules of genes, mRNAs, and proteins. Virtual experiments are possible on a simulation model, which can lead to hypotheses of molecular interactions in a target biological model much easier and faster than actual biological experiments. Therefore, it is promising to apply the computer simulation technique for the circadian clock models. Indeed, several preceding studies have provided interesting demonstration of the usefulness of the simulation approach. Leloup and Goldbeter [5] presented the computational model of mammalian circadian clock with the *Per*, *Cry*, *Bmal1*, *Clock*, and *Rev-Erbα* genes. These authors accounted for autonomous, sustained circadian oscillations in conditions corresponding to continuous darkness, and for entrainment by LD cycles. They extended the study in the subsequent paper [4], showing that the small changes in the parameters governing CRY levels can shift the peak in *Per* mRNA from the light to the dark phase, or in some case, entirely prevent entrainment. Kurosawa and Goldbeter [3] used similar models for circadian rhythms in *Neurospora* and *Drosophila* and studied the dependence of free-running period and amplitude of the external LD cycles on the entrainment of these rhythms. Rand [11] *et al.* discussed on the source of extraordinary stability found in the circadian clocks based on the system analysis of a computer model.

Hybrid functional Petri nets (HFPNs) [6] have successfully been employed in order to model many biological processes including apoptosis induced by Fas [6], Notch signaling pathway [7], and p53 with related genes [2]. Constructed HFPN models can be simulated with Cell Illustrator [16]. Its excellent user interface and the ease of modification to an HFPN model makes it possible to smoothly examine the effects of a manipulation, such as gene disruption, on a target biological system. We have applied HFPN model to a mammalian circadian clock model comprised of the five genes *Per*, *Cry*, *Rev-Erbα*, *Bmal1*, and *Clock*. As reported in our previous paper [8], the feedback loop model of Figure 1 have provided important insight on a possible mechanism responsible for the phase difference between *Per* and *Bmal* mRNAs. Comparison between the simulation results and the observation from biological experiments [8] enabled us to predict the presence of an unidentified interaction among the clock genes. In the present paper, we extended the previous approach to the dynamical behavior of the molecular circadian clock in response to environmental light. As an application of this model, we analyzed the recovery process from jet lag when traveling from Tokyo to New York. The results demonstrated the transient shifting process of *Per* mRNA oscillation from stable oscillation in Tokyo to the one in New York, showing that this case requires 6 or 7 days for recovering from jet lag.

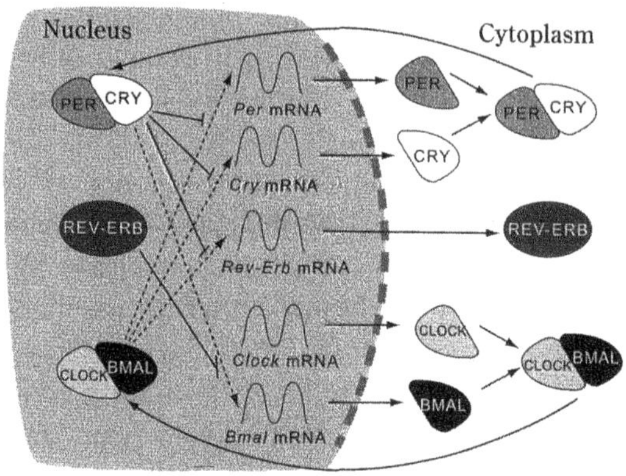

Fig. 1. A model of the circadian system of the clock genes in the mouse.

2. Gene Regulatory Network of Circadian Clock

2.1. *Feedback loops of genes and their products*

Recent molecular biological studies have disclosed that the circadian rhythm of the SCN is generated at the level of the gene expression, protein synthesis and transport of transcription factors across nuclear membrane. The genes that are involved in this intracellular system are called clock genes. They include *Per (Per1, Per2, Per3), Cry (Cry1, Cry2), Rev-Erb (Rev-Erbα), Clock* and *Bmal (Bmal1)*. The transcription of these clock genes and its regulation by their product proteins constitute a negative feedback loop, effectively to generate an oscillation in the activity of SCN cells. In the present mathematical model, *Per1, Per2 and Per3* are combined into single *Per* so as to make the model simple. Similarly, *Cry1 and Cry2* are treated together as *Cry*. So, the model is made of 5 genes, namely, *Per, Cry, Rev-Erb, Clock* and *Bmal*. Each mRNA produces a corresponding protein, PER, CRY, REV-ERB, CLOCK or BMAL. Once proteins are synthesized, they start to interact with each other. PER and CRY bind to form a protein complex, PER/CRY, and CLOCK and BMAL also form a complex, CLOCK/BMAL. Then these complexes enter into the nucleus together with REV-ERB, CLOCK and BMAL. PER/CRY in the nucleus interferes CLOCK/BMAL that is activating transcriptions of *Per, Cry* and *Rev-Erb* genes. So the PER/CRY protein complex effectively represses the transcriptions of *Per, Cry* and *Rev-Erb*. This completes the negative feedback loop, giving rise to an oscillation. In addition, *Bmal* transcription is activated by PER/CRY, and repressed by REV-ERB. This system of complicated feedback loops is responsible for circadian rhythm to be generated in the SCN of the brain.

2.2. *Phase shift of circadian rhythm by light*

The circadian clock keeps on running even when the time cues in the environmental are totally removed. However, the period of this free-running rhythm is a little longer or shorter than 24hr. The organism under the natural environment adjusts their running of the clock by external signals so as to synchronize or entrain the organism with environmental cycles. Those entraining agents are alteration of light and dark (LD cycle), temperature cycle, eating time, social contact and so on. The strongest of those is light. Physiological mechanisms of the entrainment have partly been discovered. *Per* gene have the non-coding DNA sequence with which transcription is transiently induced by light and can mediates entrainment. The level of *Per* is known to be higher during the day and lower during the night. A light stimulus at night induces *Per* mRNA level in the SCN [13]. This rise of *Per* mRNA triggers the change in the state of the circadian feedback loop, and eventually leads to the phase shift of the circadian clock.

3. Light Induced Phase Response Simulation

3.1. *Hybrid functional Petri net*

We employed hybrid functional Petri net (HFPN) to model the circadian gene regulatory mechanism. HFPN consists of three types of elements, places, transitions, and arcs whose symbols are illustrated in Figure 2. The HFPN has two kinds of places; continuous place and discrete place. A continuous place holds a real number as a concentration of a substance such as mRNA and protein. A discrete place holds a number of tokens. This paper uses discrete places to express the day time or the night time as shown in Figure 9. Continuous and discrete types are also availabile for transitions of HFPN. Continous transition is used to represent a biological reaction such as transcription and translation, at which the reation speed is assigned as a parameter. At a discrete transition, delay time is assigned as a parameter. The delay time of each discrete transition in Figure 9 is 12hr that represents the period of day or night time. Arcs are classified into three types; normal, test, and inhibitory arcs. Normal arc connects a place to a transition or vice versa. Test or inhibitory arc represents a condition and is only directed from a place to a transition. Each of normal arc from a place, test arc, and inhibitory arc has a threshold by which the parameter assinged to the transition at its head is controlled. A normal arc from a place or a test arc (an inhibitory arc) can participate in activating (repressing) a transition at its head, as far as the content of a place at its tail is over the threshold. For either of test and inhibitory arcs, no amount is consumed from a place at its tail. Formal definition of HFPN is found in the paper [6].

3.2. *HFPN model under free-running conditions*

Figure 3 shows the HFPN model for the molecular circadian clock without external disturbances, described in Figure 1. With proper choice of parameters in the Figure 3, computer simulation yielded stable rhythms in mRNAs of 4 clock genes, *Per, Cry, Bmal, Rev-Erb* with the same period, while *Clock* level stayed constant, as shown in Figure 4. Parameters of transition speeds and arc thresholds have been detemined so that the phase relations of the product concentrations of the five genes

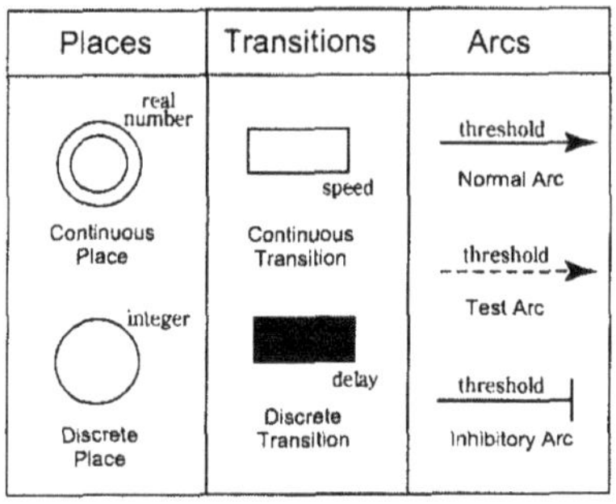

Fig. 2. The symbols of hybrid functional Petri net.

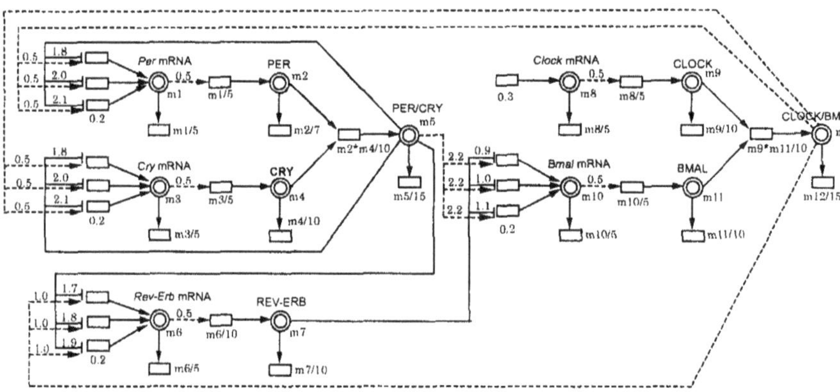

Fig. 3. HFPN representation of the circadian gene clock mechanism in Figure 1. The accompanying variable at a place (doubled circle) represents the concentration of the corresponding mRNA, protein or protein complex. For example, the variable m1 indicates the concentration of *Per* mRNA. Reaction speed (the rate of transcription, translation, complex formation or degradation) is expressed by a simple formula at a transition (rectangle). For example, the formula $m1/5$ indicates the translation rate of PER protein that depends on the variable m1 for the *Per* mRNA concentration. The real number besides an arc is the threshold for the content of the place attached to this arc. For example, the translation of Clock mRNA occurs during the period that the place value of Clock mRNA exceeds 0.5.

match the biological observations written in the literature. The simple structure of the formulas in the HFPN model enables easy detemination of these parameters through the Cell Illustrator GUI. This common period, thereafter, is regarded as the free-running period of the clock. (Cell Illustrator file of this model is available from the webpage [14].)

3.3. *Response of HFPN model to a light pulse*

After confirming the stable oscillation in the model, we explored the effects of a light stimulus on the rhythm of the mRNAs in this molecular clock system. In order to incorporate the effects of light, we assigned circadian time. Period length of the oscillation was set to a whole circadian day and divided into 24 circadian

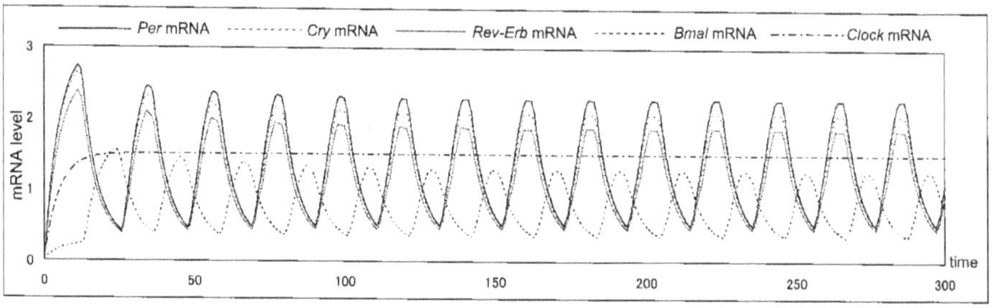

Fig. 4. Oscillations of *Per, Cry, Rev-Erb, Bmal,* and constant *Clock* mRNAs in the HFPN model of Figure 3.

times(CTs). Since neurophysiological experiments [9] showed that peak time of *Per1* mRNA in the SCN occurs at CT8, the time when *Per* mRNA reaches the highest in our simulation was assigned as CT8. Other CTs were determined accordingly. Subjective day corresponds the first half of a circadian day between CT0 and CT12, and subjective night from CT12 to CT0. Light exposure transiently increases *Per* mRNA only if exposure occurs during the subjective night. Referring the results of these biological experiments [9], we assumed that *Per* mRNA level transiently increases in response to light, the increment of which depends on the CT in the way shown in Figure 5. The consequence of this induction of *Per* mRNA in our HFPN model was computed and illustrated in Figure 6. It is clear that an instantaneous increase of *Per* mRNA at the time of light exposure, after some transient duration lasting several cycles, brought about a permanent phase shift in the subsequent free-running rhythms. Dependency of the phase shift amount on the time of light imposition is presented in Figure 7. This phase-response relation seems quite similar to the well-known phase response curve to light of Figure 8, which has been known from animal experiments [10]. This result demonstrates that our model system of the molecular clock can successfully and quantitatively simulate the behavioral phase shifts of the circadian clock, only if the light induction of one kind of clock gene (*Per*) was taken into consideration [12]. Given no observation about the realization of phase response relation has been made, this result gives the first suggestion that the present five genes feedback mechanism is essential for phase response behavior of the mammals.

4. Entrainment to LD Cycles

4.1. *Entrainment by light with the extended HFPN model*

We further explored dynamical behavior of the present molecular clock model under repeating light exposures, simulating the effect of LD cycles. In order to take into account the periodic and phase-dependent increases in *Per* level on light exposures, we introduced in the HFPN model the gate component before light affect *Per* mRNA in the circadian clock. The gate is closed during subjective day and open at subjective night. Note that the Goldbeter's group did not incorporate this

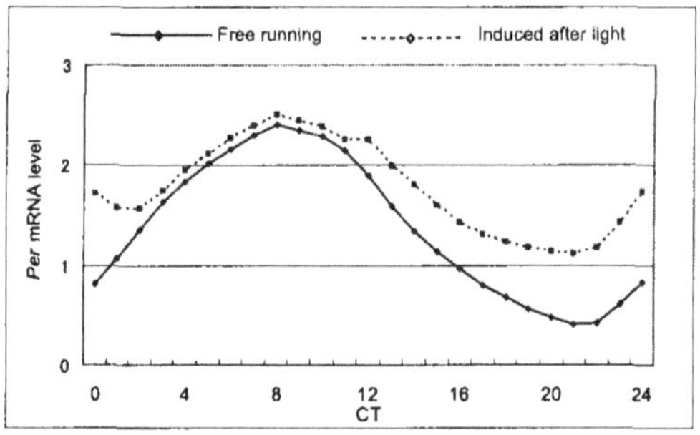

Fig. 5. *Per* mRNA levels in free-running and induced after light.

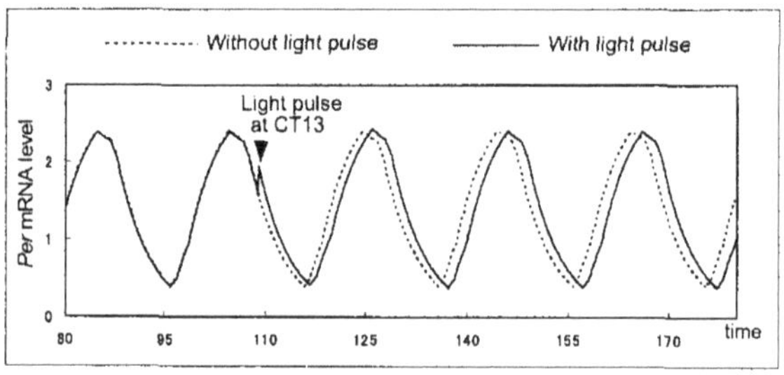

Fig. 6. Simulation result of the phase shift by a light at CT13.

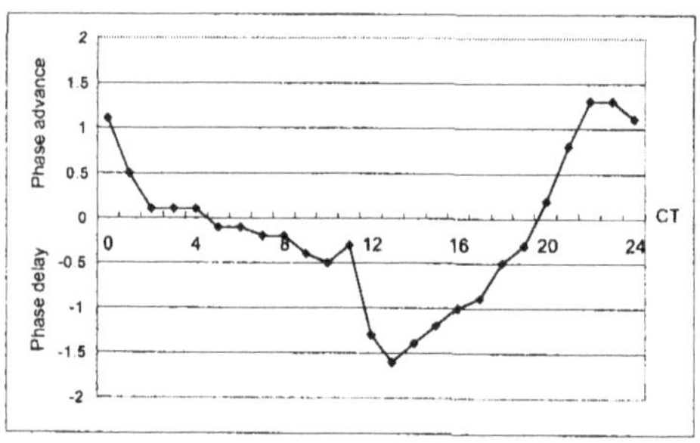

Fig. 7. Phase response curve obtained from the simulation.

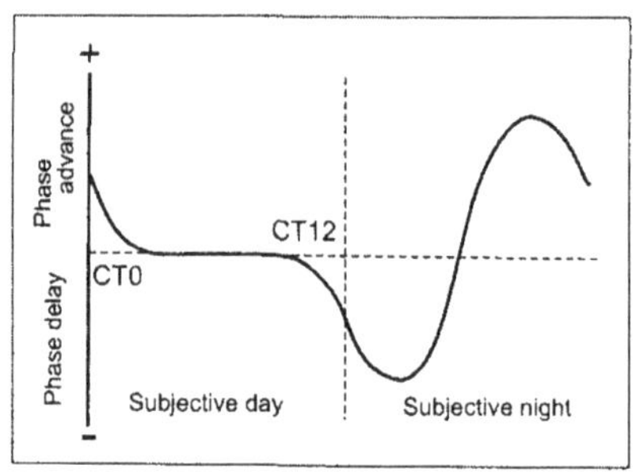

Fig. 8. Phase response curve described in biological literature [10].

gating system into their models [4, 5]. To implement the gate into our model simulation, simple two components shown on the gray background in Figure 9 were added between light stimulus and the endogenous oscillator. The place Day (m18) of LD cycle component yields 1 during the day and 0 at night. The Gate component on the right is to increase *Per* mRNA level according to the levels of *Cry* and *Bmal* mRNAs, PER and CLOCK/BMAL proteins at the time of light imposition. *Per* mRNA may not be used because this value would be changed by light. Dependencies on these internal levels were given at four continuous transitions in Gate component, and adjusted so as to reproduce the increases shown in Figure 5.

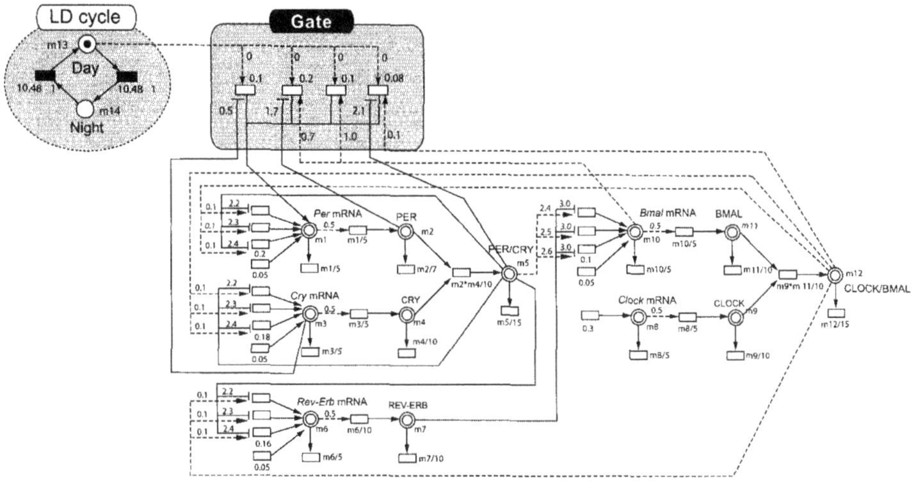

Fig. 9. Extended HFPN model with **LD cycle** and **Gate** components. Four continuous transitions in **Gate** component serve to increase the amount of *Per* mRNA at flow speeds assigned to them. Activation of each transition is controlled based on the presence of token in the place **Day** in **LD cycle** component and the condition regulated by test/inhibitory arc directed to the transition. Cell Illustrator file of this model is available on the webpage [14].

Simulation on the Cell Illustrator of the gate model system in Figure 9 showed the rhythms in mRNAs in the mouse circadian clock, as shown in Figure10. The Figure 10-A shows *Per* mRNA levels under free-running and entrained conditions. Black and white portions of black-white bar indicate dark and light periods, respectively. Solid line in Figure 10-A shows that, upon imposition of 12 hr light and 12 hr dark cycles, the phase of the rhythm gradually shifted before being entrained to the external LD cycles. Comparison with the dotted line with which a stable free-running rhythm was depicted indicates that phases of the circadian oscillation keep on delaying relative to the free-running rhythm. This reflects the fact that mouse free-running period is shorter than 24 hr. Periodic light exposures, in effect, caused phase delays so as to compensate the difference between free-running period and LD cycles and eventually entrain the endogenous rhythm to the external LD cycles. When the external LD cycles were significantly longer or shorter relative to the endogenous circadian period, the circadian clock could not entrain to those cycles. This biological observation was successfully reproduced in our model simulations, as shown in Figure 10-B and Figure 10-C. Figure 10-B shows the case where external cycles are 20hr (10hr light / 10hr dark) and Figure 10-C where the cycles are 26hr (13hr light / 13hr dark). In both cases, Per mRNA level did not faithfully follow the external LD cycle. These results of deviation from external cycles demonstrate the capability of our model to faithfully simulate the entrainment of the circadian clock observed in many animals. Simulation with the modified HFPN model with LD cycle generator confirmed the biological fact that the entrainment to LD cycle is only achieved when the environmental period is close to 24hr.

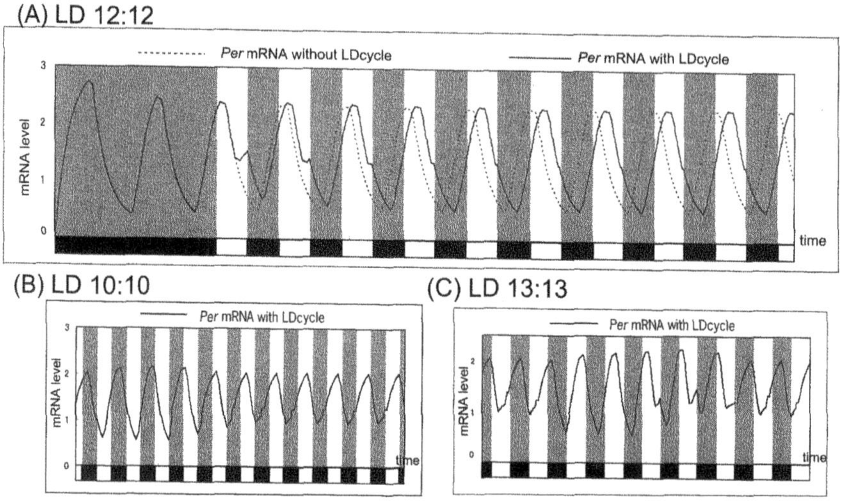

Fig. 10. (A) *Per* mRNA increases only during the late subjective day when the gate for light is open, whose amount of increase is determined by the timing of light exposure (Figure 5). Comparison between curves in solid and dotted lines clearly shows that this increase delays the phase of *Per* mRNA and entrains the rhythm to 24hr LD cycle. (B) and (C) show the cases of 20hr (10:10) and 26hr (13:13) LD cycles, respectively. In both cases, *Per* mRNA rhythms are unable to be entrained by the LD cycles.

4.2. *Jet Lag Simulation: Flying from Tokyo to New York*

As an application of LD cycle synchronization, we have simulated recovery process from the jet lag using the HFPN model of Figure 9. To take free-running cycle 24.2h of human [1] into account, delay times 10.48 of discrete transitions in mouse model of Figure 9 have been modified to 10.26. We examined the case of a traveler who takes the flight for the flight NH10 (All Nippon Airways) which departs from Tokyo at 11:00AM and arrives at New York at 9:30AM (flight time is 12.5hr) [15]. The time difference from Tokyo to New York is -14h. Figure 11 illustrates *Per* mRNA oscillation before and after the flight. Upper and lower black-white bars at the bottom of the figure represent LD cycles in Tokyo and New York, respectively. Flight time period of 12.5hr, which is indicated by the gray bar in this figure, is treated as dark period in the simulation. Since the LD cycles in New York is almost inverted to that in Tokyo, a large phase shift is requiered in the rhythm of the travelers. In fact, on the first day in New York, they advanced the phase as much as 3 hrs by exposing themselves to the light at the time when the maximum phase advance is attained as shown in Figure 11. Figure 12 is an actgram representation of the numerical data produced from the HFPN model in Figure 9. Actgram is an expression of animal locomotor activity. Gray bar in each row shows the period when an animal (human in this case) is in action. In this figure, gray color is applied to the duration when the level of *Per* mRNA is higher than 1.07. A black bold horizontally long rectangle in the 5th row indicates the 12.5hr period in a cabin flying from Tokyo

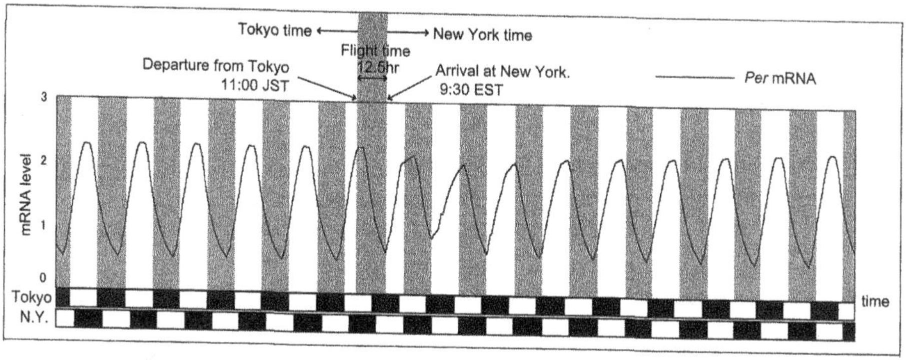

Fig. 11. Level of *Per* mRNA before, during and after the travel from Tokyo to New York. Gray bars indicate dark periods including 12.5hr flight time. Two black-white bars show LD cycles at Tokyo and New York.

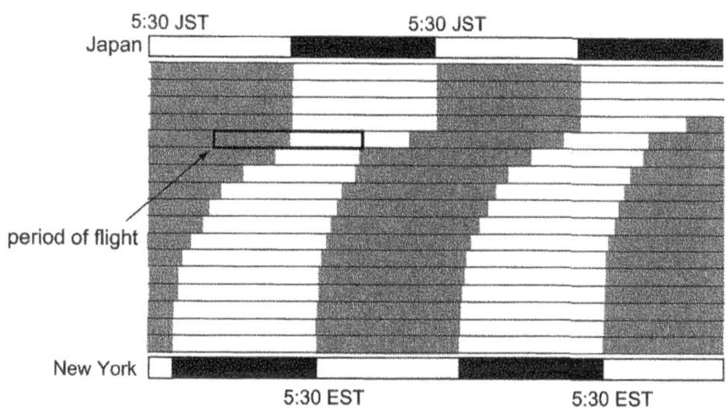

Fig. 12. Actgram obtained from the jet lag simulation for the flight which leaves Tokyo at 9:30AM and arrives at New York at 11:00AM. Horizontal gray bars indicate the high expression periods of *Per* mRNA, i.e, the period when human is in action. This diagram shows that the period keeps shifting to the earlier time until being adjusted to the LD cycles in New York.

to New York.

5. Discussion and Future works

In the present paper, we have applied Hybrid Functional Petri net (HFPN) techniques to the molecular system of clock genes that is responsible for the generation of circadian rhythms, and extended the analysis to the response to light and entrainment to LD cycles. Computer simulation reproduced a phase response curve similar to that reported in biological literature, suggesting that the behavioral phase response properties are a manifestation of the molecular clock. When the gate was introduced before the oscillating system, which is closed during the subjective day, oscillation of mRNAs in the HFPN model responded so as to entrain itself to external LD cycles. Furthermore, entrainment was found possible only when periodicity of the external LD cycle was close to 24 hr. This reproduction of the characteristic behavior of entrainment found in the biological observation strengthens the usefulness of our HFPN model. We also studied the mRNA rhythms in this model under the situation mimicking the jet lag caused by inter time-zone flight. Our model system with the gate considered successfully reproduced unstable transition period corresponding physiological symptoms of malaise during the jet lag. It is interesting that the model learned from the molecular mechanisms responsible for rhythm generation was able to predict the behavior of the circadian clock under LD cycles. It also suggested activity patterns of individual persons often experienced on such an occasion as a jet lag. Computer simulation in this paper may provide a scientific insight to the molecular machinery of the gene regulatory system of the circadian clock. Future studies will help to find a way to alleviate health problems derived from various types of sleep disorder, as well as a remedy to jet lag.

Acknowledgements

This work was partially supported by Grant-in-Aid for Scientific Research on Priority Areas "Systems Genomics" (17017008) and Grant-in-Aid for Scientific Research (B) (19300103) from the Ministry of Education, Culture, Sports, Science and Technology of Japan.

References

[1] Czeisler, C.A., *et al.*, Stability, precision, and near-24-hour period of the human circadian pacemaker., *Science*, 284:2177–2181, 1999.

[2] Doi, A., *et al.*, Simulation-based validation of the p53 transcriptional activity with hybrid functional Petri net., *In Silico Biol.*, 6(1):1–13, 2006.

[3] Kurosawa, G. and Goldbeter, A., Amplitude of circadian oscillations entrained by 24-h light-dark cycles., *J. Theor. Biol.*, 242:478–488, 2006.

[4] Leloup, J.C. and Goldbeter, A., Modeling the mammalian circadian clock: Sensitivity analysis and multiplicity of oscillatory mechanisms., *J. Theor. Biol.*, 230:541–562, 2004.

[5] Leloup, J.C. and Goldbeter, A., Toward a detailed computational model for the mammalian circadian clock., *Proc. Natl Acad. Sci. USA*, 100(12), 7051–7056, 2003.

[6] Matsuno, H., *et al.*, Biopathways Representation and Simulation on Hybrid Functional Petri Net., *In Silico Biol.*, 3(3):389–404, 2003.

[7] Matsuno, H., *et al.*, Boundary formation by notch signaling in Drosophila multicellular systems: experimental observations and a gene network modeling by Genomic Object Net., *Pac. Symp. Biocomput.*, 8:152–1632, 2003.

[8] Matsuno, H., Inouye, S.T., Okitsu, Y., Fujii, Y., and Miyano, S., A new regulatory interaction suggested by simulations for circadian genetic control mechanism in mammals., *J. Bioinf. and Comput. Biol.*, 4(1):139–157, 2006.

[9] Miyake, S., *et al.*, Phase-dependent responses of Per1 and Per2 genes to a light-stimulus in the suprachiasmatic nucleus of the rat., *Neurosci. Lett.*, 294(1):41–44, 2000.

[10] Pittendrigh, C.S. and Daan, S., A functional analysis of circadian pacemakers in nocturnal rodents. V. Pacemaker structure: A clock for all seasons., *J. Comp. Physiol.*, 106:223–355, 1976.

[11] Rand, D.A, Shulgin, B.V., Salagar, D. and Millar, A.J., Design principles underlying circadian clocks., *J. R. Soc. Interface*, 1:119–130, 2004.

[12] Shigeyoshi, Y., *et al.*, Light-induced resetting of a mammalian circadian clock is associated with rapid induction of the mPer1 transcript., *Cell*, 91:1043–1053, 1997.

[13] Takahashi, J.S., DeCoursey, P.J., Bauman, L., and Menaker, M., Spectral sensitivity of a novel photoreceptive system mediating entrainment of mammalian circadian rhythms., *Nature*, 308:186–188, 1984.

[14] `http://genome.ib.sci.yamaguchi-u.ac.jp/~ISMB2008/`

[15] `http://www.ana.co.jp/`

[16] `http://www.cellillustrator.org/`

PART B
Keynote Addresses

SEQUENCING THE TRANSCRIPTOME *IN TOTO*

SEAN M. GRIMMOND
s.grimmond@imb.uq.edu.au

Expression Genomics Laboratory
Institute for Molecular Bioscience
University of Queensland, AUSTRALIA

Abstract

Since the sequencing of the mouse and human genomes, there has been a concerted effort to define their complete transcriptional output. EST, full length cDNA sequencing, and transcriptome annotation efforts by FANTOM, ENCODE and other consortia surveyed mammalian expression space, revealing that loci on average generate 6-10 transcripts. Alternative promoters, splicing and 3'UTRs are commonplace.

While these data have provided an excellent atlas of what can be generated from mammalian genomes, we have not had, until recently, the right genomic tools to place this transcriptional complexity into a biological context. Array based profiling has been an excellent tool for assessing overall gene activity, but lacks the sensitivity and resolution required to study complete transcriptome content

RNA sequencing (RNAseq) has recently been demonstrated in several eukaryotic species and is redefining our understanding of mRNA transcriptome content and mRNA dynamics, all at a single nucleotide resolution. We have developed methods for performing multi-gigabase shotgun sequencing of human and mouse transcriptomes and have developed approaches to assess locus activity and demonstrated its improved sensitivity relative to the current "gold standard" array platforms. We also use RNAseq to assess the expression levels of variant transcripts via diagnostic sequences. Thirdly, we are able to perform genome-wide transcriptome discovery. Finally we have also established approaches to identify alternations to the reference sequence content, allowing us to search for expressed polymorphisms, mutations or events such as RNA editing.

These data are combined with RNAseq surveys of other fractions of the transcriptome (i.e. small RNA and polysome-associated RNAs) to gain a fuller picture of coding and functional RNA content. This is being used to define, at unprecedented resolution, the transcriptional networks driving specific biological states.

References

[1] Cloonan N, Forrest ARR, Kolle G, Gardiner BBA, Faulkner GJ, Brown MK, Taylor DF, Steptoe AL, Wani S, Bethel G *et al.*: Stem cell transcriptome profiling via massive-scale mRNA sequencing. *Nat Meth* 2008, 5(7):613–619.

MODERN HOMOLOGY SEARCH

MING LI
mli@cs.uwaterloo.ca

School of Computer Science
University of Waterloo
Waterloo, CANADA

Abstract

Dynamic programming [1] has full sensitivity, but too slow for large scale homology search. FASTA / BLAST type of heuristics [2] trade sensitivity for speed. Can we have both sensitivity and speed?

We present the mathematical theory of optimized spaced seeds which allows modern homology search to achieve high sensitivity and high speed simultaneously. The spaced seed methodology is implemented in our PatternHunter software [3, 4], as well as many other modern homology search software, serving thousands of queries daily.

The theory is then extended and implemented in ZOOM [5] to do fast genome scale reads mapping for the second generation sequencers.

Joint work with Bin Ma, John Tromp, D. Kisman, Hao Lin, and Zefeng Zhang.

References

[1] S.F. Altschul, W. Gish, W. Miller W, E.W. Myers, D.J. Lipman. Basic local alignment search tool. *J Mol Biol* 215:3(1990), 403–410.

[2] T.F. Smith, M.S. Waterman, Identification of Common Molecular Subsequences. *Journal of Molecular Biology*, 147(1981), 195–197.

[3] B. Ma, J. Tromp, M. Li, PatternHunter: Faster and more sensitive homology search. *Bioinformatics*, 18:3(2002), 440–445.

[4] M. Li, B. Ma, D. Kisman and J. Tromp. PatternHunter II: highly sensitive and fast homology search. *J. Bioinformatics and Computational Biology*, 2:3(2004), 417–440.

[5] H. Lin, Z. Zhang, M.Q. Zhang, B. Ma, M. Li. ZOOM! Zillions of oligos mapped. *Bioinformatics*. In press. 2008.

MODELING HUMAN GENOME-WIDE COMBINATORIAL REGULATORY NETWORKS INITIATED BY TRANSCRIPTION FACTORS AND MICRORNAS USING FORWARD AND REVERSE ENGINEERING

YI-XUE LI
yxli@sibs.ac.cn

Shanghai Center for Bioinformation Technology
and
Shanghai Institutes for Biological Sciences, Chinese Academy of Sciences, CHINA

Abstract

MicroRNAs are short endogenous non-coding transcripts which regulate their target mRNAs by translational inhibition or mRNA degradation. Recent microRNA transfection experiments show strong evidence that microRNAs influence not only their target but also non-target genes, but how the regulatory signals are transduced from microRNAs to the downstream genes remains to been elucidated. We suspect that primary and secondary regulatory mechanisms, initially triggered by microRNAs, form refined local networks in the cell. In light of this hypothesis, a comprehensive strategy was developed to reconstruct combinatory networks of primary and secondary microRNA regulatory cascades, using microRNA's target and non-target gene expression profiles and information of microRNA-regulated transcription factors (TF) and TF regulated genes. This strategy was then applied to 53 microRNA transfection expression datasets and led to discovery of combinatorial regulatory networks triggered by 20 microRNAs. Many of these networks were enriched with genes whose functional roles were consistent with known regulatory roles of microRNAs. More importantly, a tumor-related regulatory network and related pathways were discovered, in which novel discoveries were integrated with existing knowledge on the regulatory mechanisms of four microRNAs. In the network, by activating mir-34 family, the tumor suppressor gene p53 can inhibit five target oncogenes, four of which have never been reported. Our approach was carried out on a sizeable number of public microRNA transfection experiment datasets, enabling a global view of combinatory regulatory networks triggered by microRNAs. Through reconstructing microRNA-triggered combinatory regulatory networks, the work help identify the true degradation targets of mammal microRNAs, and more importantly, aid in fundamental understanding of microRNA related biological functional processes.

RECONSTRUCTING THE CIRCUITS OF DISEASE: FROM MOLECULAR STATES TO PHYSIOLOGICAL STATES

ERIC E. SCHADT
eric_schadt@merck.com

Department of Genetics
Rosetta Inpharmatics, LLC/Merck Research Labs, USA

Abstract

Common human diseases and drug response are complex traits that involve entire networks of changes at the molecular level driven by genetic and environmental perturbations. Efforts to elucidate disease and drug response traits have focused on single dimensions of the system. Studies focused on identifying changes in DNA that correlate with changes in disease or drug response traits, changes in gene expression that correlate with disease or drug response traits, or changes in other molecular traits (e.g., metabolite, methylation status, protein phosphorylation status, and so on) that correlate with disease or drug response are fairly routine and have met with great success in many cases. However, to further our understanding of the complex network of molecular and cellular changes that impact disease risk, disease progression, severity, and drug response, these multiple dimensions must be considered together. Here I present an approach for integrating a diversity of molecular and clinical trait data to uncover models that predict complex system behavior. By integrating diverse types of data on a large scale I demonstrate that some forms of common human diseases are most likely the result of perturbations to specific gene networks that in turn causes changes in the states of other gene networks both within and between tissues that drive biological processes associated with disease. These models elucidate not only primary drivers of disease and drug response, but they provide a context within which to interpret biological function, beyond what could be achieved by looking at one dimension alone. That some forms of common human diseases are the result of complex interactions among networks has significant implications for drug discovery: designing drugs or drug combinations to impact entire network states rather than designing drugs that target specific disease associated genes.

THE EMERGING GENERALIZATIONS OF PROKARYOTIC GENOMICS

EUGENE V. KOONIN
koonin@ncbi.nlm.nih.gov

National Center for Biotechnology Information
National Library of Medicine
National Institutes of Health, Bethesda MD, USA

Abstract

The first bacterial genome was sequenced in 1995, and the first archaeal genome in 1996. Soon after these breakthroughs, an exponential rate of genome sequencing was established, with a doubling time of approximately 18 months for bacteria and approximately 34 months for archaea. Comparative analysis of the hundreds of sequenced bacterial and dozens of archaeal genomes leads to several generalizations on the principles of genome organization and evolution. A crucial finding that enables functional characterization of the sequenced genomes and evolutionary reconstruction is that the majority of archaeal and bacterial genes have conserved orthologs in other, often, distant organisms. However, comparative genomics also shows that horizontal gene transfer (HGT) is a dominant force of prokaryotic evolution, along with the loss of genetic material resulting in genome streamlining. A crucial component of the prokaryotic world is the mobilome, the enormous collection of viruses, plasmids and other selfish elements which are in constant exchange with more stable chromosomes and serve as HGT vehicles. Thus, the prokaryotic genome space is a tightly connected, although compartmentalized, network, a new notion that undermines the "Tree of Life" model of evolution and requires a new conceptual framework and tools for the study of prokaryotic evolution.

A NEW UNDERSTANDING OF THE HUMAN GENOME

JOHN MATTICK
j.mattick@imb.uq.edu.au

Institute for Molecular Bioscience
University of Queensland, AUSTRALIA

Abstract

It appears that the genetic programming of mammals and other complex organisms has been misunderstood for the past 50 years, because of the assumption – largely true in prokaryotes, but not in complex eukaryotes – that most genetic information is transacted by proteins. The numbers of protein-coding genes do not change appreciably across the metazoa, whereas the relative proportion of non-protein-coding sequences increases markedly. Moreover, while only a tiny fraction encodes proteins, it is now evident that the majority of the mammalian genome is transcribed in a developmentally regulated manner, and that most complex genetic phenomena in eukaryotes are RNA-directed. Evidence will be presented that (i) regulatory information scales quadratically with functional complexity and hence the majority of the genomes of the higher organisms comprises regulatory information; (ii) there are thousands of non-protein-coding transcripts in mammals that are dynamically expressed during differentiation and development, including in embryonal stem cell and neuronal cell differentiation, and T-cell and macrophage activation, among others, many of which show precise expression patterns and subcellular localization in the brain; (iii) many 3'UTRs are not only linked to but are also expressed in a regulated manner separately from their associated protein-coding sequences to transmit genetic information *in trans* (iv) there are large numbers of small RNAs, including new classes, expressed from the human and mouse genomes, that may be discerned from bioinformatic analysis of genomic and deep sequencing transcriptomic datasets; and (v) much, if not most, of the mammalian genome may not be evolving neutrally, but rather is composed of different types of sequences (including transposon-derived sequences) that are evolving at different rates under different selection pressures and different structure-function constraints. There is also genome-wide evidence of editing of noncoding RNA sequences, especially in the brain and especially in humans (Alu elements), which may constitute a key part of the molecular basis of memory and cognition. Taken together, these and other observations suggest that the majority of the human genome is devoted to an very sophisticated RNA regulatory system that directs developmental trajectories and mediates gene-environment interactions via the control of chromatin architecture and epigenetic memory, transcription, splicing, RNA modification and editing, mRNA translation and RNA stability.

AUTHOR INDEX

www.ingramcontent.com/pod-product-compliance
Lightning Source LLC
LaVergne TN
LVHW082257150926
848178LV00006B/98